QM Library

23 1359459 6

QL495 mon

**WITHDRAWN
FROM STOCK
QMUL LIBRARY**

Biosynthesis in Insects
Advanced Edition

Dedication

To Sir John and Lady Cornforth (Kappa and Rita), pioneers of biosynthesis, who taught me so much, this book is fondly and respectfully dedicated

Biosynthesis in Insects
Advanced Edition

E. David Morgan
Chemical Ecology Group, Keele University, Keele, UK

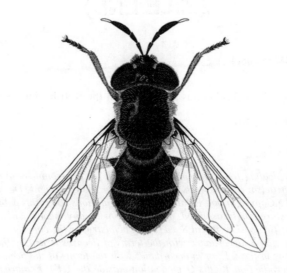

RSCPublishing

ISBN: 978-1-84755-808-4

A catalogue record for this book is available from the British Library

© E. David Morgan 2010

All rights reserved

Apart from fair dealing for the purposes of research for non-commercial purposes or for private study, criticism or review, as permitted under the Copyright, Designs and Patents Act 1988 and the Copyright and Related Rights Regulations 2003, this publication may not be reproduced, stored or transmitted, in any form or by any means, without the prior permission in writing of The Royal Society of Chemistry, or the copyright owner, or in the case of reproduction in accordance with the terms of licences issued by the Copyright Licensing Agency in the UK, or in accordance with the terms of the licences issued by the appropriate Reproduction Rights Organization outside the UK. Enquiries concerning reproduction outside the terms stated here should be sent to The Royal Society of Chemistry at the address printed on this page.

The RSC is not responsible for individual opinions expressed in this work.

Published by The Royal Society of Chemistry,
Thomas Graham House, Science Park, Milton Road,
Cambridge CB4 0WF, UK

Registered Charity Number 207890

For further information see our web site at www.rsc.org

Foreword

Biosynthesis as a subject for study is less than a century old. It began in almost total darkness lit by one guiding light: a tested theory of molecular structure. When the new methods of examination arrived (spectroscopy in the ultraviolet, visible and infrared; mass and NMR spectrometry) their power was amplified by the existence of thousands of compounds with accurately known structures, their functional groups in different settings shining out from the spectra like jewels. The new methods, too, required much less material; this in turn made it easier to apply the various newly available techniques of chromatography, which were more powerful and convenient with smaller samples. Add to all this the newly emerging availability of isotopic labelling, and you have an armoury to attack the problems of biosynthetic pathways The attack has been highly successful. The results show, as few other techniques could, the adaptability of living cells and their power to conjure up, from a limited range of starting points, agents with such variety and effectiveness.

Insects offer an unusually wide variety of specific chemically mediated reactions with their environment and within their own species and their own bodies. This book, by an expert in the field, reviews the progress made and tests all readers who want to know how much they assimilated. If they gain no more than a conviction that chemistry is the lifeblood of biology, their time will not have been wasted.

John Warcup Cornforth

Preface

Our knowledge of the substances that insects make, for what purpose they make them and how they make them is growing rapidly and, because of the enormous number of insect species, this subject is potentially vast. Since the first edition of *Biosynthesis in Insects* appeared, the subject has advanced in many directions.

The first edition had a more didactic approach for students, since it grew out of a series of lectures to research students at the Universidade Federal de Alagoas in Brazil, expanded later through lectures to research students at Keele. It seemed more appropriate now to speak directly to research workers for the expanding subject, while retaining a little explanatory material. The subject stretches across the boundaries of mechanistic organic chemistry through biochemistry, entomology and, as it grows, elements of endocrinology, genetics and molecular biology. Not everyone is equally familiar with these diverse subjects.

It is not a textbook of biochemistry, nor is it intended to cover the energetics of producing metabolites, nor the enzymology of insects. Too little is known about the last. The endocrinology of production of insect hormones and pheromones has been well covered elsewhere. Nevertheless, the wide range of substances made by insects can be reduced to some kind of order and system by considering them from their biosynthetic origins; and by considering their biosynthesis, we can understand better what structures and types might be expected in studying new compounds isolated in studies of insect behaviour and physiology. Where information is available about biosynthesis in related arthropods, like millipedes, centipedes, spiders, mites, ticks and opilionids (Phalangida), it is included so comparisons can be attempted across the Arthropoda.

Biosynthesis in Insects, Advanced Edition
By E. David Morgan
© E. David Morgan 2010
Published by the Royal Society of Chemistry, www.rsc.org

Contents

Acknowledgements		xv
Chapter 1	**Introduction**	**1**
	1.1 Studying Biosynthetic Pathways	4
	1.2 Plant *versus* Insect Biosynthesis	5
	1.3 Arthropods and Insects	7
	1.4 Chemical Communication	8
	1.5 The Structures of Natural Products	9
	Background Reading	10
	References	10
Chapter 2	**Enzymes and Co-enzymes**	**12**
	2.1 The Chemical Function of Enzymes	12
	2.1.1 Lysozyme	14
	2.1.2 Metallo-enzymes	15
	2.1.3 Cytochromes	17
	2.2 Co-enzymes	18
	2.2.1 Adenosine Triphosphate	19
	2.2.2 Co-enzyme A	20
	2.2.3 Nicotinamide Adenine Dinucleotide	21
	2.2.4 Flavin Adenine Dinucleotide	23
	2.2.5 Thiamine Diphosphate	23
	2.2.6 Pyridoxal Phosphate	25
	2.2.7 Tetrahydrofolic Acid	26
	2.2.8 *S*-Adenosyl Methionine	28
	2.2.9 Vitamins	28
	2.3 Biosynthesis of Formic Acid	29

Biosynthesis in Insects, Advanced Edition
By E. David Morgan
© E. David Morgan 2010
Published by the Royal Society of Chemistry, www.rsc.org

	2.4	Pyruvic Acid	29
	2.5	Nitrogen Fixation	31
	2.6	Chirality	31
		2.6.1 Asymmetric Induction	32
	Background Reading		35
	References		35

Chapter 3 Experimental Methods 37

    3.1 Tracing Biosynthetic Pathways 37
    3.2 Radio-isotope Labelling 39
          3.2.1 Specific Incorporation 40
          3.2.2 Locating the Site of Synthesis 41
          3.2.3 Radio-labelling Examples 42
    3.3 Heavy Isotope Labelling 43
          3.3.1 Kinetic Isotope Effects 45
          3.3.2 Other Isotope Effects 45
          3.3.3 Examples of Isotopic Labelling 46
          3.3.4 Virginae Butanolide 50
          3.3.5 ^{13}C-^{13}C Coupling 51
    3.4 Analytical Aspects 53
          3.4.1 Chirality 53
    3.5 Molecular Biology in Biosynthetic Studies 55
          3.5.1 Using the Polymerase Chain Reaction 55
          3.5.2 Functional Genomics 58
          3.5.3 Model Organisms 60
    Background Reading 62
    References 62

Chapter 4 Fatty Acids and Cuticular Hydrocarbons 66

    4.1 Introduction 66
    4.2 Fatty-acid Biosynthesis 66
          4.2.1 The Synthetic Enzyme Complex 67
          4.2.2 Unsaturated Acids and Desaturase Enzymes 72
          4.2.3 Eicosanoids 76
          4.2.4 Branched Fatty Acids 79
    4.3 Cuticular Hydrocarbons 82
          4.3.1 Hydrocarbon Biosynthesis 83
          4.3.2 Branched-chain Hydrocarbons 85
          4.3.3 Alkenes 90
          4.3.4 Physical State of Hydrocarbons 91
    Background Reading 92
    References 92

Chapter 5 Aliphatic Compounds from Fatty Acids — 96

- 5.1 Lepidopteran Sex Pheromones — 96
  - 5.1.1 Hydrocarbons — 97
  - 5.1.2 Epoxides — 99
  - 5.1.3 Other Oxygenated Compounds — 100
- 5.2 Lepidopteran Defence — 110
- 5.3 Coleopteran Compounds — 110
  - 5.3.1 Hydrocarbons, Acids and Lactones — 110
- 5.4 Coleopteran Defence — 114
  - 5.4.1 Coccinellinae — 115
  - 5.4.2 Chilocorinae — 117
  - 5.4.3 Scymninae — 118
  - 5.4.4 Epilachninae — 119
- 5.5 Dipteran Pheromones — 122
  - 5.5.1 *Musca* — 122
  - 5.5.2 *Drosophila* — 123
  - 5.5.3 *Glossina* — 125
- 5.6 Hymenoptera — 125
  - 5.6.1 Bees — 126
  - 5.6.2 Wasps — 129
  - 5.6.3 Ants — 130
- 5.7 Isoptera – Termites — 135
- 5.8 Blattodea – Cockroaches — 137
- 5.9 Green Leaf Volatiles — 138
- Background Reading — 141
- References — 141

Chapter 6 Aceto-propiogenins — 146

- 6.1 Acetogenins — 146
- 6.2 Polyketides — 147
- 6.3 Butyric Acid Compounds — 148
- 6.4 Acetogenin Pheromones — 150
- 6.5 Aceto-propiogenins — 151
  - 6.5.1 Ants — 152
  - 6.5.2 Coleoptera — 154
  - 6.5.3 Other Arthropods — 159
- 6.6 Pyrones and Lactones — 160
- 6.7 Carbocyclic Compounds — 162
- 6.8 Spiroacetals — 166
- 6.9 Pederin — 173
- Background Reading — 176
- References — 176

Chapter 7 Lower Terpenes — 179

- 7.1 Introduction — 179
- 7.2 Monoterpene Biosynthesis — 180
  - 7.2.1 The Mevalonate Pathway — 180
  - 7.2.2 The Methylerythritol Phosphate Pathway — 185
- 7.3 Monoterpene Pheromones — 186
  - 7.3.1 Bark Beetles — 186
  - 7.3.2 Weevils — 194
  - 7.3.3 Other Pheromones — 195
- 7.4 Monoterpene Defensive Compounds — 196
  - 7.4.1 Iridoids — 196
  - 7.4.2 Termite Defensive Secretion — 201
- 7.5 Sesquiterpenes — 202
  - 7.5.1 Sesquiterpene Pheromones — 202
  - 7.5.2 Sesquiterpene Defences — 204
  - 7.5.3 Lac Insects — 208
- 7.6 Homoterpenes — 209
  - 7.6.1 Juvenile Hormone — 212
- Background Reading — 214
- References — 215

Chapter 8 Higher Terpenes and Steroids — 220

- 8.1 Introduction — 220
- 8.2 Diterpenes — 220
  - 8.2.1 Termites — 221
- 8.3 Sesterterpenes — 224
- 8.4 Triterpenes and Steroids — 224
  - 8.4.1 Sterols in Insects — 227
  - 8.4.2 Phytosterol Dealkylation — 228
- 8.5 Insect Moulting Hormone – Ecdysteroids — 231
  - 8.5.1 Ecdysteroid Biosynthesis — 232
  - 8.5.2 Ecdysteroids in Adult Insects — 236
  - 8.5.3 Ecdysteroid Inactivation and Excretion — 238
- 8.6 Sterol Pheromones — 239
- 8.7 Steroid Defensive Substances — 240
  - 8.7.1 Saponins from Triterpenes — 242
- 8.8 Mammalian Hormones in Insects — 243
- 8.9 Tetraterpenes — 245
  - 8.9.1 Insect Vision — 249
- Background Reading — 250
- References — 251

Contents xiii

Chapter 9 Aromatic Compounds — 255

9.1 Aromatic Compounds in Nature — 255
9.2 The Shikimic Acid Pathway — 255
9.3 Phenyl-C_3 Compounds — 256
    9.3.1 Aromatic Pheromones — 260
    9.3.2 Compounds from Chorismic Acid — 261
    9.3.3 Aromatic Amines — 263
9.4 Phenols — 265
9.5 Quinones — 269
    9.5.1 Bombardier Beetles — 270
9.6 Insect Pigments — 271
    9.6.1 Melanin — 272
    9.6.2 Naphthoquinones and Anthraquinones — 275
    9.6.3 Aphins — 276
    9.6.4 Pterins — 277
    9.6.5 Tetrapyrroles — 281
    9.6.6 Ommochromes and Ommins — 284
    9.6.7 Anthocyanins and Flavones — 285
9.7 Chitin, Cuticle and Sclerotisation — 286
References — 289

Chapter 10 Alkaloids and Compounds of Mixed Biosynthetic Origin — 294

10.1 Alkaloids — 294
10.2 Insect Alkaloids — 295
    10.2.1 Ant Venoms — 296
    10.2.2 Myrmicarins — 298
    10.2.3 Other Examples — 299
    10.2.4 Amino Acid Derivatives as Pheromones — 302
10.3 Compounds of Mixed Biosynthetic Origin — 303
10.4 Luciferin — 306
10.5 Plant Volatile Elicitors — 309
Background Reading — 311
References — 311

Chapter 11 Plant Substances Altered and Sequestered by Insects — 315

11.1 Introduction — 315
    11.1.1 Nuptial Gifts — 316
11.2 Cardiac Glycosides — 316
11.3 Pyrrolizidine Alkaloids — 318
    11.3.1 Pyrrolizidines in Lepidoptera — 319
    11.3.2 Pyrrolizidines in Coleoptera — 320
11.4 Cyanogenic Glucosides — 322

11.5	Veratrum Alkaloids	325
11.6	Other Examples	326
11.7	Origins of Sequestration	327
	Background Reading	328
	References	328

Looking Ahead   332

Trends in Research	332
Agendum	333
References	334

Appendix 1   Common Abbrevations   336

Appendix 2   Glossary of Terms   338

Subject Index   344

Acknowledgements

I am particularly grateful to Ralph Howard, one of the pioneers of insect chemistry, for generously reading all of my manuscript, and giving me many helpful comments. Professor Gary Blomquist, a leading light in insect biosynthesis studies, has kindly read Chapters 4, 5 and part of 7. I want to thank Professor William Kitching for reading my section on spiroacetals, Dr Jörn Piel for checking what I have written on pederin and Dr David Stanley for checking the section on eicosanoids and immunity. Neil Oldham now knows more than I ever taught him about experimental methods, and has been helpful with that. His first-hand knowledge of iridoid biosynthesis has been helpful. Gary Blomquist, Claus Tittiger and Frédèric Tripet have guided my infant steps in the difficult subject of molecular biological methods, and I am greatly indebted to them. I am grateful to Professor Wittko Francke for his help and advice. Dr Gordon Hamilton helped me with the confused subject of sandfly pheromones and Prof. Chris Ramsden with some difficult rearrangement reactions. I thank Professor René Lafont for kindly sending me a number of his papers and reviews on ecdysteroids. Dr S. O. Andersen has been kindly helpful with the subject of melanin formation, and I thank Dr M. Sundaram with copies of his papers on the subject.

I am grateful to Dr John Shanklin of Brookhaven National Laboratory and Dr Ed Cahoon of the Danforth Plant Sciences Center for again letting me use the diagrams of castor oil desaturase, Prof. John Mann for permission to use a figure from his work and Dr Tittiger for diagrams from his publications. I am obliged to Dr Jonathan Banks in Australia and Dr Keith S. Brown in Brazil for helping me concerning the final position of aphinin research. Dr Merlin Fox at the Royal Society of Chemistry has been a constant help and guided me on difficult publishing issues.

Those of us who are laboratory-bound see relatively few of the insects we read about, and still fewer of them alive. I am grateful to various friends for contributing their photographs of (biosynthetically) interesting and useful insects. First of all, I must thank Mark Crocker for his beautiful picture of the honeybee on the cover and Robert Whyte of Save our Waterways Now, Queensland, who has generously given several pictures. Professors J. Holopainen, J. K. Lindsey and Andrei Lobanov; Drs R. G. Vogt, Alex Wild, Kevin Wanner, Jack Kelly Clark, P. J. Bryant, D. B. Fenolio and Arthur Anker have also provided beautiful pictures from their work. There are some excellent insect photographers, professional and amateur, who have allowed me to use their pictures here. I am indebted to Lynette Schimming, François Bonneton, Tim Ransom, Seabrooke Leckie, Sebastian Lübcke, Henrik Hemplemann, Nigel Jones, Judy Burris and Wayne Richards, Charles Schurch Lewallen, N. Sloth, Ashley Bradford, Shane Farrell, Josef Dvorák, Alessio Di Leo, Steve Ogden, Simon Knott, Jean Hutchins and Roger Rittmaster, Todd Burrows, Jan Kaspar, Kristin Viglander, Colin Halliday, B. Hamers, Matt Edmonds, Bruce Marlin and José Luis Calleiras, all of whom have contributed photos. There are several organisations that have allowed me to use their copyrighted or openly available pictures – US Department of Agriculture, EPPO, Paris, Entomart, University of California IPM Project, The Botanic Garden Trust, Sydney, Laspilitas – and I have used three pictures from Wikimedia Commons, with appropriate acknowledgements. Dr Norman J. Fashing has kindly given me the use of his electron micrograph of *Rhyzoglyphus robini*, and the Canadian Government the use of a drawing of *Carpophilus hemipterus*.

I am indebted to Dr Martin Bay Hebsgaard of Edinburgh University for permission to use his beautiful drawing of the hoverfly *Callicera aurata* (an endangered species in the British Isles) used on the title page.

CHAPTER 1
Introduction

It is now over 420 million years since land plants first appeared on our Earth, at the end of the Silurian Period, and the first insects appeared about the same time. The oldest known fossil insect, *Rhyniognathus hirsti*, is between 396 and 407 million years old, from the beginning of the Devonian Period. There was a massive wipe out of insects in the Permian-Triassic extinction about 252 million years ago, but when flowering plants developed 200 million years ago, many Hymenoptera, Lepidoptera, Diptera and some Coleoptera evolved with them (Grimaldi and Engel, 2005). Insects have therefore branched off from the rest of the animals a long time ago, and have been evolving for much longer than mammals (Figure 1.1). Therefore a much greater diversity in genetic inheritance can be expected among insects than among mammals. Estimated numbers of insect species vary from one million to thirty million, making up 80% of all world species. New species are discovered at the rate of about 5000 per year. At any time, it is further estimated there must be some 10^{18} individual insects alive.

Potentially, the subject of this book is gigantic but, at present, knowledge of insect biochemistry is but fragmentary.

All living organisms share much of primary or basic metabolism. The production of nucleic acids, amino acids, sugars, lipids and macromolecules made from these are processes common to all. Secondary metabolism is carried out by all organisms, but differs with different groups or species. Primary metabolism is sometimes defined as comprising all the reactions and compounds essential to the life of the cell, while secondary metabolism is the system of processes and compounds necessary to maintain the life of the whole organism. Secondary

---

Biosynthesis in Insects, Advanced Edition
By E. David Morgan
© E. David Morgan 2010
Published by the Royal Society of Chemistry, www.rsc.org

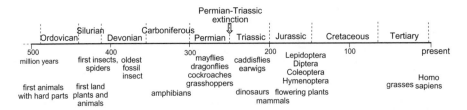

Figure 1.1 The antiquity of insects. The scale is in millions of years. Details vary considerably between authors. This figure is based on information in Grimaldi and Engel (2005).

metabolites may be structural material like bone, chitin or hair, antibacterial or antifungal protectants, they may give protection from predators, they may produce warning or camouflage pigments, they may be used in signalling, such as hormones or pheromones, or they may have some function not yet identified. The division into primary and secondary metabolites is rather arbitrary and is rejected by some modern authors. It stresses the old idea of one enzyme-one metabolite, whereas one enzyme may catalyse changes in several or even many compounds (Firn and Jones, 2009).

The range of secondary metabolites is enormous and presents a never-ending source of curiosity and challenge. Yet this great array of substances is produced from relatively few basic building blocks that are part of primary metabolism. Figure 1.2 attempts to summarize, very briefly, the basic groups of compounds from which secondary metabolites are made. The carbon atoms of all substances, whether from plant or animal, are ultimately derived from carbon dioxide *via* photosynthesis. It is remarkable that most compounds we consider here come through a very limited number of precursors. Figure 1.2 shows that many groups of compounds are formed *via* relatively few biosynthetic paths. Biosynthesis is the building up of chemical compounds through the physiological processes that take place in living animals, plants and micro-organisms. Biosynthesis requires starting materials, energy and catalysts. The starting materials may be any compounds from primary or secondary metabolism. All the compounds down the centre of Figure 1.2 are important starting materials for insect biosynthesis. The energy needed comes ultimately from sunlight, but is often immediately supplied through the breakdown of adenosine triphosphate, and the reactions are catalysed by enzymes.

Insects share some biochemical characteristics with all living organisms, others with all animals, but others are peculiar to insects alone, or to a few species or even a single caste of a single species (like males or

Introduction

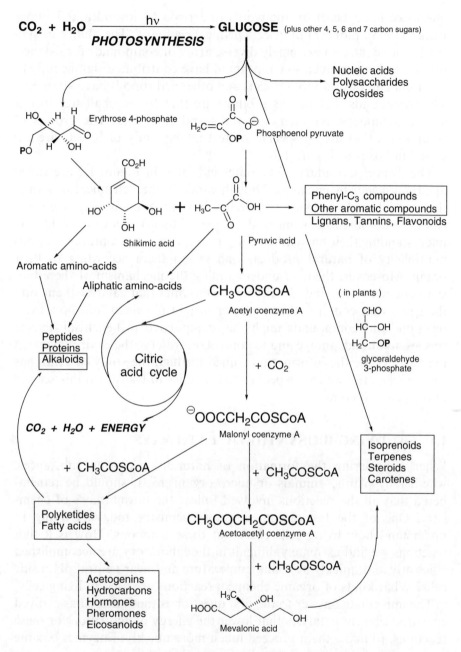

Figure 1.2 A summary of the chief biosynthetic routes to primary and secondary metabolites. Those of particular interest here are enclosed in boxes. Notice that glucose, phosphoenol pyruvate and erythrose-4-phosphate are the key intermediates for all these classes of compounds. **P** indicates a phosphate ester. Adapted from a figure in J. Mann, *Chemical Aspects of Biosynthesis*, 1994, by permission of Oxford University Press.

queens of honeybees). In the words of Jerrold Meinwald and Thomas Eisner (1995), pioneers in insect chemical ecology, "The ability to synthesize or acquire an extremely diverse array of compounds for defence, offence and communication appears to have contributed significantly to the dominant position that insects and other arthropods have attained". The compounds that insects produce are therefore a challenge to our ability to understand their structures and functions. The groups of compounds that are of special interest in the study of insects are enclosed in boxes in Figure 1.2.

The diverse secondary metabolites indicated in Figure 1.2 are called *natural products* by chemists. Though varied in their chemical structure, they are all made by one of these few biosynthetic pathways (in some cases, a combination of more than one of them) (Hanson, 2003). By understanding their biosynthetic origins one can make some sense of this multiplicity of natural products and group them according to their origin. Moreover, through understanding the mechanism of these biosynthetic reactions, and identifying the enzymes that catalyse them, and the genes that code for them, greater insight is gained into how these reactions and compounds might be manipulated in beneficial or pest species, to our advantage and to control or preserve the environment. In the past decade the information gained through the study of genes has leapt ahead, and we can expect genetic studies to transform this science in the years to come.

1.1 STUDYING BIOSYNTHETIC PATHWAYS

When considering the formation of naturally occurring substances, whether by plants, animals or micro-organisms, it should be remembered that all the reactions involved follow the normal laws of chemistry. One of the fascinating areas of chemistry today is trying to understand how living cells carry out these reactions. How is it that reactions we find extremely difficult in the laboratory are accomplished efficiently and quickly at room temperature and near neutral pH inside cells? What kinds of organic chemical reactions are used in living cells?

The immediate answer to the first question is that nature has evolved enzymes, efficient catalysts that lower the energy of activation for these reactions, to make them proceed much more quickly. Enzymes became active catalysts through repeated accidental, evolutionary changes over time, so that now almost every step in biosynthesis and metabolism is catalysed by an enzyme. Whatever the apparent "magic" effect of enzymes, the reactions must still obey the laws of thermodynamics; the reaction will still be explicable in terms of electron push and pull, bond

Introduction

and charge movements, proximity effects, pH, concentration of reactants and catalysts, and will follow the rules of organic chemistry. No biochemical reaction can go forward if the overall energetics are unfavourable to formation of the product. To give a fuller explanation, it is helpful to consider the nature and function of enzymes and some of the co-enzymes that often function with them (Chapter 2).

Emil Fischer, at the beginning of the twentieth century, had isolated two enzymes called invertin and emulsin. Invertin hydrolysed only α-D-glucosides (sucrose is an example) while emulsin hydrolysed β-D-glucosides. From this and his knowledge of sugars he correctly deduced that these enzymes were asymmetrically constructed molecules; in modern terms, they are chiral. Biochemical reactions take place on the chiral *surface* of an enzyme, which makes an important distinction from solution chemistry. The first enzyme obtained in a pure, crystalline state was urease, in 1926. It soon became clear that it and other enzymes were proteins.

The energetics and kinetics of enzymic reactions are important to the biochemist, but are not essential to our understanding of *what kinds* of compounds are produced by insects. Nevertheless, living systems are not static. Schoenheimer and Rittenberg (1936) showed that when an animal was allowed to drink heavy water (D_2O or ^2H_2O) for a few days, the fatty acids of that animal became labelled with deuterium. When normal water then replaced heavy water, the deuterium again disappeared from the fatty acids, showing that cells, and whole animals, are in a state of dynamic equilibrium.

Organic chemical reactions that take place in living systems can be divided into five simple types, which are illustrated in Figure 1.3. Enzymes are known that catalyse all these types of reactions, but that simplicity is hidden in the great variety of enzymes and the individual reactions they catalyse.

1.2 PLANT *VERSUS* INSECT BIOSYNTHESIS

Plants have the ability to make a much greater diversity of compounds than animals. Above all, plants can use photosynthesis, splitting water in the light reaction, and in the dark reaction creating carbohydrates from carbon dioxide and hydrogen (Figure 1.4). Plants (and micro-organisms) have exclusive access to the shikimic acid pathway (Chapter 9), the aromatic amino acids and the methylerythritol pathway to terpenes. The case of the polyunsaturated acids (Section 4.1.2) may be unclear, since at least 12 insects have been shown to be able to make linoleic acid, but linolenic acid remains a substance made only by plants. The sterols

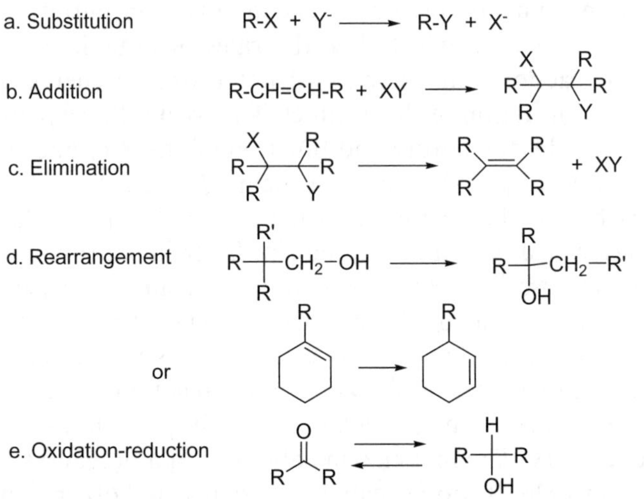

Figure 1.3 The essential types of organic biochemical reactions.

Light reaction:

$$H_2O + NADP^+ \xrightarrow{h\nu} NADPH + H^+ + \tfrac{1}{2}O_2$$

$$ADP + H_2PO_4^- \longrightarrow ATP$$

Dark reaction:

$$6\ CO_2 + 12\ NADPH \longrightarrow C_6H_{12}O_6 + 12\ NADP^+$$

Figure 1.4 An outline of the reactions of photosynthesis in green plants. For NAD^+, NADH, ADP and ATP see Chapter 2.

(Chapter 8) can be made by plants and higher animals, but not by insects. The formation of carotenes (Chapter 8) by insects is doubtful. Compounds such as chlorophyll, starch, cellulose, lignin, tannins, anthocyanins, flavones and triterpenes can be made only by plants.

On the other hand, all the general biosynthetic methods used by insects, discussed in this book, are available to plants. The fatty acids and their derivatives, such as hydrocarbons (Chapter 4), the acetogenins (Chapter 6) and especially the terpenes (Chapters 7 and 8) and aromatic compounds (Chapter 9), can all be made by plants. Acetogenins are not as prominent among plant products as the others, except in the special case of formation of anthocyanins and flavones. Only special areas are left to insects alone. It is surprising to find, when compounds can be made by both plants and insects, that often they have found similar or the same way to biosynthesize them. This is sometimes attributed to parallel evolution, but the future may reveal other ways in which they share processes.

Plants, in their constant struggle with insects, have produced both physical defences (hairs, spines and thick waxy surfaces) and chemical defences (stinging trichomes, alkaloids, toxins and feeding deterrents) against insects, while insects are constantly evolving ways to overcome them. An interesting example of plant counter-attack are the phytoecdysteroids, made by plants, which mimic the natural moulting hormone of insects and may disrupt normal development of the insect feeding on them (Chapter 7). There are plant anti-juvenile hormone compounds too. Nevertheless, there is probably not a single plant species without at least one insect that has found a way to overcome its defences. A subject of growing interest is the production of volatile compounds by plants that have been attacked by insects that signal this to other plants and to parasitoids of the attacking insects (Section 5.9).

1.3 ARTHROPODS AND INSECTS

The arthropods were probably the first organisms to emerge from the sea, and insects were the first invertebrates to fly. Arthropods, nematodes and some smaller, less familiar groups are sometimes known as the Ecdysozoa, or moulting animals. The arthropods can be sub-divided into Crustacea (crabs, lobsters, shrimp, barnacles and woodlice), Chelicerata (spiders, ticks, mites, scorpions and others), Hexapoda or Insecta, and Myriapoda (millipedes, centipedes and other minor groups), although different authors use different classifications (see Kendall, 2009). These classes separated a long time ago, so they have developed quite differently, but there are interesting parallels in development. Spiders and millipedes have sometimes developed chemical defences or communication chemicals similar to those of insects. It is therefore useful always to seek comparisons.

The insects are divided into the Apterygota, primitive wingless insects (springtails and silverfish) that have as yet received little chemical study, and the Pterygota, or winged insects, which form the great majority. The latter in turn are divided into the Exopterygota or Hemimetabola, which hatch from eggs to *nymphs* and closely resemble their final adult form or *imago* (grasshoppers, cockroaches, termites, bugs, stick insects, *etc.*) (Figure 1.4), and Endopterygota or Holometabola, which hatch from eggs to *larvae*, which may have a very different form and habitat from the adult. Larvae proceed to a resting form called the *pupa* as the tissues are completely remodelled and from the pupa emerges the adult form. The Holometabola include beetles, butterflies and moths, flies, fleas, bees, wasps and ants. Almost half of all the insect species are beetles.

The isolation of insect chemicals began slowly. Kermesic acid or venetian red, a pigment from beetles (Chapter 9), has been known and used from ancient times. Wray, in 1670, reported formic acid by distillation of formicine ants. It was not until the 1930s that it began to be recognized that some Lepidoptera males were chemically attracted to females, and only in 1959 was the first sexual attractant (bombykol, from the silk moth *Bombyx mori*) isolated and identified (Butenandt *et al.*, 1959). From that time onward, aided by development of chromatographic and sensitive mass spectrometric techniques, the study of insect natural products has grown to be a major discipline of science.

1.4 CHEMICAL COMMUNICATION

Communication by chemicals must be a sense developed early in evolution, because it is found from bacteria (where it is referred to as quorum sensing) (Waters and Bassler, 2005) to higher animals. Humans are such unsuitable experimental animals, it is difficult to know if they use chemical communication. Chemicals used for communication are called **semiochemicals**, those used between different species, *e.g.* between plant and animal, or animal and animal, are called collectively **allelochemicals**, and these are further sub-divided in a system depending upon whether they benefit the sender (**allomones**), the receiver (**kairomones**) or both (**synomones**), and there are other categories. Those used between individuals of one species are called **pheromones**. Many of the compounds from insects considered in this book are pheromones (Greek, *phero* = carry or convey), defensive or offensive substances (allomones, Greek *allos* = other) or hormones (Greek, *hormao* = excite or impel). Hormones are used for chemical communication *within* the individual. The types of compounds used as pheromones and allomones are highly variable and they appear to have evolved many times, while the hormones are relatively conserved, and the same hormones serve many or all insects and can be common to many invertebrate classes.

Pheromones are the group of insect compounds that have found greatest application in agriculture and forestry. For example, a large number of lepidopteran species are important agricultural pests. Their females use sexual pheromones to attract males for mating. The pheromones can be used to aid control of pests in one of several ways. Traps baited with synthetic pheromone can be used to detect the arrival of a pest, or to assess the build-up of the species in a crop, so that

Introduction

insecticides can be used more sparingly and at the correct time. In a few cases, trapping alone can be effective in removing enough of the males to prevent mating and so control the pest. Sometimes the pheromone is scattered throughout the crop so that males are unable to locate females (mating disruption). Sometimes a wrong isomer can completely inhibit the response to a pheromone, so a lure containing some of the inhibitor can disrupt mating. Both Coleoptera and Lepidoptera can be pests in forestry and there, too, pheromone traps have been found effective. Pests in stored products are particularly suitable for pheromone trapping, where use of insecticides is undesirable. Sales of pheromones worldwide still represent only a few percent of the total value of sales of insecticides, which are of the order of billions of US dollars, but pheromones sales are steadily growing.

Insect defensive compounds are usually effective at short distance and their toxicity or repellency is not sufficient for them to have found any industrial application. Venoms can be powerful, but usually require injection. Of the insect hormones, ecdysteroids (Chapter 7) have not yet found practical application, but there are several examples, in special circumstances, of very effective use of juvenile hormone mimics in pest control. The study of semiochemicals is therefore an important part of insect biosynthesis.

1.5 THE STRUCTURES OF NATURAL PRODUCTS

Knowing the probable biosynthetic origin of a new compound can be a great help to decide what is its likely structure, what is an improbable structure and what may be its structural formula. A candidate structure may be difficult to eliminate completely, because nature is full of

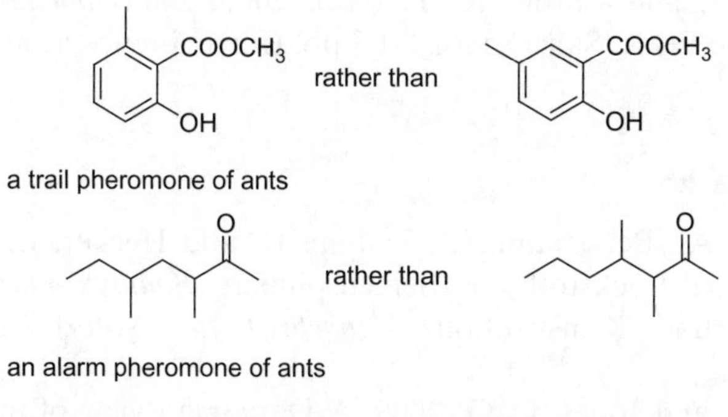

Figure 1.5 On the left are two structures with simple biosynthetic origins, while on the right are two structures for which a simple biosynthetic route cannot be given.

surprises. This book should help the reader to decide which among some alternative structural possibilities is the more likely. In Figure 1.5, the compounds on the left are insect pheromones where the likely biosynthetic origin can be easily deduced from the structures, while it is very difficult to see how those on the right can be made by known routes. One of those on the right has been found in at least one insect, the other has not. Comparison of a sample of the synthesized compound with the natural compound in physical and biological properties is an important final stage in identification.

BACKGROUND READING

Harborne, J. B. 1994. *Introduction to Ecological Biochemistry*, 4th edition, Academic Press, London (Insect feeding preferences, Chapter 5, pp. 128–161; Animal pheromone and defence substances, Chapter 8, pp. 211–242).

Howse, P., Stevens, I. and Jones, O. 1998. *Insect Pheromones and Their Use in Pest Management*, Chapman and Hall, London, 369 pp. (Insect semiochemicals and communication, Chapter 1, pp. 3–37; The role of pheromones in insect behaviour and ecology, Chapter 2, pp. 38–68; Factors controlling responses of insects to pheromones, Chapter 3, pp. 69–104).

Mann, J. 1994. *Chemical Aspects of Biosynthesis*, Oxford University Press, Oxford (Secondary metabolism and ecology, Chapter 2, pp. 303–328).

Metabolism, Wikipedia; en.wikipedia.org/wiki/Metabolism.

Natural Products, Wikipedia; en.wikipedia.org/wiki/Natural_products.

Torssell, K. B. G. 1997. *Natural Product Chemistry*, Swedish Pharmaceutical Society, Stockholm (Introduction and general considerations, Chapter 1, pp. 12–41; Chemical ecology, Chapter 2, pp. 42–79).

Resh, V. H. and Cardé, R. T. (ed.), 2003. *Encyclopedia of Insects*, Academic Press, San Diego, 1266 pp. (for reference at any point).

REFERENCES

Butenandt, A., Beckmann, R., Stamm, D. and Hecker, E. 1959. Über den Sexual-Lockstoff des Seidenspinners *Bombyx mori*. Reindarstellung und Konstitution. *Zeitschrift für Naturforschung*, **14b**, 283–284.

Firn, R. D. and Jones, C. G. 2009. A Darwinian view of metabolism – molecular properties determine fitness. *Journal of Experimental Botany*, **60**, 719–726.

Grimaldi, D. and Engel, M. S. 2005. *Evolution of the Insects*, Cambridge University Press, Cambridge, 755 pp.
Hanson, J. R. 2003. *Natural Products: The Secondary Metabolites*, Royal Society of Chemistry, Cambridge, Chapter 1, pp. 1–34.
Kendall, D. A. 2009. www.kendall-bioresearch.co.uk/class.htm
Meinwald J. and Eisner, T. 1995. The chemistry of phyletic dominance. *Proceedings of the National Academy of Science, USA*, **92**, 14–18.
Schoenheimer, R. and Rittenberg, D. 1936. Deuterium as an indicator in the study of intermediary metabolism. I. *Journal of Biological Chemistry*, **111**, 163–168.
Waters, C. M. and Bassler, B. L. 2005. Quorum sensing: cell-to-cell communication in bacteria. *Annual Review of Cell and Developmental Biology*, **21**, 319–346.

CHAPTER 2
Enzymes and Co-enzymes

2.1 THE CHEMICAL FUNCTION OF ENZYMES

To make any compound, an organism requires an energetically favourable chemical reaction, a starting material, usually a catalyst to help the reaction proceed in the required direction and at a useful rate, and possibly an energy source to make the reaction go forward. The catalyst is an enzyme. The **substrate**, the substance that is being altered, becomes attached to the enzyme in some way, and at a particular part, called the **active site**, where the reaction takes place. An enzyme can have one or more active sites. Many enzymes require a cofactor to make the reaction go forward. If that cofactor is attached firmly to the enzyme, it is called a prosthetic group, as is the zinc atom in carboxypeptidase (Section 2.1.2) or the haem group in cytochromes (Section 2.1.3). If the cofactor is loosely held, it is called a co-enzyme. The prosthetic group or co-enzyme, if one is involved, is held close to the substrate at the active site. "An enzyme first binds its substrate in a particular orientation by using a variety of weak binding forces (hydrogen bonding, electrostatic attraction, dipole-dipole interaction, hydrophobic attraction, and so on), and then uses a variety of strategically placed functional groups and controlled conformational changes to induce reaction between them" (Cornforth, 1984).

The essential effect of the enzyme is to lower the activation energy of the reaction by altering the transition state so that it can proceed with less energy requirement and therefore faster (Bugg, 2004). Enzymes

---

Biosynthesis in Insects, Advanced Edition
By E. David Morgan
© E. David Morgan 2010
Published by the Royal Society of Chemistry, www.rsc.org

typically accelerate a reaction up to 10^6 times. The rate is only limited by the rate of diffusion of the substrate to the catalytic site. In a reaction in which A and B can be inter-converted, but at equilibrium the concentration of A is much greater, the reaction can be made to produce B by rapid removal of B through its conversion to C. In this way many reactions with unfavourable equilibria can be used in biosynthesis (see Section 7.2.1). Enzymes are selective and must recognize specific elements of structure. They can have structural specificity, that is, they only operate with molecules of a certain type or shape. They can have regiospecificity, that is, they affect only certain parts of the molecules, for example, catalyse esterification of hydroxyls in one part of a sterol but not another. They can have stereospecificity, for example, catalyse the oxidation of (*R*)-alcohols but not (*S*)-alcohols, or produce (*Z*)-alkenes but not (*E*)-alkenes.

The control of biosynthesis is achieved by the control of enzyme activity. This control requires induction or activation of an enzyme, and repression or shut down of activity. There are three sorts of control recognized: allosteric regulation, covalent modification and synthesis and degradation of the enzyme. Allosteric enzymes have sites additional to the active site, allosteric sites (*allo* = other), which can accommodate the final product of the synthesis. As the concentration of this product rises, it becomes bound to this allosteric site and shuts down the activity of the enzyme. Covalent modification is a means whereby the enzyme can be locked into a stable state of either activity or inactivity. Sometimes adding a phosphate group to the enzyme can put it into this on-or-off state. Thirdly, the induction of synthesis of the enzyme can be caused by activation of the genes that code for the enzyme, causing the formation of messenger RNA (mRNA) and synthesis of more molecules of the enzyme. Often the accumulation of the end product of the synthesis can repress gene activation, so no mRNA is produced and no more enzyme molecules synthesized.

Three examples to illustrate the way enzymes function follow. The first example, lysozyme, has no prosthetic group and does not require a co-enzyme. The second illustrates an enzyme with a metal ion as prosthetic group. The third is an example from the all-important cytochrome enzymes, encountered frequently in reactions of many insect compounds, including the formation of the juvenile and moulting hormones.

The three-dimensional shapes of enzymes can be observed with the computer programme RasMol (www.openrasmol.org). Proteins can be rotated and observed from all angles. The Protein Data Bank (PDB) provides nearly 60,000 protein structures (www.rcsb.org/pdb).

2.1.1 Lysozyme

Lysozymes are a group of enzymes found in many places, such as tears, saliva and the white of an egg. They break down bacterial cell walls, hence the name: an enzyme that causes bacterial lysis. They provide a simple example of how an enzyme facilitates a chemical reaction. Lysozymes function by breaking down cell wall polymers called peptidoglycans by breaking the bond between *N*-acetylmuramic acid and N-acetylglucosamine (Figure 2.1). Lysozyme from egg whites has a relatively small molecule for a protein, with 129 amino acid residues linked in a single protein chain, and molecular mass of 13,930. It was the first enzyme to have its total structure determined by X-ray analysis (Blake *et al.*, 1965), and to have its active site identified. The protein chain of lysozome is twisted and folded into a shape like a ball with a cleft down one side (rather like an apple with a large bite out of it, Figure 2.1). The polysaccharide network on the surface of the bacteria fits into the cleft. The point in the structure of the polysaccharide

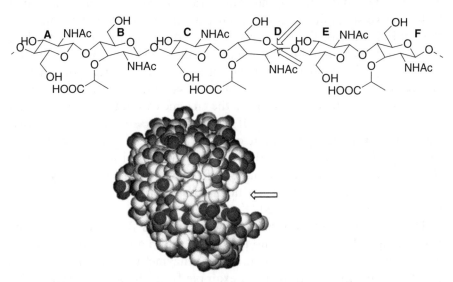

Figure 2.1 The polysaccharide molecule found in the walls of certain bacterial cells is the substrate broken by the lysozyme molecule. The polysaccharide consists of alternating residues of two kinds of amino sugar: N-acetylglucosamine and N-acetylmuramic acid. In the portion of polysaccharide chain shown here A, C and E are N-acetylglucosamine residues; B, D and F are N-acetylmuramic acid residues. The shape of ring D is a distorted boat form when adsorbed on the enzyme. The position attacked is indicated by the arrows. Below is a molecular model of the lysozyme molecule showing the cleft containing the active site.

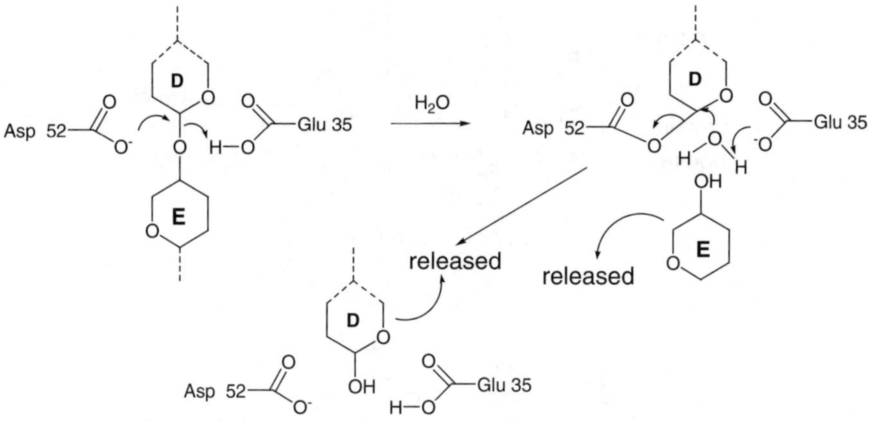

Enzymes and Co-enzymes 15

Figure 2.2 A detailed portion of the active site of lysozyme, showing how the aspartic acid number 52 and glutamic acid number 35 of lysozyme work together to break the polysaccharide chain.

(the substrate) that is attacked is shown in Figure 2.1 and the mechanism of the enzyme reaction is shown in Figure 2.2. Six rings of the polysaccharide fit into the cleft of the lysozyme molecule, and are held firmly in position by hydrogen bonds and other interactions. Ring D is held in a flattened conformation, so the bond to ring E is strained and prepared for reaction. Adjacent to it are two carboxylic acid groups on amino acids which are part of the enzyme; that of aspartic acid number 52 is in polar surroundings and is in the ionized form; that of glutamic acid number 35 is in non-polar surroundings and is in the unionized form (Vocadlo *et al.*, 2001). These two groups and a water molecule take part in the reaction as shown in Figure 2.2.

2.1.2 Metallo-enzymes

In the example of lysozyme, the catalytic effect is entirely due to the C-, H-, O- and N-containing groups of the protein. In many enzymes there is a prosthetic group that often contains a metal ion bound to the protein. For example, magnesium is found in many enzymes that manipulate phosphate and diphosphate, because magnesium chelates well with diphosphate. Other metals used in enzymes are Zn, Fe, Cu, Mn, Co and Mo. Some 400 enzymes are known that contain zinc atoms.

Carboxypeptidase A is a metallo-enzyme from the pancreas of mammals (insects have a similar enzyme) that contains a zinc atom at its active site, and one that has been well studied (Rees and Lipscomb, 1981). It hydrolyses amino acids from the carboxylate end of proteins,

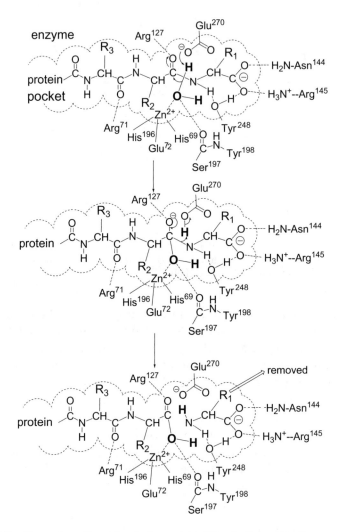

Figure 2.3 The carboxylic acid end of a protein sitting in the active site, a pocket in the enzyme carboxypeptidase A, showing how it is held in place by bonding to the Zn^{++} atom and various amino acids. The probable mechanism, based on the hydrolysis of a known peptide, is shown, with the water molecule used for the hydrolysis shown in bold type. Once the amino acid on the right is cleaved, it can diffuse away and the protein moves up into the pocket for the next amino acid to be cleaved in the same way.

one by one (Figure 2.3). The zinc atom holds together the three-dimensional shape of the active site and co-ordinates a water molecule that is used in hydrolysing the peptide bond. These examples help to show how various changes in the electronic environment and shape of the substrate caused by the closeness of certain parts of the enzyme reduce

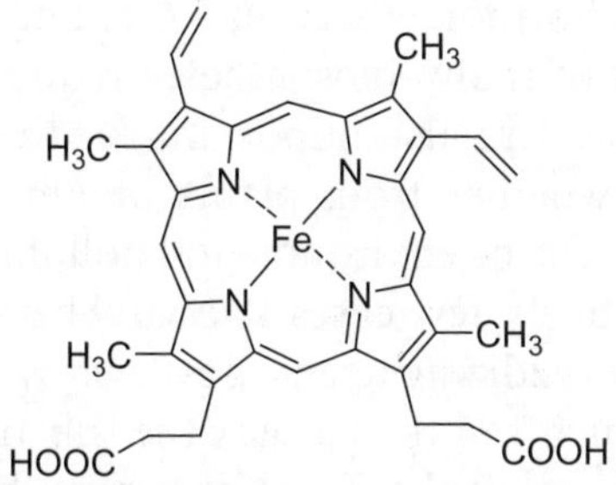

Figure 2.4 The structure of haem. The tetrapyrrole without the iron atom is known as protoporphyrin IX (see also Chapter 8).

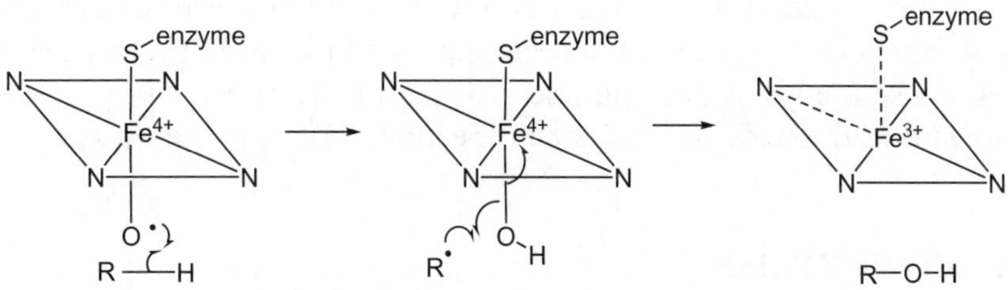

The overall reaction is:
$$R\text{-}H + O_2 + NAD(P)H + H^+ \xrightarrow{P450} R\text{-}OH + H_2O + NAD(P)^+$$

Figure 2.5 Schematic drawing of the centre of a haem group of a cytochrome enzyme catalysing an oxidation of a hydrocarbon. This is a free radical reaction, as indicated by the arrows with a single barb.

the activation energy necessary for the reaction to proceed (Casiday and Frey, 1998).

2.1.3 Cytochromes

There is in all living cells a family of membrane-bound enzymes known as cytochromes, so called because they contain a haem group (Figure 2.4) attached covalently, and therefore have visible spectra, *i.e.* are coloured. Haem represents another kind of prosthetic group. Cytochromes catalyse oxidation-reduction reactions through the changing valency of the central iron atom. A distinct group of them are the cytochrome P450 enzymes, or mixed function oxidases. They have a visible absorption band near 450 nm when the haem iron is in the reduced state and complexed to carbon monoxide. For insects, they are important in the oxidizing of alkanes to alcohols, alkenes to epoxides, in introducing an OH group into aromatic rings, for carbon-carbon bond cleavage and for the formation of juvenile hormone (Section 7.6.1) and ecdysteroids (Section 8.5). The later stages of many biosyntheses require hydroxylation by a cytochrome P450. They are also important for the detoxifying of many ingested substances, whether from plants or animals, or are synthetic substances, like pesticides or environmental pollutants (Feyereisen, 1999; 2005). There are relatively few cases known where oxygen acts directly with an enzyme as an oxidizing agent, however cytochromes do activate molecular oxygen, which can then attack the substrate of whatever kind, as in Figure 2.5. Notice that the Fe^{3+} atom must be re-oxidized to Fe^{4+} before the enzyme can be used again. It accepts another molecule of O_2, splits off OH^- and is restored to Fe^{4+}-$O^{\cdot}$ for re-use. Other substances that contain haem groups are haemoglobin and catalase, but see also porphyrins (Section 9.6.5). There is now a large number of cytochromes P450 for which the genes have been sequenced (for example, see Section 3.5). These are given code numbers, *e.g. CYP9T2* is a P450- producing ipsdienol in the bark beetle *Ips pini* (Section 7.3).

2.2 CO-ENZYMES

Co-enzymes are relatively small organic molecules, which are required with the enzyme to perform many chemical reactions or to carry a substrate between enzymes. The co-enzyme may itself be reversibly changed during the reaction, and must be continually recycled. It is then a chemical reagent in the enzyme-catalysed reaction. Some co-enzymes are covalently attached to the substrate; in other cases, the co-enzyme and the substrate are adsorbed together onto the active site of the enzyme. Wikipedia contains useful entries on some, but not all, co-enzymes (en.wikipedia.org/wiki).

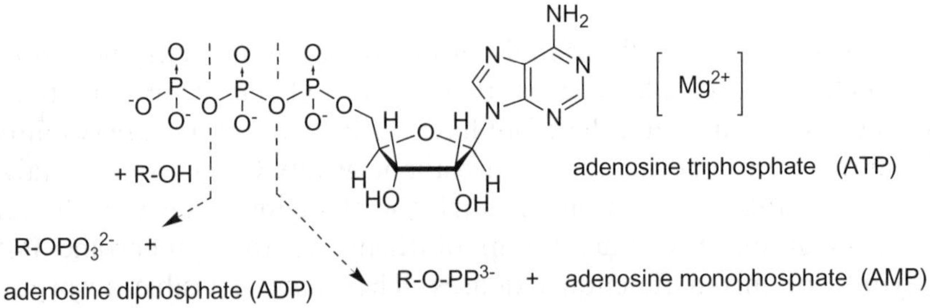

Figure 2.6 Adenosine triphosphate (ATP) and its cleavages to adenosine diphosphate and adenosine monophosphate. In cells ATP is normally complexed with magnesium ions which partly neutralize the large negative charge of the phosphates.

2.2.1 Adenosine Triphosphate

Adenosine triphosphate (ATP) is both a co-enzyme and a ubiquitous source of energy in cells (Figure 2.6). As a co-enzyme it transfers a phosphate or diphosphate group to many substrates. Enzymes catalysing this reaction are known as kinases or phosphotransferases. The very first step in using an acetate ion requires an equivalent of ATP (see co-enzyme A).

As an energy source or energy carrier, ATP takes advantage of the fact that thermodynamically it is very unstable, but kinetically it is very stable. At standard temperature, and 1 M concentrations, the hydrolysis of ATP to give adenosine diphosphate (ADP) and inorganic phosphate (usually abbreviated to P_i) yields (ΔG^0) $-30.5\,kJ\,mol^{-1}$. Under typical cellular conditions, ΔG is approximately $-57\,kJ\,mol^{-1}$.

Similarly, hydrolysis of ATP to adenosine monophosphate (AMP) and inorganic diphosphate (PP_i) gives about $45.6\,kJ\,mol^{-1}$. With an enzyme, the kinetic barrier is lowered and hydrolysis of ATP occurs quickly and the energy released can be used to overcome the activation energy barrier of some other enzyme reaction. ATP is associated with Mg^{2+}, which partly neutralizes the high negative charge.

About 38 molecules of ATP are synthesized for each molecule of glucose oxidized to CO_2 and H_2O, or 130 ATP molecules for each molecule of palmitic acid oxidized. To give some indication of the amount of ATP used, a much-quoted example gives the amount per day turned over in a human body. Assuming an adult person of 70 kg weight consumes 11,700 kJ (2800 kcal) per day, they will produce 117 moles or 65 kg of ATP per day, so almost one's body weight. Since the body contains only about 50 g (0.1 mole) of ATP-ADP mixture, it means that each molecule of ATP must be recycled about 1300 times a day. Other co-enzymes similarly must be continually recycled.

Cells requiring a reserve of energy, such as muscle cells, have a stored supply of high-energy phosphate that can be rapidly converted to ATP. In insects this is arginine phosphate (Figure 2.7), with an enzyme rapidly

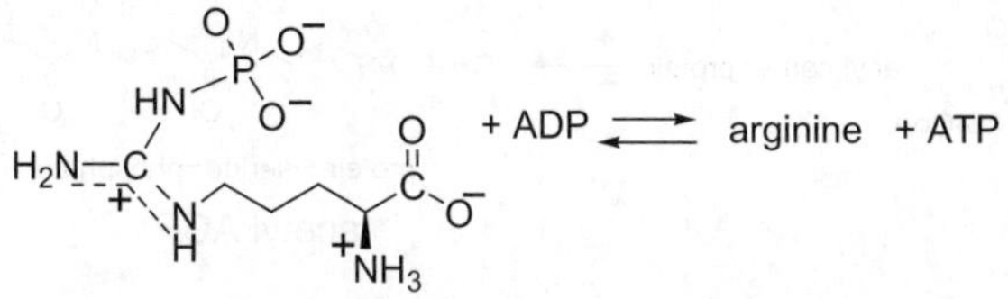

Figure 2.7 Arginine phosphate, the reserve energy supply of insects and some other animals, rapidly convertible into ATP as required.

working in either direction, reforming arginine phosphate in a quiescent period.

The production of bioluminescence by fireflies is an example of a reaction requiring ATP (Section 10.4). Dried firefly tails are commonly used to photometrically assay very small amounts of ATP.

2.2.2 Co-enzyme A

Co-enzyme A can be described as a "handle" for carboxylic acid groups (Figure 2.8). It picks up and transports acetyl groups, or acyl groups in general. It is particularly important in the degradation of fatty acids, in the oxidation of pyruvic acid and in the early stages of terpene synthesis (Chapter 7). The reactive part of the molecule is the thiol group, so the name is usually abbreviated to CoA-SH. The thiol gives with, *e.g.* acetic acid, a thioester, co-enzyme A thioacetate, or briefly CH_3COS-CoA or AcSCoA. Note that acetic acid must first be phosphorylated with ATP to render it sufficiently reactive to condense with CoA-SH. Aldehydes and ketones have relatively reactive carbonyl groups, with reactivity slowed chiefly by bulky R groups. In esters, reactivity at $C=O$ carbon is decreased through orbital overlap from the OR group. Absence of

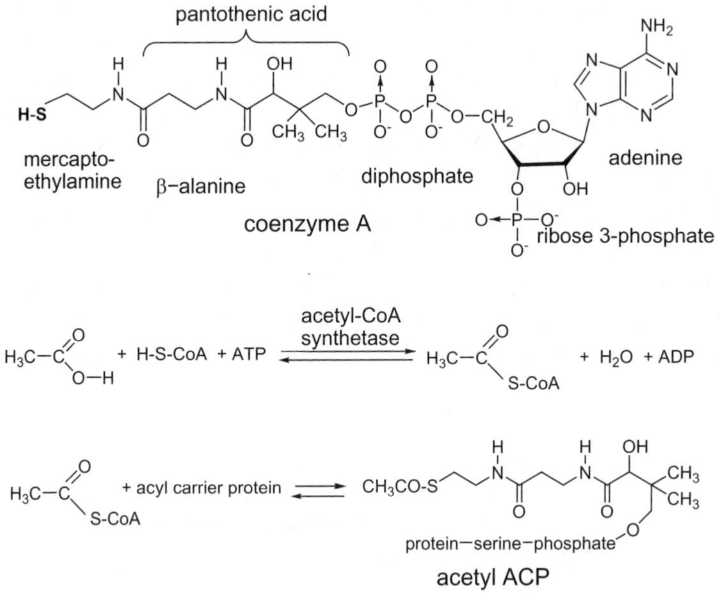

Figure 2.8 Co-enzyme A, showing its constituent parts, and the reaction between co-enzyme A and an acid, assisted by ATP. For the synthesis of fatty acids, co-enzyme A is replaced by acyl carrier protein (ACP).

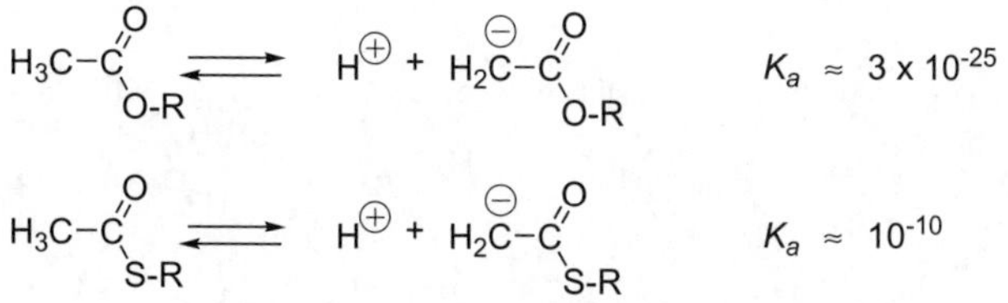

Enzymes and Co-enzymes

Figure 2.9 There is a large increase in the ease of dissociation of an α-proton in a thioester compared with a normal ester, which is important for biosynthetic condensation reactions.

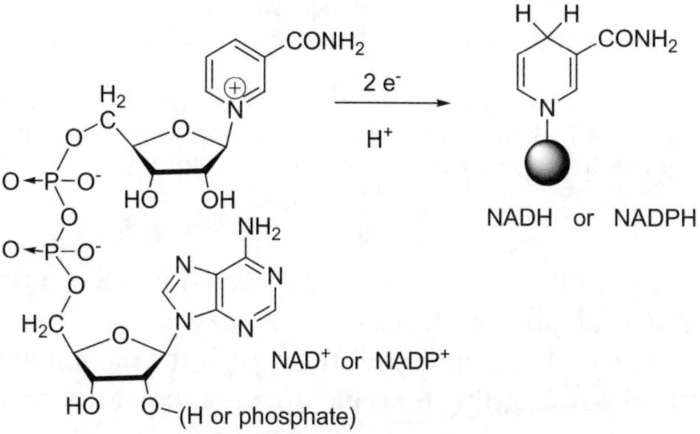

Figure 2.10 Nicotinamide adenine dinucleotide NAD^+ (or $NADP^+$ with an extra phosphate on C-2 of ribose) oxidized and reduced forms. The sphere in the right-hand structure represents the remainder of the molecule.

orbital overlap from C-S with the larger sulfur atom means that reactivity at the thioester C=O is more like that of a ketone. There is consequently a large increase in the acidity constant for the removal of an α-proton from a thioacetate compared with an acetate (Figure 2.9). The importance of thioesters is evident when considering the biosynthesis of fatty acids (Chapter 4) and terpenes (Chapter 7).

For the special case of synthesis of fatty acids (Section 4.1.1) co-enzyme A is exchanged through an equilibrium reaction for acyl carrier protein (ACP) (Figure 2.8), which has the same phosphopantetheinyl group as co-enzyme A.

2.2.3 Nicotinamide Adenine Dinucleotide

Nicotinamide adenine dinucleotide and nicotinamide adenine dinucleotide phosphate (NAD^+ and $NADP^+$, respectively) are co-enzymes for

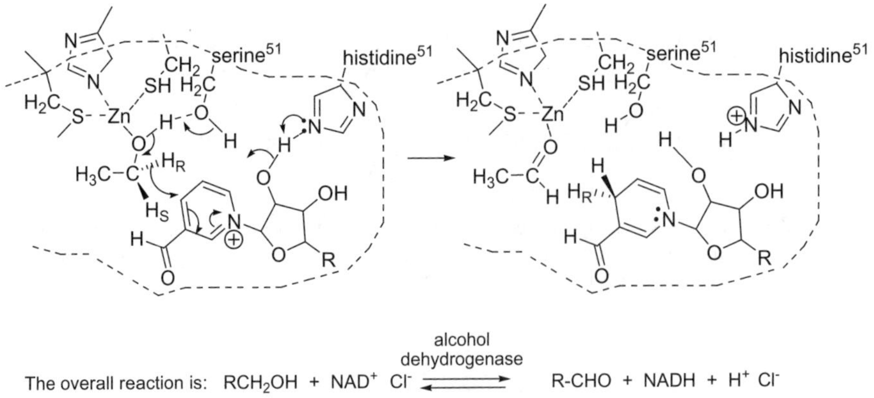

The overall reaction is: RCH$_2$OH + NAD$^+$ Cl$^-$ $\xrightleftharpoons{\text{alcohol dehydrogenase}}$ R-CHO + NADH + H$^+$ Cl$^-$

Figure 2.11 The active site of liver alcohol dehydrogenase, showing how the oxidation of ethanol and the reduction of NAD$^+$ are co-ordinated, and the *pro-R* hydrogen of ethanol is transferred to NAD$^+$.

many enzymes catalysing oxidations and reductions. Since oxidation-reductions remove or add electrons, the co-enzyme can be considered a carrier of electrons. The essential reactive part is the nicotinamide portion (Figure 2.10). The rest of the molecule orients it correctly on the enzyme active site.

Nature provides mechanisms to separate metabolic or catabolic processes from synthesis or anabolic processes. Here NAD$^+$ is used in catabolic reactions, NADP$^+$ is used in anabolic processes. NAD$^+$ and NADP$^+$ are oxidizing agents and NADH and NADPH are reducing agents (Pollak *et al.*, 2007).

The reactivity of liver alcohol dehydrogenase has been studied in detail. The reaction of an ethanol molecule with NAD$^+$ on the active site of this enzyme is illustrated in Figure 2.11. The acidity of the OH group of ethanol is increased considerably by co-ordination to zinc, and the negative charge on zinc stabilizes the developing negative charge on oxygen as the O-H bond is broken. The way the NAD$^+$ is held above the ethanol molecule distinguishes between the pro-chiral hydrogen atoms (see Section 2.4.1). The *pro-R* hydrogen atom of ethanol becomes attached to the lower side (as drawn) of the reduced nicotinamide molecule. When NADH is used for reduction, the same hydrogen, from the lower side of the molecule is transferred to the back of the carbonyl group (as drawn in Figure 2.11). This has been established by replacing either the *pro-S* or the *pro-R* hydrogen of ethanol by deuterium, and seeing whether the deuterium is retained by acetaldehyde or is taken up by NADH.

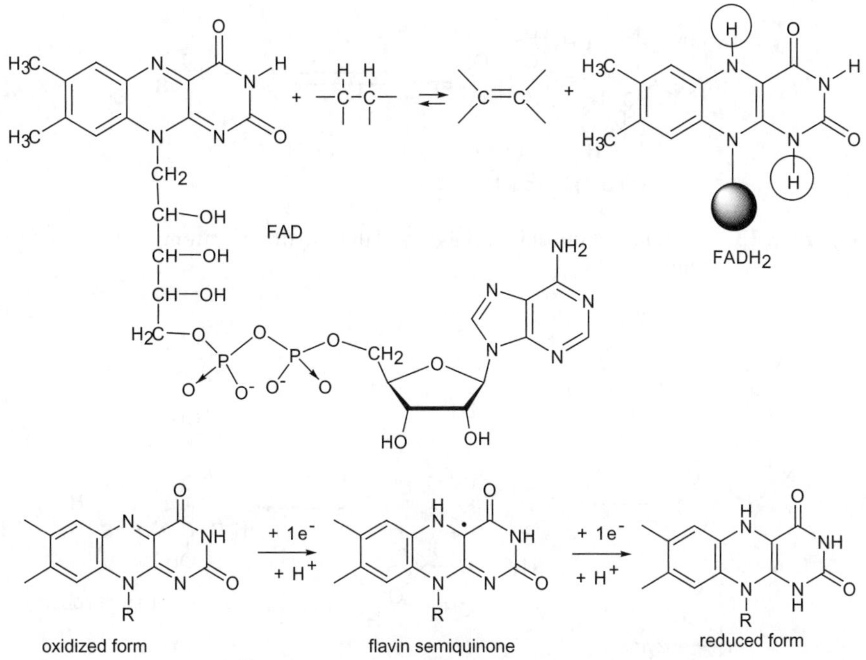

Figure 2.12 Flavin adenine dinucleotide FAD, oxidized and reduced forms. The two circled hydrogen atoms in FADH$_2$ are those removed from carbon. Below are the three forms of flavin, including the intermediate free radical semiquinone.

2.2.4 Flavin Adenine Dinucleotide

Flavin adenine dinucleotide (FAD) is also an oxidation-reduction reagent, but with two reduced forms, FADH and FADH$_2$, and with a more diverse function with different enzymes. The interest here is its use in reduction of C=C bonds and removing 2H from adjacent carbon atoms. In Figure 2.12, the two hydrogen atoms added to the co-enzyme and removed from the substrate are circled. FAD also oxidizes oxy-acids, amines and some amino acids. FADH can react directly with molecular oxygen to form an intermediate free radical (Figure 2.12).

2.2.5 Thiamine Diphosphate

β-Keto-acids are easily decarboxylated, but α-keto-acids, like pyruvic acid, an important intermediate in the breakdown of glucose to acetic acid, are not. The latter decarboxylation is achieved through the co-enzyme

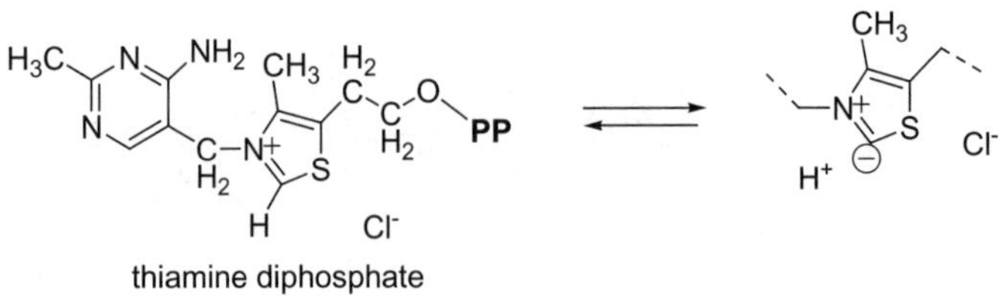

Figure 2.13 Thiamine diphosphate has one labile hydrogen atom at the reactive centre.

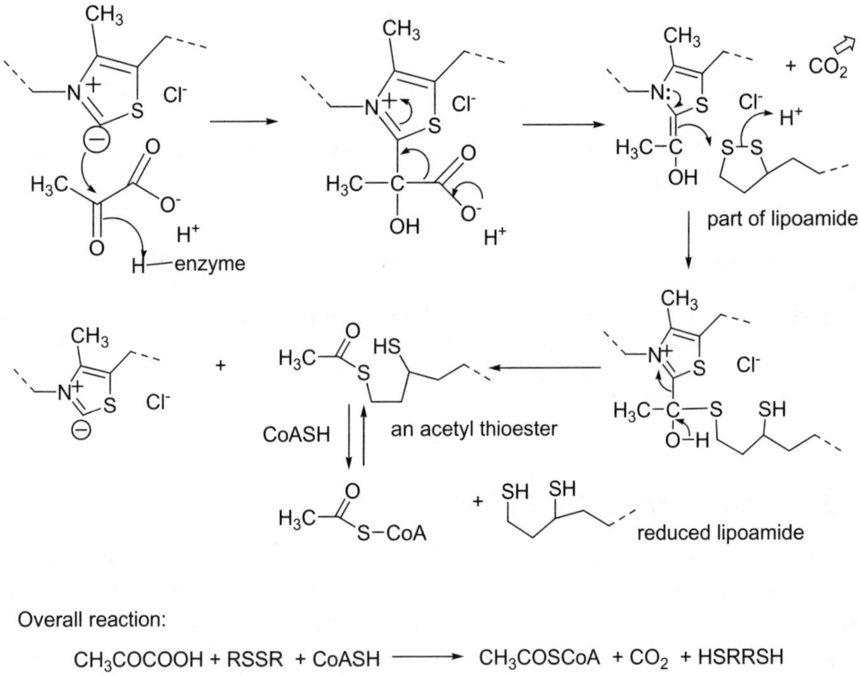

Overall reaction:

$$CH_3COCOOH + RSSR + CoASH \longrightarrow CH_3COSCoA + CO_2 + HSRRSH$$

Figure 2.14 Thiamine diphosphate reacts with pyruvic acid, which loses CO_2, then the remaining two-carbon fragment reacts with the dithiolane ring of lipoamide to give a thioacetate ester, cleaved to acetyl CoA and reduced lipoamide.

thiamine diphosphate (Figure 2.13). The hydrogen atom next to nitrogen in the thiazole ring of thiamine diphosphate is acidic and easily removed. Shaking thiamine diphosphate with D_2O gives rapid exchange of this H for D. Thiamine diphosphate converts pyruvic acid, from glucose metabolism, into the equivalent of a β-keto-acid (with $C=N^+$ instead of $C=O$), which then easily loses CO_2 by decarboxylation.

Higher organisms oxidize pyruvic acid to acetic acid (as AcSCoA) and CO_2, with lipoamide acting as another co-enzyme. Approximately half the CO_2 exhaled by animals is produced by decarboxylation with thiamine. The lipoamide inserts itself after the decarboxylation step (Figure 2.14). The AcSCoA produced in the last step either is ultimately oxidized to CO_2 (through the citric acid or Krebs cycle) or serves as the vital starting material for the biosynthesis of fatty acids, acetogenins and terpenes (Figure 1.2). The reduced lipoamide is regenerated with FAD and, in turn, the $FADH_2$ is regenerated with NAD^+ and the NADH produced is oxidized in the electron transport chain, ultimately producing more energy.

2.2.6 Pyridoxal Phosphate

Pyridoxal phosphate is a multi-purpose reagent that can accomplish reactions catalysed by several enzymes. Because it remains attached to the enzyme, it is better regarded as a prosthetic group. Its reactivity depends upon the positively charged nitrogen atom of the pyridinium ring acting as an electron sink. Chiefly, it is responsible for transamination, that is, the conversion of α-keto-acids to α-amino acids and the reverse. This is achieved by a series of reactions while the pyridoxal phosphate is bound to a transaminase enzyme by several interactions. For de-amination, pyridoxal forms an imine with the amino acid, and that is converted to an imine of pyridoxamine and a keto-acid. Hydrolysis of the imine gives an α-keto-acid and pyridoxamine, which must be converted back to pyridoxal for re-use (Figure 2.15). In the metabolism of proteins, the individual amino acids are de-aminated in this way. The α-keto-acid is passed to the citric acid cycle (Figure 1.2) for energy production, or is sometimes used in biosynthesis. The ammonia ultimately produced from deamination is potentially toxic; it can be excreted as a dilute solution directly into the water by fish, and some other aquatic animals. In higher animals it is converted into harmless products. In insects, uric acid is the important one. In mammals the amine group is transferred to the urea cycle and is excreted as urea.

Pyridoxal phosphate is also used for decarboxylation of amino acids to produce biologically active amines (Figure 2.16) (see also Chapter 10), and for the inter-conversion of the amino acids serine and glycine *via* an aldol reaction (Section 2.2.7). Note that with pyridoxal, the bond being broken is the one perpendicular to the pyridine ring. The enzyme catalysing the particular reaction ensures the amino acid is held in the correct configuration (Figure 2.16).

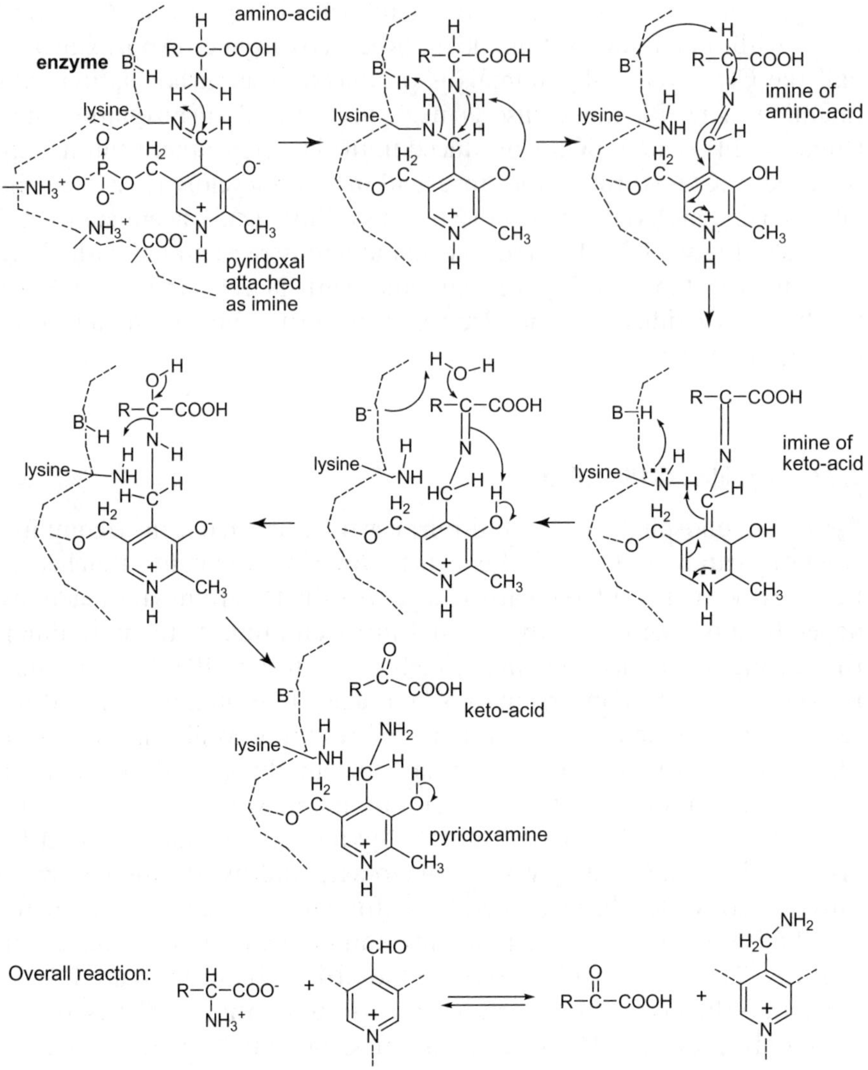

Figure 2.15 The removal of ammonia from an amino acid by pyridoxal phosphate (PLP) and a transaminase enzyme. The PLP is held tightly to the enzyme by a lysine and ionic bonding of the phosphate group. **B** indicates some general base.

2.2.7 Tetrahydrofolic Acid

Two co-enzymes are used in the manipulation of single methyl groups: tetrahydrofolic acid and *S*-adenosyl methionine. The ultimate sources of the one-carbon fragment are the amino acids serine and glycine (sometimes methionine and histidine also) aided by the co-enzyme

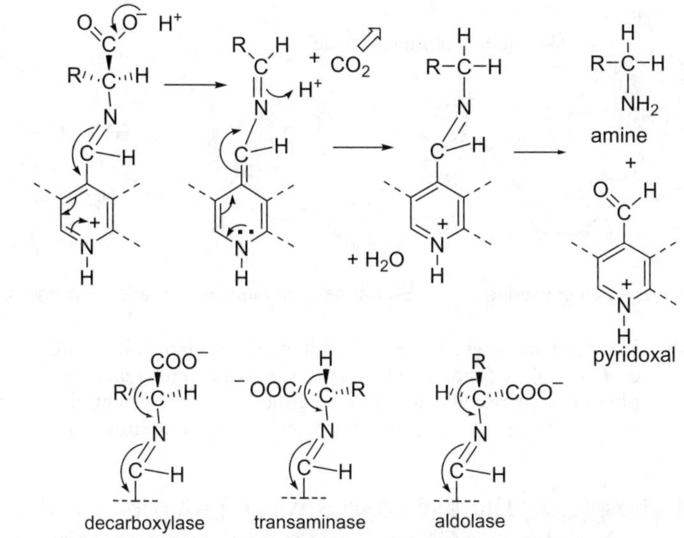

Enzymes and Co-enzymes 27

Figure 2.16 The sequence of steps by which an amino acid attached to pyridoxal and a decarboxylating enzyme is converted to an amine and CO_2. Enzymes using pyridoxal as cofactor hold the amino acid in different configurations to break different bonds. The bond broken is always perpendicular to the pyridine ring.

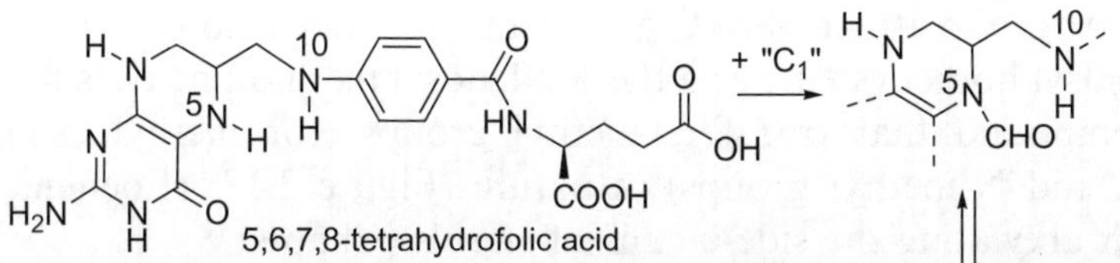

Figure 2.17 The reduction of a one-carbon fragment attached to tetrahydrofolic acid to give the source of formyl, methyne, methylene and methyl groups. This sequence of reactions can run in either direction, to give a methyl group from a formyl group, or the reverse as required.

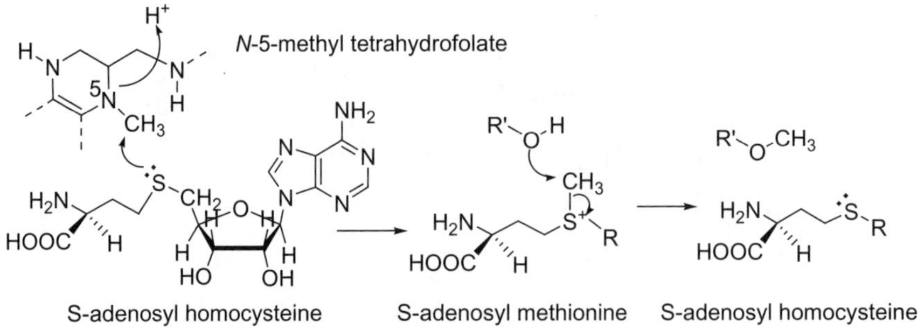

Figure 2.18 Reaction between 5-methyltetrahydrofolic acid and *S*-adenosyl homocysteine gives *S*-adenosyl methionine, which can react with an alcohol, phenol or carboxylic acid to give a methyl ether, a phenolic methyl ether or a methyl ester, respectively, regenerating *S*-adenosyl homocysteine.

pyridoxal phosphate. The hydroxymethylene group of serine is transferred to tetrahydrofolic acid as a methylene group, catalysed by the enzyme serine hydroxymethyltransferase. Once the single carbon fragment becomes attached to N-5 of tetrahydrofolic acid it can be reduced to methyl with NADH (Figure 2.17), or converted to a methyne attached to N-5 and N-10, or hydrated to a formyl group.

2.2.8 *S*-Adenosyl Methionine

A methyl group attached to C-5 of tetrahydrofolic acid is transferred to *S*-adenosyl homocysteine and the *S*-adenosyl methionine thus formed is the compound that transfers methyl groups (for methyl esters and ethers, and N-methyl groups) in nature (Figure 2.18) (Loenen, 2006), and for alkylating the side-chains of sterols (Chapter 8).

2.2.9 Vitamins

There are other known co-enzymes (*e.g.* cobalamine, ascorbic acid), but those mentioned are the ones most involved in biosynthesis. It should be noted that higher animals have lost the ability to synthesize some of these co-enzymes or parts of them. By definition, a vitamin is an essential substance that the body cannot make for itself and must acquire through its food. Humans cannot make pantothenic acid (vitamin B_5) needed for co-enzyme A, nicotinamide (vitamin B_3) for NAD^+, riboflavin (vitamin B_2) for FAD, thiamine (vitamin B_1), folic acid (vitamin B_9 or M) or pyridoxal (vitamin B_6). Insects also require several of the B-group vitamins for larval development, sometimes in very low amounts, and these can be passed down through eggs (Dadd, 1985). Other compounds, *e.g.*

p-aminobenzoic acid, are often added to artificial diets for insects but it is not certain that they are true vitamins.

2.3 BIOSYNTHESIS OF FORMIC ACID

All ants of the subfamily Formicinae have lost the ability to sting but can spray a concentrated solution of formic acid (up to 65% in some species) from their venom glands. Many species of ground beetles (Carabidae) spray up to 70–75% formic acid from their pygidial glands. Millipedes, arachnids, the red-humped caterpillar of *Schizura concinna* (Lepidoptera: Notodontidae) (Plate 1) and two species of *Oxytrigona* stingless bees also use formic acid in defence. Hefetz and Blum (1978a; b) studied the biosynthesis in the carpenter ant *Camponotus pennsylvanicus*, and showed by using radio-labelled compounds that the formic acid can be formed from serine, glycine or histidine with the help of tetrahydrofolic acid (Figure 2.19). Apparently any compound capable of contributing a C_1 fragment can be a potential source of formic acid. The four enzymes necessary for these steps were all shown to be present in the venom gland.

2.4 PYRUVIC ACID

Pyruvic acid and its derivative phosphoenol pyruvate have already appeared in Figure 1.1, and pyruvic acid in discussion of the action of

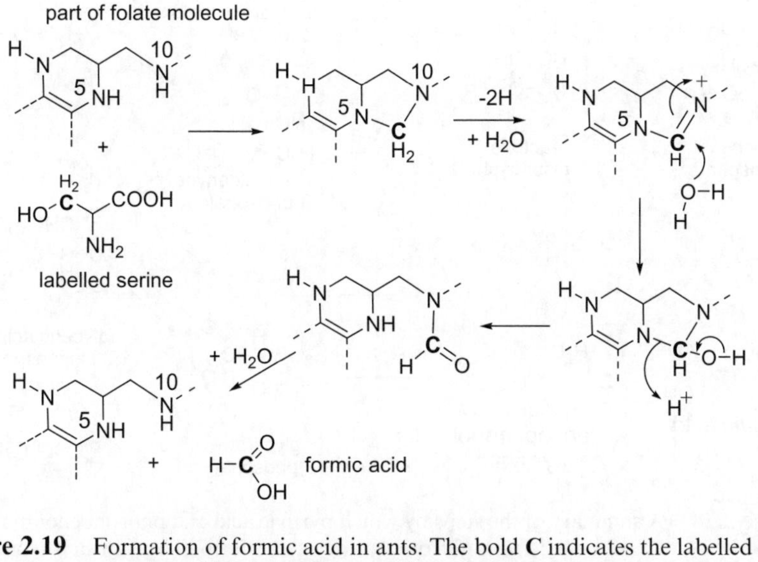

Figure 2.19 Formation of formic acid in ants. The bold **C** indicates the labelled atom of serine which is incorporated into formic acid.

thiamine diphosphate (Figure 2.14). Pyruvic acid and phosphoenol pyruvate are important intermediates in metabolism and synthesis.

When glucose is metabolized, the first steps are the conversion of glucose to glucose 6-phosphate, followed by isomerization to fructose 6-phosphate and then conversion to fructose 1,6-bisphosphate (Figure 2.20). This is cleaved by an aldol reaction running in reverse, and catalysed by the enzyme aldolase. The products are dihydroxyacetone 1-phosphate and glyceraldehyde 3-phosphate, two compounds interconvertible through a common enol form. Glyceraldehyde 3-phosphate can be de-phosphorylated and reduced to glycerol (see below) by NADPH, but most of it is phosphorylated again and oxidized to glyceric acid 1,3-bisphosphate, linked to the conversion of one molecule of NAD^+ to NADH. Glyceric acid bisphosphate loses one phosphate to ADP, forming ATP while the 3-phosphate is isomerized to glyceric acid 2-phosphate. Loss of water from this compound gives phosphoenol pyruvate. Transfer of the phosphate to ADP gives another molecule of ATP, while free enolpyruvate isomerizes to the keto-form, pyruvic acid. The summary in Figure 2.20 does not give all the stages, nor does it consider the mechanisms or the energetics of this important process. For that the reader is referred to a standard textbook of biochemistry.

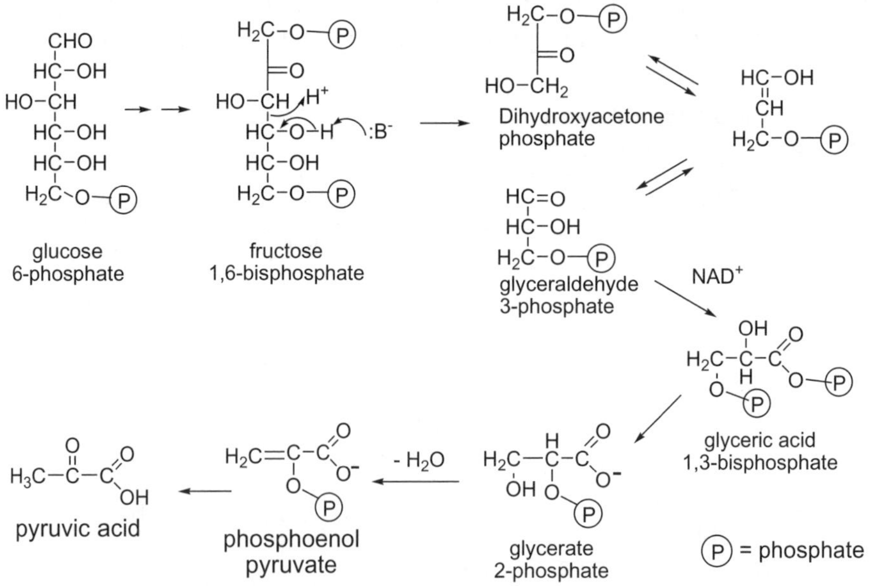

Figure 2.20 A summary of the steps by which pyruvic acid and phosphoenol pyruvate are obtained from glucose during metabolism. The key step is a retro-aldol reaction on fructose 1,6-bisphosphate.

Some insects are cold hardy, that is, they have evolved ways of surviving very low temperatures at which their blood might freeze and the ice-crystals pierce their cell walls. One way is for insects to produce large quantities of glycerol, which lowers the freezing point of their blood. Glycerol concentration can reach 2M or more, representing 20% of the fresh body weight of the insect in winter months. It has been shown that the larvae of the cold-hardy goldenrod gall moth *Epiblema scudderiana* produces a special aldolase that converts fructose 1,6-bisphosphate into glycerol. The enzyme is more active at 5 °C than at 20 °C (Holden and Storey, 1994).

2.5 NITROGEN FIXATION

One of the remarkable abilities of living cells is the conversion of inert, atmospheric nitrogen into ammonia when catalysed by the enzyme nitrogenase. The equivalent industrial reaction, the Haber process, requires extreme conditions. Only some prokaryotic uni-cellular organisms possess nitrogenase. Nitrogen fixation was first recog;nized as occurring in the Formosan subterranean termite *Coptotermes formosanus* (Isoptera: Rhinotermitidae) (Plate 2) (Breznak *et al.*, 1973). Many termites feed on dry wood, which is very deficient in nitrogen. Symbiotic nitrogen fixation in their gut has been shown to make a significant contribution to their nitrogen economy. The life spans of higher termites are reduced from 250 days to about 13 days by feeding them with certain antibiotics that kill both bacteria and spirochetes.

It is now known that a number of insects, essentially those living on wood, have symbiotic bacteria in their gut that can fix nitrogen. As well as many termite species, they include the bark beetle *Dendroctonus terebrans*, the cockroach *Cryptocerus punctulatus*, the scarabaeid beetle *Cetaria* sp., the fruit fly *Ceratatis capitata* and the stag beetle *Dorcus rectus*. The complete genome of a nitrogenase bacteroid living in the cells of a bacterium *Pseudotrichonympha grassii* in the gut of *Coptotermes formosanus* has recently been established (Hongoh *et al.*, 2008).

2.6 CHIRALITY

The great majority of compounds made by living organisms are chiral. A substance is chiral when it and its mirror images are not identical (the mirror image of the molecule cannot be superimposed upon it). The two mirror-image forms of a chiral substance, called enantiomers, are different substances. Although most of their chemical properties are identical, they can have very different biological properties. The big difference between solution chemistry and enzyme chemistry is that enzyme-catalysed reactions take place on a surface, as shown in Figure 2.11. Enzyme-catalysed reactions are usually stereospecific, that is, if the product is chiral, only one enantiomer is formed. They are also stereoselective, that is, if two isomers are available for reaction, only one is usually selected for reaction. Chirality in natural products is a direct consequence of being produced by enzymes. Since our own taste receptors (on our tongues) and odour receptors (in our nasal passage) are themselves chiral, we can detect some chiral differences. For example the amino acid L-asparagine, first isolated from asparagus juice, has a bitter taste, while its enantiomer, the unnatural D-asparagine, tastes sweet (Figure 2.21). Our odour receptors can distinguish between (S)-$(+)$-carvone and (R)-$(-)$-carvone (Figure 2.21), and we have a much lower threshold for detecting the first form, while the different flavours of lemons and oranges are due to our ability to detect the enantiomeric forms of the monoterpene limonene (Figure 7.1). Chirality in insect perception is discussed in Section 3.4.1. The natural juvenile hormone of insects $(+)$-JH I (Chapter 7) is about 10,000 times more active than its enantiomer $(-)$-JH I. Cholesterol (Chapter 8) has nine asymmetric centres, yet only one enantiomer exists naturally.

The usual cause of chirality in organic molecules is an asymmetric carbon atom, one with four different groups attached (Figure 2.22). The rules for describing the arrangement of groups about an asymmetric carbon atom, the Cahn–Ingold–Prelog rules, can be found in any textbook of organic chemistry. Priority is assigned to atoms or groups in descending order of atomic mass (Figure 2.22).

2.6.1 Asymmetric Induction

Asymmetric induction is the formation of a chiral substance from an achiral starting material. Glycerol has a symmetric molecule, but when it

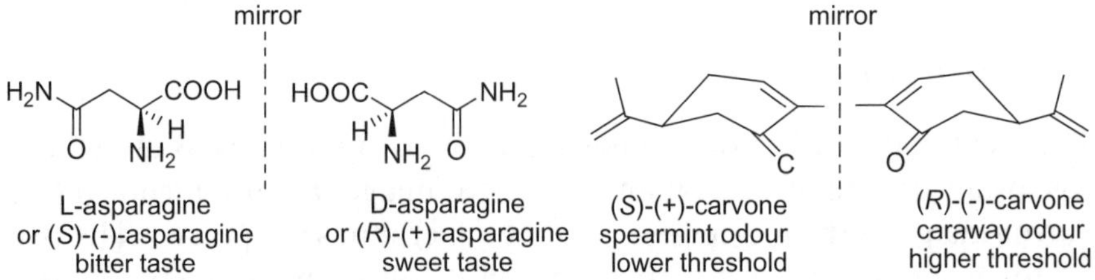

Figure 2.21 Examples of pairs of enantiomers that we are able to distinguish by taste and odour.

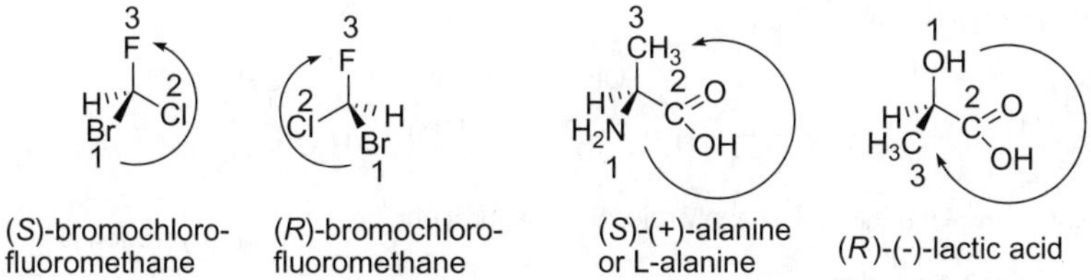

Enzymes and Co-enzymes

Figure 2.22 Illustration of use of the Cahn–Inglod–Prelog rules. (S)-(+)-Alanine and (R)-(−)-lactic acid are naturally occurring compounds. The symbols (+) and (−) here indicate optical rotation, but there is no relation between rotation and assignment of S and R to a chiral centre.

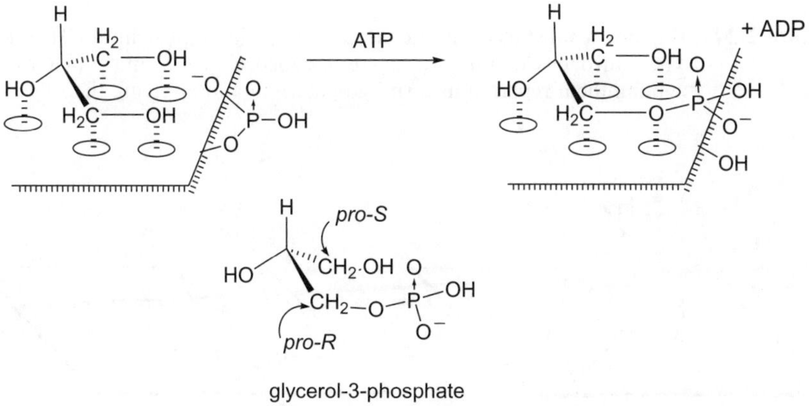

Figure 2.23 Glycerol held on an imaginary enzyme surface, being selectively phosphorylated. (The numbering of glycerol phosphate presents a problem, because L-glycerol-3-phosphate is the same compound as D-glycerol-1-phosphate. To avoid confusion, for glycerol, the *pro-S* CH_2 group is numbered 1 and the *pro-R* CH_2 numbered 3. Therefore the compound shown is the 3-phosphate.)

is held on the surface of an enzyme and one hydroxyl group is selectively phosphorylated, that symmetry is destroyed and chiral glycerol-3-phosphate is formed (Figure 2.23).

If a molecule is not chiral, it is designated as *pro-chiral* if a single substitution will make it chiral. Ethanol is an achiral compound; its CH_2 group is described as *pro-chiral*. To distinguish between the two hydrogen atoms of this group, the Cahn–Ingold–Prelog rules are applied to describe the group, after each hydrogen in turn is notionally replaced by deuterium, and the priority rule applied (Figure 2.24). If the new chiral centre has the *S* configuration, then the hydrogen that has been replaced is *pro-S*, and, conversely, if the new centre with deuterium is *R*,

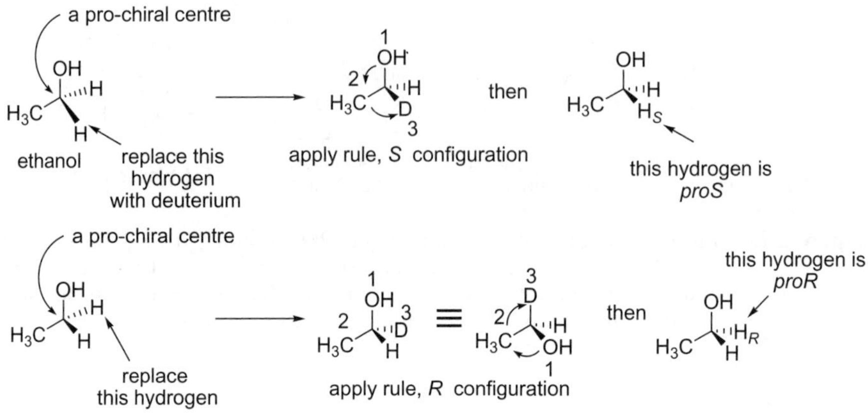

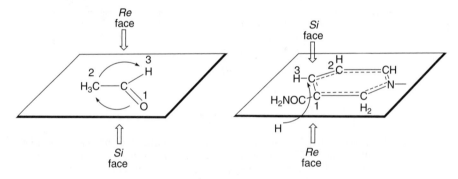

Figure 2.24 Ethanol, a substance with no chiral centres, has a pro-chiral CH_2 group. The hydrogens of this group are assigned as *pro-S* and *pro-R* by imagining them replaced in turn by deuterium and applying the rules.

Figure 2.25 Pro-chiral planar groups are assigned *Re* and *Si* faces by the Cahn–Ingold–Prelog chirality rules, illustrated by acetaldehyde and the nicotinamide portion of NAD^+.

then the hydrogen that was replaced is *pro-R*. The enzyme alcohol dehydrogenase and NAD^+ stereospecifically remove the *pro-R* hydrogen attached to the carbon bearing the O–H group in the oxidation of ethanol to acetaldehyde (Figure 2.11).

Planar groups or molecules may be *pro-chiral*. The acetaldehyde produced by oxidation of ethanol with NAD^+ (Figure 2.11) is pro-chiral. Applying the priority rules, one side is assigned the *Re* face and the other the *Si* face (Figure 2.25). In the biosynthesis of fatty acids (Section 4.1.1), the keto-group of acetoacetyl ACP (Figures 4.3 and 4.5) is reduced by NADPH from the *Si* face to give the (*R*) enantiomer of β-hydroxybutyryl ACP.

Methanol is not *pro-chiral*. Inserting one deuterium atom gives CH_2D-OH, which is *pro-chiral*, and inserting a tritium (or ^3H) in the latter gives CHDT-OH, which is chiral, and a useful biosynthetic tool.

Examples of chiral induction are considered in Chapter 7 in the biosynthesis of terpenes, where isopentenyl pyrophosphate is held on the enzyme surface and the *pro-R* hydrogen is selectively removed (Figure 7.4). The same *pro-R* hydrogen is removed in terpene chain extension, for example, in the formation of geranyl pyrophosphate (Figure 7.4).

BACKGROUND READING

Garrett, R. H. and Grisham, C. M. 1995. *Biochemistry, International Edition*, Saunders, Fort Worth, Chapter 14, Coenzymes and Vitamins, pp. 462–510.

Torssell, K. B. G. 1997. *Natural Product Chemistry*, Swedish Pharmaceutical Society, Stockholm, Chapter 3, Carbohydrates and primary metabolites, pp. 80–116; Chapter 7, Amino-acids, peptides and proteins, pp. 313–347.

REFERENCES

Blake, C. C., Koenig, D. F., Mair, G. A., North, A. C., Phillips, D. C. and Sarma, V. R. 1965. Structure of hen egg-white lysozyme. A three-dimensional Fourier synthesis at 2 Ångstrom resolution. *Nature*, **206**, 757–761.

Breznak, J. A., Brill, W. J., Mertins, J. W. and Coppel, H. C. 1973. Nitrogen fixation in termites. *Nature*, **244**, 577–579.

Bugg, T. D. H. 2004. *Introduction to Enzyme and Coenzyme Chemistry*, 2nd edition, Blackwell Publishing, Oxford, pp. 292.

Casiday, R. and Frey, R. 1998. (www.chemistry.wustl.edu/~edudev/LabTutorials/Carboxypeptidase/carboxypeptidase.html).

Cornforth, J. W. 1984. Stereochemistry of life. *Interdisciplinary Science Reviews*, **9**, 393–398.

Dadd, R. H. 1985. Nutrition: Organisms. In: *Comprehensive Insect Physiology, Biochemistry and Pharmacology*, vol. 4 (Kerkut, G. A. and Gilbert, L. I., ed.), Pergamon Press, Oxford, pp. 313–390.

Feyereisen, R. 1999. Insect P450 enzymes. *Annual Review of Entomology*, **44**, 507–533.

Feyereisen, R. 2005. Insect cytochromes P450. In: *Comprehensive Molecular Insect Science*, vol. 4 (Gilbert, L. I., Iatrou, K. and Gill, S. S., ed.), Elsevier, Oxford, pp. 1–77.

Hefetz, A and Blum, M. S. 1978a. Biosynthesis of formic acid by poison glands of formicine ants. *Biochimica et Biophysica Acta*, **543**, 484–496.

Hefetz, A and Blum, M. S. 1978b. Biosynthesis and accumulation of formic acid in poison gland of carpenter ant *Camponotus pennsylvanicus*. *Science*, **201**, 454–455.

Holden, C. P. and Storey, K. B. 1994. Purification and characterization of aldolase from the cold hardy insect *Epiblema scudderiana* – enzyme role in glycerol biosynthesis. *Insect Biochemistry and Molecular Biology*, **24**, 265–270.

Hongoh, Y., Sharma, V. K., Prakash, T., Noda, S., Toh, H., Taylor, T. D., Kudo, T., Sakaki, Y., Toyoda, A., Httori, M. and Ohkuma, M. 2008. Genome of an endosymbiont coupling N_2 fixation to cellulolysis within protist cells in termite gut. *Science*, **322**, 1108–1109.

Loenen, W. A. M. 2006. S-Adenosyl methionine: jack of all trades and master of everything? *Biochemical Society Transactions*, **34**, 330–333.

Pollak, N., Dolle, C. and Ziegler, M. 2007. The power to reduce: pyridine nuleotides – small molecules with a multitude of functions. *Biochemical Journal*, **402**, 205–218.

Rees, D. C. and Lipscomb, W. N. 1981. Binding of ligands to the active site of carboxypeptidase A. *Proceedings of the National Academy of Sciences USA*, **78**, 5455–5459.

Vocadlo, D. J., Davies, G. J., Laine, R., Withers, S. G. 2001. Catalysis by hen egg-white lysozyme proceeds *via* a covalent intermediate. *Nature*, **412**, 835–838.

CHAPTER 3
Experimental Methods

3.1 TRACING BIOSYNTHETIC PATHWAYS

To study a biosynthetic reaction or scheme, one must form an hypothesis, suggesting how the target compound might be formed, and what the starting material or enzyme substrate should be. This should be guided by known chemical reactions and examples of other enzyme-catalysed reactions. One way the hypothesis can be tested is by preparing an isotopically labelled quantity of the substrate, and supplying it to the synthesizing organism, tissue, gland or cells. After an incubation period, the metabolite must be isolated, purified and identified, guided by the isotopic label. The amount of labelled compound converted into labelled product may be very small when using intact animals or plants. Micro-organisms give better incorporation.

It is becoming possible through gene technology to recognize the possible RNA required to make a particular enzyme, and so to create by reverse transcription the DNA of that gene and express it in a bacterium, yeast or cell culture. The labelled substrate can then be fed to this culture, and the products analysed. This method has been particularly useful in recent work on the enzymes used in the biosynthesis of lepidopteran and coleopteran pheromones (Section 5.1 and Section 7.3).

It is important when using live insects to use enough labelled compound to obtain enough material to work with at the end of the experiment, and yet not so much that normal metabolism of the insect is altered, and abnormal products are formed. A further problem with whole organisms is that the product may not be formed because the

---

Biosynthesis in Insects, Advanced Edition
By E. David Morgan
© E. David Morgan 2010
Published by the Royal Society of Chemistry, www.rsc.org

enzyme is not active. Sometimes the enzymes required for a particular biosynthesis have to be switched on by feeding, or a hormone or neuropeptide. A moth only emits its sex pheromone at a particular time of day, when it is mature and ready for mating. The enzymes making that pheromone, or its final stages, will only be active at that time.

A biosynthetic pathway is only certain when we know (a) the ultimate source in primary metabolism from which the compound of interest derives (for example, a fatty acid, sugar, amine or others described in the following chapters) and (b) how the atoms of that primary precursor are distributed in the final product. The first task of finding the ultimate source, or starting substance, is usually not so difficult. The second objective may be very difficult and subject to all sorts of pitfalls and misleading clues.

A labelled compound is required to know that the substrate supplied and the product are related. This is usually achieved by incorporating some isotopic element, by which the change can be followed. It may be a radioactive isotope, such as ^{3}H, ^{14}C, ^{32}P or ^{35}S, that can be followed by its radiation; or it can be a stable heavy isotope, such as ^{2}H, ^{13}C, ^{15}N or ^{18}O, that can be traced by mass spectrometry or nuclear magnetic resonance (NMR) spectroscopy (Table 3.1). Another possible strategy, much used with micro-organisms, but also with the fruit fly *Drosophila melanogaster*, is to use mutant strains of an organism that lack the

Table 3.1 Some isotopes used in biosynthesis studies and how they are detected.

Isotope	Natural abundance	Detection method	Nuclear spin
^{1}H	99.98	^{1}H NMR, ^{13}C NMR coupling	1/2
^{2}H	0.015	NMR, MS, ^{13}C NMR shift	1
^{3}H	(half-life 12.26 y)	Radioactivity	–
^{12}C	98.9	^{1}H NMR shift	0
^{13}C	1.1	^{13}C NMR, MS	1/2
^{14}C	Age and source dependent (half-life 5730 y)	Radioactivity	–
^{14}N	99.63	NMR, MS	1
^{15}N	0.37	^{15}N NMR, MS	–1/2
^{16}O	99.8	MS	0
^{17}O	0.037	^{17}O NMR, MS	–5/2
^{18}O	0.20	MS, ^{13}C NMR shift	0
^{31}P	100	^{31}P NMR, MS	1/2
^{32}P	(half-life 14.28 d)	Radioactivity	–
^{32}S	95.02	MS	0
^{34}S	4.21	MS	0
^{35}S	(half-life 87.2 d)	Radioactivity	0

enzymes to complete a particular synthesis, or to add a specific enzyme inhibitor (for example, 2-octynoic acid is an inhibitor of fatty-acid synthesis), so that intermediates in the pathway accumulate and can be identified. A mutant strain of yeast was important in discovering mevalonic acid and its place in terpene biosynthesis (Chapter 7) and a number of mutants of the bacterium *Escherichia coli* helped to understand the steps of the shikimic acid pathway (Chapter 9).

Caution is necessary to ensure that the labelled substrate is not broken down to acetate and then re-built into other compounds. This was the case initially with the labelling study of dendrolasin (Section 7.5.2). The amount of label that needs to be incorporated into the starting material depends upon the method of detection. This can be a minute fraction with radioactive isotopes. More of a stable isotope would be required for identification by ^{13}C NMR spectroscopy than by mass spectrometry, for example.

3.2 RADIO-ISOTOPE LABELLING

Radioactive isotopes were formerly much used in biosynthetic studies, but stable isotopes are more used now because they can avoid the difficult degradation steps required with radio-labels. Some common precursor compounds *e.g.* ^{14}C-labelled glucose can be purchased but, frequently, isotopically enriched compounds have to be synthesized from low-molecular-mass compounds such as ^{14}CO$_2$, ^{3}H$_2$O, K^{14}CN and CH$_3$^{35}SH and the synthesis of the desired labelled compounds alone may be a major research task.

Compounds can be either uniformly labelled or specifically labelled at known atoms. By growing a plant or green alga in the presence of ^{14}CO$_2$, uniformly labelled glucose can be obtained, in which all the ^{12}C carbon atoms have an equal probability of being replaced by ^{14}C. More usual is to use a specifically labelled compound, such as [2-^{14}C]-acetic acid, in which carbon atoms only in the methyl group atoms are labelled, or [1-^{14}C]-acetic acid, where only the carboxylate carbon atoms are labelled. Specific labelling is used in experiments in which a biosynthetic route is suggested, a labelled compound is fed to the system and a plan of degradation of the product compound is made that will permit showing that the radio-labelled atoms are in the predicted places. For this, all the common degradative reactions of organic chemistry can be used, *e.g.* ozonolysis, decarboxylation, double-bond cleavage, *etc.* and the Barbier–Wieland degradation of fatty acids, by which one carbon atom at a time is removed from the chain.

The amount of radio-labelled compound obtained at the end of the biosynthesis experiment may be so little that more "cold" or unlabelled final compound may have to be added to have enough material to manipulate. In the studies of the biosynthesis of cholesterol (Chapter 7) unlabelled cholesterol for dilution is available in abundance. That is rarely so with insect substances, and a sample of the "cold" target compound may have to be synthesized too.

Sometimes it is useful to watch for the appearance of radioactivity in a particular gland or in a compound at periods after injection, to obtain information about the rate or site of biosynthesis. A hormone or pheromone may only be formed, or a gland activated, at a certain stage of development. A technique not much used but potentially valuable is to take sections through an insect after it has imbibed a radioactive compound and expose the sections on a photographic plate to see in which organs the activity accumulates (whole body autoradiography).

3.2.1 Specific Incorporation

If intact plants or animals are used, incorporation of a labelled intermediate can be extremely low (0.01 to 1%) because many processes and organs are competing for use of the labelled compound. The *specific incorporation* of the isotope is defined as the specific activity of the desired product divided by the specific activity of the starting material expressed as a percentage (Campbell, 1974). It is an important measure in biosynthesis studies. The specific activity of a radioactively labelled compound is in turn defined as the amount of radioactivity in a given amount of material (expressed as counts per minute per mole (cpm/mol) or cpm/g or decompositions per minute, dpm). Radioactivity is generally expressed in Becquerels ($1\,\text{Bq} = 1$ disintegration sec^{-1}) or in Curies ($1\,\text{Ci} = 3.7 \times 10^{10}\,\text{Bq}$).

Sir Derek Barton is said to have suggested a scale of label incorporation by which 1% is considered excellent, 0.1% is good, 0.01% is positive and 0.001% or less is dubious, probably negative. Better incorporation of the intermediate can be achieved with micro-organisms, tissue culture of insect cells or excised glands (see coccinellines, Chapter 4), or by having a cell-free system of partially purified enzymes. Alternatively, genes for making certain enzymes are now available, and an insect gene may be inserted into a micro-organism (bacterium), or insect cell culture, which can then be grown to produce more of the required labelled compound (Section 3.5). This will become increasingly useful as more genes are sequenced.

3.2.2 Locating the Site of Synthesis

Radio-labelling can be a very convenient way to locate where in the insect body the compound of interest is being produced. Possibly, as in the case of cuticular hydrocarbons, the compound or compounds are being synthesized in one place and deposited in another. To prove the site of synthesis, it is necessary to show that the excised tissue is capable of synthesizing the compound.

(Z)-10-Heptadecen-2-one is the major component of the aggregation pheromone of the fruit fly *Drosophila buzzatii* while 2-tridecanone from the same insect has an inhibiting effect on the aggregation. To find where and how these compounds were synthesized, insect bodies were separated into heads, thoraces and abdomens. Each of these body parts was incubated with sodium [1-^{14}C]acetate. No radioactive pheromone was recovered from the heads or thoraces, but 1.1% of the recovered label was incorporated into heptadecenone by the abdomens. In addition, 0.2% of label was incorporated into 2-tridecanone (Figure 3.1). Since the site had been traced to the abdomen, in a second series of experiments various organs of the abdomen were separated and incubated with labelled acetate. It was now found that only the ejaculatory bulb (part of the genital organs) of mature male flies synthesized the compounds. This time 3.3% of the recovered label (0.4% of the applied label) was recovered in the heptadecenone, and 1.0% of the recovered label (0.1% of the applied label) in tridecanone (Skiba and Jackson, 1993).

As an example of one of their experiments, 0.2 µl of sodium [1-^{14}C]acetate solution containing 440,000 cpm (counts per minute) was incubated with the ejaculatory bulbs at pH 6.8 in phosphate buffer. The products were extracted with hexane and subjected to radio-gas chromatography. The total label recovered was 54,000 cpm, but most of this was in fatty acids, 1162 cpm were recovered in the heptadecenone peak (2.15% of recovered label, 0.26% of applied label) and 625 cpm in

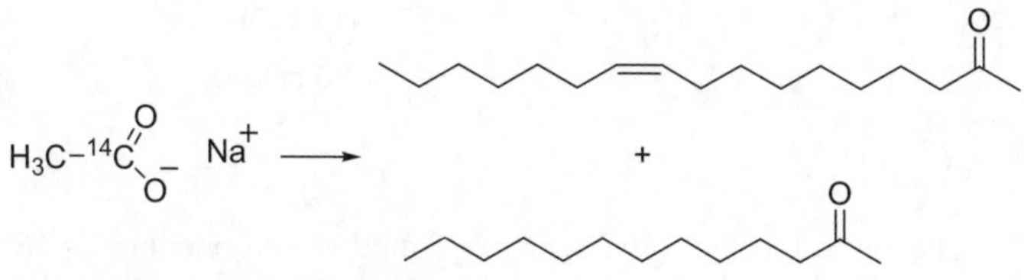

Figure 3.1 The production of radio-labelled (Z)-10-heptadecanone and 2-tridecanone from radio-labelled sodium acetate by the ejaculatory bulbs from the abdomens of *Drosophila buzzatii*. Note that the position of labelled atoms was not known, only the total incorporation of label, and therefore the details of the route of synthesis were not known.

tridecanone (1.15% of recovered label, 0.14% of applied label). The experiments effectively located the site of biosynthesis and showed the compounds were being synthesized from acetate units, but it did not show the position in the compound of the labelled atoms, so the full biosynthetic story was not revealed.

3.2.3 Radio-labelling Examples

One starts from a hypothesis for the biosynthetic route, which will predict which atoms should be labelled from a known, labelled starting material. The hypothesis predicts which atoms of the final product should be labelled, and the final labelled product must be degraded to show the label is on the predicted atoms. Such degradation studies can require great skill and are time consuming.

The biosynthesis of 2-hydroxy-6-methylacetophenone by the ant *Rhytidoponera aciculata* has been studied by injecting sodium [2-^{14}C]acetate into mealworm (*Tenebrio molitor*) larvae and feeding these to the ants (Tecle *et al.*, 1986) (Figure 3.2). The radioactive extract of ants was diluted with unlabelled 2-hydroxy-6-methylacetophenone, which was then purified and isolated as its 2,4-dinitrophenylhydrazone, to give a solid with which the radioactivity could be counted. The

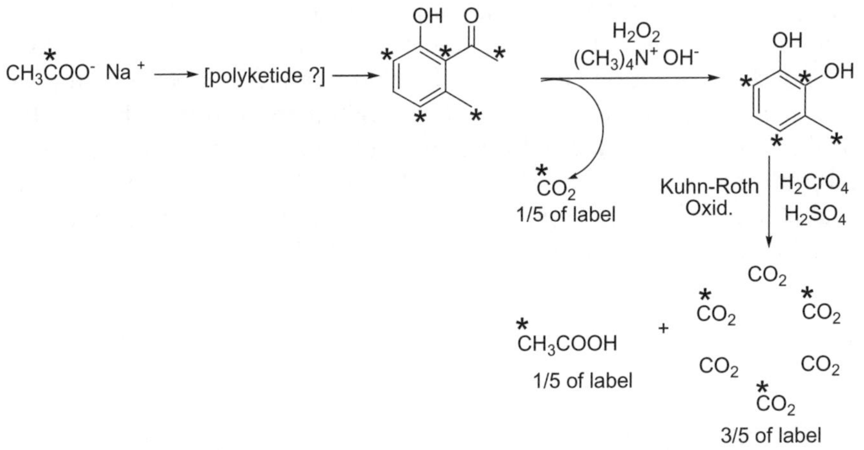

Figure 3.2 The radio-labelling of 2-hydroxy-6-methylacetophenone by *R. aciculata* ants. The hypothesis predicted that the compound was made from a string of acetate units joined head-to-tail, and therefore the carbon atoms bearing an asterisk should be radio-labelled. Note that the method did not locate the position of all the labelled carbon atoms, but did confirm the exact position of two labelled atoms, and showed that the remaining three were in the benzene ring.

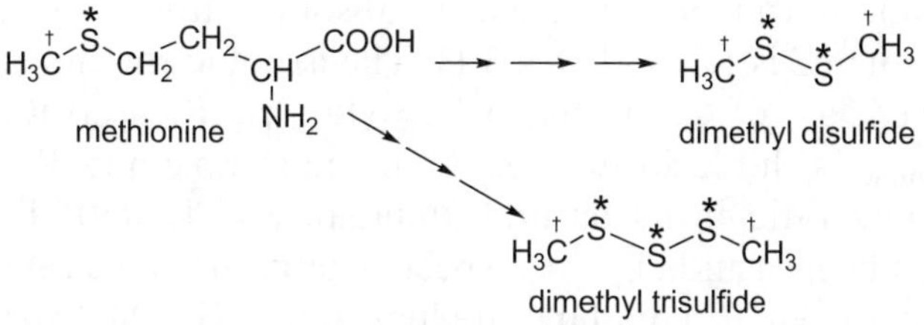

Experimental Methods

$$H_3C\overset{*}{\underset{\dagger}{S}}\text{-}CH_2\text{-}CH_2\text{-}\underset{NH_2}{CH}\text{-}COOH \longrightarrow \longrightarrow \longrightarrow H_3C\overset{*}{\underset{\dagger}{S}}\text{-}\overset{\dagger}{\underset{*}{S}}\text{-}CH_3$$

methionine → dimethyl disulfide

$$H_3C\overset{*}{\underset{\dagger}{S}}\text{-}\overset{*}{S}\text{-}\overset{\dagger}{S}\text{-}CH_3$$

dimethyl trisulfide

Figure 3.3 To confirm a hypothesis that dimethyl disulfide and dimethyl trisulfide were made from S-adenosyl methionine, methionine labelled as shown with ^{14}C and ^{35}S indicated that the methylthio-group was incorporated into both compounds in ants. Methionine labelled in the carboxylate group was not incorporated.

compound was oxidized with hydrogen peroxide to remove the acetyl group, which removed one unit of radioactivity. The 3-methylcatechol resulting was oxidized by the powerful Kuhn–Roth method, which gives acetic acid from methyl groups (it too was labelled with one unit), while the rest of the molecule was oxidized to CO_2 (containing three units of activity).

Sometimes double or even triple labelling can be useful. Some ponerine ants have dimethyl disulfide and dimethyl trisulfide in their mandibular glands. Crewe and Ross (1975) used doubly labelled methionine to show that the methyl and sulfur were incorporated together into these compounds by *Paltothyreus tarsatus* ants (Figure 3.3). The ants were injected with L-[^{35}S]-methionine and L-(methyl-[^{14}C])methionine. Both labels were detected in the dimethyl disulfide and trimethyl disulfide, but when methionine labelled in the carboxyl group was injected, there was no labelling of the sulfides. Whatever the actual intermediates through which the methylthio-group is converted, clearly methionine can be used for the production of both compounds.

3.3 HEAVY ISOTOPE LABELLING

Developments in mass spectrometry and NMR spectroscopy have made it more convenient in many cases to study biosynthetic processes with heavy isotopes, with more information obtainable from a combined use of these methods. The position of labelling, particularly for ^{13}C, can be deduced from intact molecules by NMR spectroscopy, without the degradation process needed with radio-labelling. Labelling with heavy isotopes, however, requires rather greater incorporation of the label for secure identification. Here the important measure is isotopic enrichment, that is, the degree by which a particular isotope is raised above the natural level. It can be expressed as absolute atom %. The natural abundance of ^{13}C is 1.1% (Table 3.1). The isotopic enrichment can also be described as $\delta^{13}C$, where $\delta^{13}C = (R_{sample} - R_{natural})/R_{natural} \times 100$, where R_{sample} is the ratio of $^{13}C:^{12}C$ in the sample and $R_{natural}$ is the corresponding ratio in its natural abundance (1.1/98.9). The ratio of isotopes can be measured by mass spectrometry or, in the case of ^{13}C, by the intensity of ^1H-^{13}C coupling satellites in the ^1H NMR spectrum of a singlet hydrogen atom that is not overlapped by other signals. NMR experiments are available for more complex overlapping signals (Nolis et al., 2009).

Where total light and heavy isotope is required, isotope ratio mass spectrometry is used. For ^{13}C and ^{15}N the sample is combusted in oxygen, and after clean-up and reduction of NO_x to N_2 the mixture of CO_2 and N_2 is separated on a packed GC column and analysed in a magnetic sector mass spectrometer, with the ion current measured with a Faraday cup. For nitrogen the ions at masses 28, 29 and 30 are measured and for carbon masses 44, 45 and 46. For ^2H and ^{18}O, the elements are converted to H_2 and CO, similarly, separated by GC and measured by MS, using masses 2 and 3 for H_2 and 28, 29 and 30 for CO. Sulfur can be measured as SO_2.

The use of ^{18}O in the initial stages of the study of photosynthesis provides a simple example of how labelling can be informative. When $C^{18}O_2$ was supplied to a plant in sunlight, none of the ^{18}O was found in the O_2 evolved, but was recovered as $H_2^{18}O$. When $H_2^{18}O$ was used in the experiment, $^{18}O_2$ was produced. The oxygen in the atmosphere is therefore derived from splitting water, not CO_2 (Figure 3.4).

Deuterium labelling is the most useful for mass spectrometry. A single ^2H atom incorporated increases the intensity of the M+1 ion. There must be sufficient ^2H present to make the ion intensity greater than that due to the natural abundance of ^{13}C (1.1%) and any other isotopes. Natural ^2H (0.015%), ^{15}N (0.36%) and ^{17}O (0.04%) add little to the strength of the M+1 ion, and ^{18}O (0.2%) adds little to the M+2 ion. Incorporation of an intact C^2H_3 group will produce a new ion at M+3 that can be readily seen and its intensity measured to give the degree of incorporation.

$$C^{16}O_2 + 2H_2^{18}O \xrightarrow[\text{chlorophyll}]{h\nu} (CH_2^{16}O)_n + {}^{18}O_2 + H_2^{16}O$$

Figure 3.4 The oxygen released in photosynthesis is from the splitting of water, demonstrated by using water containing the heavy isotope of oxygen. The formula $(CH_2O)_n$ represents carbohydrates.

Experimental Methods

A thorough knowledge of the capabilities of mass spectrometry and NMR spectroscopy are necessary to make good use of them in biosynthetic studies with heavy isotopes. It is very useful that ^{13}C, ^{15}N and ^{18}O all have nuclear spins of n/2 (^{2}H has a spin of 1, see Table 3.1), which means they can all be studied by NMR methods.

3.3.1 Kinetic Isotope Effects

When the rate-limiting step in a reaction is the making or breaking of a bond bearing a heavy atom, a kinetic isotope effect is observed. From quantum mechanics it can be shown that increasing the mass of an atom bonded in a molecule increases the energy between ground and transition state. If one considers the C–H and C–D bonds, the activation energy for breaking the C–D bond is therefore greater, so the overall rates of C–D bond breaking are slower than for C–H bond breaking. Where overall rate of reaction depends upon breaking C–H bonds, after some time the material left unreacted will become richer in C–D bonds. Evidently, the effect is greater, the greater the difference in mass between the heavier and lighter isotopes. In the case of ^{1}H, ^{2}H and ^{3}H the effect is greatest, but effects can sometimes be observed between *e.g.* ^{18}O and ^{16}O. Other things being equal, C–H bonds should be broken seven times faster at room temperature than C–D bonds. The rate of reaction can vary widely, depending upon the bond type but in many reactions C–D bonds are often broken four to eight times more slowly than C–H bonds. Use can be made of this difference in studying the mechanism of biochemical reactions. For example, in the formation of a double bond in stearic acid (Figure 4.6), by labelling C-9 and C-10 with two deuterium atoms each, it is found that the rate of formation of oleic acid decreases sharply, showing that the slow, rate-determining step is the breaking of a C–H (or C–D) bond on carbon atom number 9. These are called primary kinetic isotope effects.

A secondary kinetic isotope effect can also be observed in some cases. It occurs when a heavy isotope is adjacent to the bond being broken. The isotope affects the internal vibrations of the molecule, and some zero-point energies affect the rate of reaction. The secondary kinetic isotope effect can have a positive or a negative effect on the rate of reaction.

3.3.2 Other Isotope Effects

Two further effects of replacing hydrogen by deuterium can be recognized. The first concerns volatility. The C–D bond is shorter than the

C–H bond, so that deuterated compounds are slightly more volatile and elute towards the front edge of a gas chromatographic peak. The methyl ester of fully deuterated stearic acid elutes as a separate peak before normal methyl stearate. The second example arises in NMR spectroscopy. Not only do ^{1}H and ^{2}H resonate at different radio-frequencies and have different number of spin states, but ^{13}C attached to ^{2}H has a slightly different chemical shift from ^{13}C attached to ^{1}H (Figure 3.8(c)). Other isotope effects are known. It is noteworthy that pure D_2O does not support animal or plant life; the small differences in bond energies are sufficient to make the D_2O less reactive.

Stereochemistry can similarly be studied by deuterium labelling of the *pro-R* hydrogens on C-9 and C-10 of stearic acid and showing that these are both lost by a *syn*-elimination to give oleic acid. This is how the stereochemistry given in Figure 4.9 was determined. Similarly our knowledge of the stereochemistry of alcohol oxidation with NAD^+ and carbonyl reduction with NADH comes from deuterium labelling (Figure 2.11).

3.3.3 Examples of Isotopic Labelling

Scarab beetles have defensive secretions based on formic acid or a mixture of intermediate mass fatty acids. The ground beetle *Scarites subterraneus* (Coleoptera:Scaribidae) (Plate 3) produces a defensive mixture of isobutyric and methacrylic acids with lesser amounts of other C_4 and C_5 acids. Two days after injection of $^2H_8$-L-valine, the defensive glands were removed and the acids present converted to pentafluorobenzyl esters. The pentafluorobenzyl esters were used because the acids alone are too volatile and too polar for GC, simple methyl or ethyl esters are too volatile and the benzyl esters can be made under mild conditions. About 20% of the deuterated valine was incorporated into labelled isobutyric and methacrylic acids (Attygalle *et al.*, 1991). Pentafluorobenzyl esters of deuterated isobutyric and methacrylic acids appeared as GC peaks, eluting just before the corresponding undeuterated esters. They had molecular masses seven and five units greater than normal isobutyric and methacrylic acids, respectively, showing that the carbon chains were derived intact from valine (Figure 3.5).

The ratio of deuterated to undeuterated methacrylic acid was greater than the ratio of deuterated to undeuterated isobutyric acid, indicating that a strong primary kinetic isotope effect was operating, during the dehydrogenation of isobutyric to methacrylic acid.

More recently this group showed that tiglic, 2-methylbutyric and ethacrylic acids, found in the pygidial glands of the carabid beetle

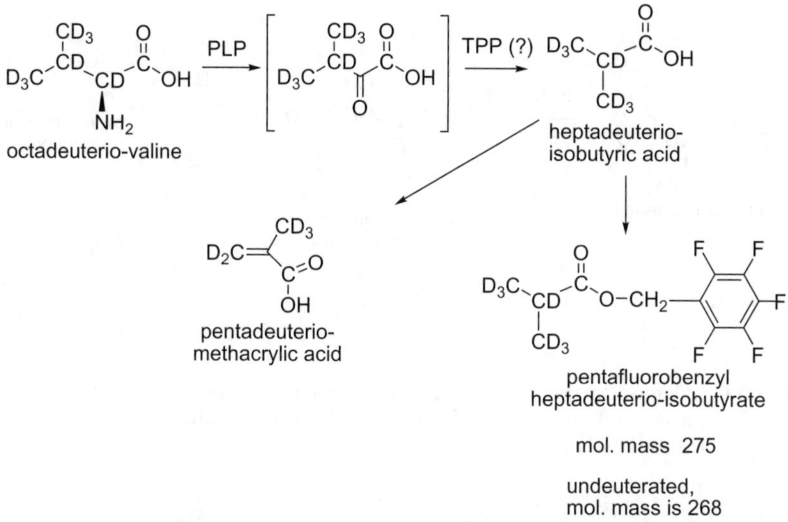

Experimental Methods

Figure 3.5 Deuterium labelling of the amino acid valine showed that isobutyric and methacrylic acids can be produced from it in the beetle *S. subterraneus*. The acids were collected and chromatographed as pentafluorobenzyl esters. PLP is pyridoxal phosphate, TPP is thiamine diphosphate.

Pterostichus californicus, are biosynthesized from isoleucine *via* 2-methylbutyric acid, using racemic [2,3,4,4-^{2}H$_4$]isoleucine and racemic allo-[2,3,4,4-^{2}H$_4$]isoleucine (Attygalle *et al.*, 2007). There are two chiral centres in isoleucine, so all isomers are present in this mixture. Their results are summarized in Figure 3.6. Again, a strong primary kinetic isotope effect was observed. The slower rate of removal of ^{2}H compared to ^{1}H left less label in ethacrylic acid, one ^{2}H being removed, and still less in tiglic acid, where two ^{2}H must be removed. The deduction of structure and the extent of labelling were all deduced from the mass spectra.

An experiment can be set up to show that the methyl ester group of methyl 6-methylsalicylate, the trail pheromone of the ant *Tetramorium impurum*, is derived from *S*-adenosyl methionine (Figure 2.18). First [^{2}H$_3$]-*S*-adenosyl methionine (synthesized from *S*-adenosyl homocysteine and [^{2}H$_3$]methyl iodide) is prepared and added to a tissue culture of the excised poison glands of the ant and incubated together. There is always a judgement to be made between adding enough labelled compound to get good incorporation of the label, and adding too much so that the metabolism of the tissue is altered. It is sometimes easier to show that a compound can serve as a precursor to the target compound than to show that it is the *normal*, natural intermediate in the

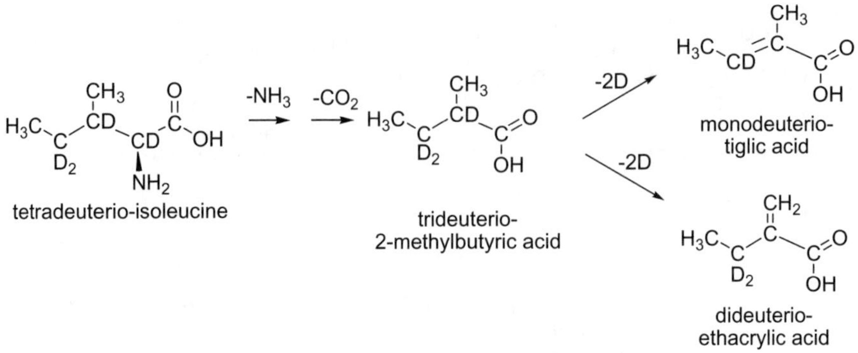

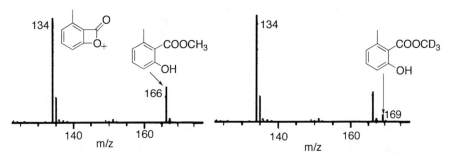

Figure 3.6 Demonstration that deuterated isoleucine is converted to 2-methylbutyric acid, and tiglic and ethacrylic acid by the beetle *P. californicus* in its pygidial gland.

Figure 3.7 Part of the mass spectrum of the ant pheromone methyl 6-methylsalicylate, and the spectrum after incubation of glands with deuterated *S*-adenosyl methionine, showing molecular ions at 166 and 169 in a ratio of 85:15.

biosynthetic path to the target. The products from the incubation can be extracted with hexane and examined by gas chromatography-mass spectrometry. Part of the mass spectrum of the normal pheromone (Figure 3.7) shows the molecular ion at m/z 166 and the daughter ion at m/z 134 formed by loss of methanol. This is known to be a concerted loss from the methyl ester and the phenol group. The product after incubation with deuterated *S*-adenosyl methionine shows a new peak at m/z 169 (M + 3), and the peak at m/z 166 is correspondingly reduced in abundance. The ion at m/z 134 is unaffected, showing that all the deuterium is lost with the methyl ester group. As noted earlier, molecules with three deuterium atoms will elute slightly earlier from the gas chromatograph. This is because the C–D bond is slightly shorter and stronger than the C–H bond, causing the deuterium-containing molecules to be slightly more volatile (an isotope effect).

The tri-deuterated S-adenosyl methionine is almost 100% pure. Essentially all the hydrogens in the methyl group have been replaced by deuterium. How much of the deuterated methyl group finds its way into methyl 6-methylsalicylate may be very small. *Isotopic enrichment* refers to the change in isotope content in the biosynthesized compound above the natural abundance and is usually expressed as atom % excess. The isotopic enrichment of the methyl group in the pheromone can be estimated from the relative abundances of the peaks at m/z 166 and 169 (Figure 3.7). The ratio is 1:0.18. The enrichment is then 15%.

The experiment can also be followed by ^{13}C NMR spectroscopy. The appearance of the total ^{13}C proton-decoupled spectrum is shown in Figure 3.8(*a*). The effect of substitution of one, two or three hydrogens in a methyl group by deuterium is shown at (*b*) and the methyl ester portion of the spectrum after incubation with deuterated S-adenosyl methionine is given at (*c*). This partial spectrum shows how the presence and number of deuterium atoms attached to ^{13}C in individual molecules can be revealed. As 85% of the molecules are unlabelled the original peak is strong with a weaker deuterium-coupled septuplet beside it, showing that three deuterium atoms are incorporated into 15% of the molecules. The NMR experiment requires rather more material because ^{13}C spectra are naturally weak (natural abundance of ^{13}C is 1.1%), but if the shifts of all the carbon atoms of the molecule are known, knowing the position of labelling can be quite simple.

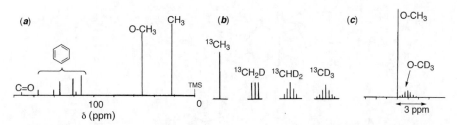

Figure 3.8 An example of the use of ^{13}C NMR spectroscopy in labelling the methyl ester of methyl 6-methylsalicylate. (*a*) The proton-decoupled spectrum of the unlabelled compound. (*b*) Substitution of one or more deuterium atoms for hydrogen on a ^{13}C-methyl group causes different multiplicity of ^{13}C-^{2}H couplings and a small progressive upfield shift (0.9 Hz for CD$_3$). (*c*) The appearance of the ester methyl of the compound after 15% incorporation of deuterium as in Figure 5.8. Note all the three deuterium atoms are incorporated intact with ^{13}C. The coupled CD$_3$ and singlet CH$_3$ almost overlap.

Cytochrome P450 enzymes can incorporate both oxygen atoms of O_2 into organic compounds. An apparatus for keeping insects in an atmosphere of $^{18}O_2$ has been described, which makes possible a decision on whether oxidation is occurring by this route (Fletcher *et al.*, 2004). The effects of ^{18}O on the chemical shifts of adjacent ^{13}C atoms are also described there.

3.3.4 Virginae Butanolide

Digressing from insects briefly, a good example of the power of ^{13}C NMR spectroscopy to solve a biosynthesis problem is found in the study of the microbiological product, virginae butanolide A (VBA) (Figure 3.9(*a*)). It is one of a series of similar signalling compounds produced by the bacterium *Streptomyces antibioticus* (Sakuda and

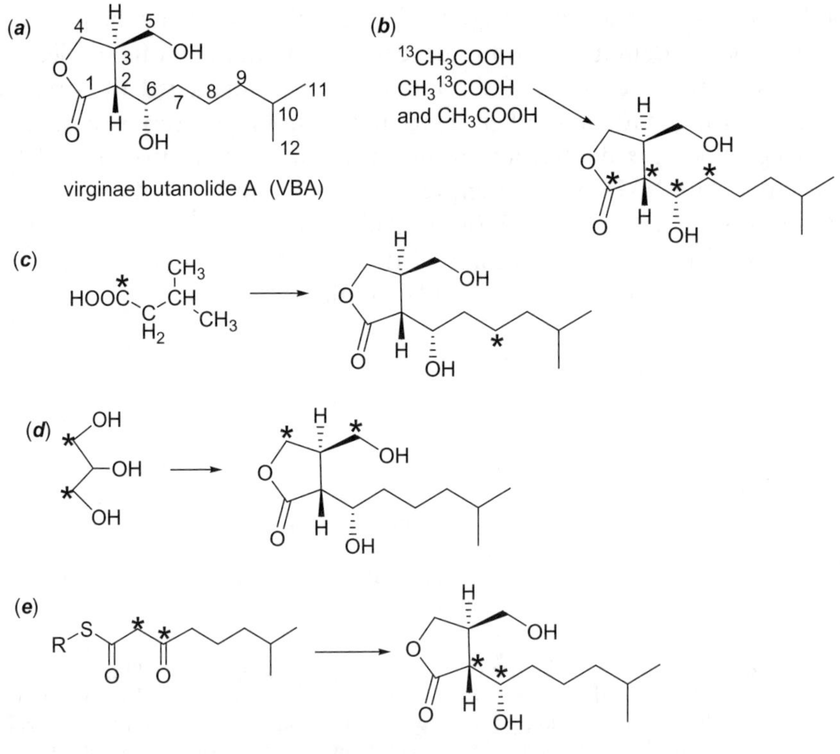

Figure 3.9 (*a*) The structure and numbering of virginae butanolide A. (*b*) Labelling introduced by supplying a mixture of $^{13}CH_3COOH$, $CH_3^{13}COOH$ and CH_3COOH. (*c*) Effect of labelling with [1-^{13}C]isovaleric acid. (*d*) Effect of feeding with [1,3-$^{13}C_2$]glycerol. (*e*) Incorporation intact of a ready-made β-keto-ester. Labelled atoms are marked with asterisks.

Yamada, 1999). First, the culture conditions for the *Streptomyces* to produce a good yield of VBA were studied, then the conditions for its isolation, and the best conditions for feeding it with sodium acetate. A mixture of sodium [1-^{13}C]acetate and [2-^{13}C]acetate and unlabelled acetate were fed to the bacterium. The ^{13}C NMR spectrum of the isolated product showed labelling in carbon atoms 1, 2, 6 and 7 only (Figure 3.9(b)). Therefore, these atoms were derived from intact acetate units. This result suggested a further experiment with [1-^{13}C]isovaleric acid, which gave an enriched NMR peak for carbon atom 8 (Figure 3.9(*c*)). Glycerol seemed a possibility for the remaining three carbon atoms, so they synthesized glycerol labelled at carbon atoms 1 and 3. As expected the NMR spectrum showed enhanced signals at carbon atoms 4 and 5 of VBA (Figure 3.9(*d*)). This time there was some labelling at carbon atoms 2 and 7 because there was some degradation of the glycerol to ^{13}CH$_3$COOH, and incorporation of this (Figure 3.9(*b*)). The ^{13}C NMR spectrum does not tell whether atoms 4 and 5 are in the same molecule, an important point for a rigorous proof, but the mass spectrum did show that molecules of VBA were either unlabelled or twice-labelled with ^{13}C. Finally to show that the isovaleric acid and the acetate units were linked into a kind of polyketide before linking to glycerol, the di-labelled thioester B (Figure 3.9(*e*)) was prepared (to resemble a CoA thioester) with two adjacent ^{13}C atoms.

The ^{13}C NMR spectrum of unlabelled VBA is shown in Figure 3.10(*a*), with the same spectrum (on a much smaller sample) after labelling with [1,3-^{13}C$_2$]glycerol (Figure 3.10(*b*)). As well as the enhanced intensity of the C-4 and C-5 atoms (about 6.2%), there was unexpected enrichment of atoms 2 and 7 (4.5 and 3.7%, respectively). This was attributed to breakdown of glycerol to acetate and incorporation of this into VBA. Two adjacent atoms, as in the β-keto-ester, gives, on incorporation intact, ^{13}C-^{13}C coupling, and in the NMR spectrum of the resulting VBA this coupling could be seen, with the doublets superimposed on the un-enriched singlets (Figure 3.10(*c*)). The researchers even went on to show that the glycerol was incorporated as dihydroxyacetone, and other finer points.

3.3.5 ^{13}C-^{13}C Coupling

In ^{13}C NMR spectra of normal abundance, or even for moderately enriched compounds, the probability of two ^{13}C atoms being next to each other in a molecule is extremely small, so no coupling between ^{13}C atoms is detectable. The ^{13}C nucleus has a spin of 1/2, so it couples like ^{1}H nuclei. Two adjacent ^{13}C atoms each appear as doublets in the (proton-decoupled) ^{13}C spectrum. Taking advantage of ^{13}C-^{13}C

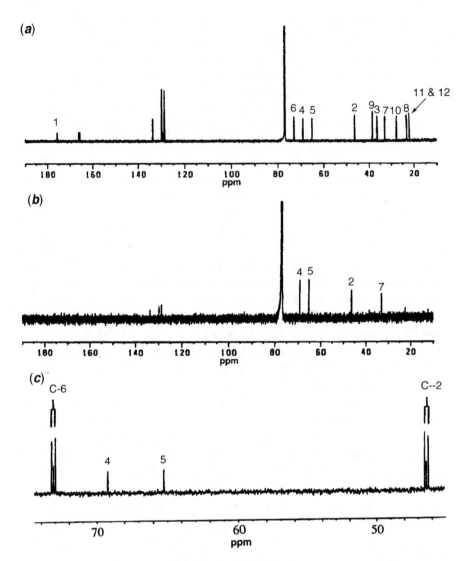

Figure 3.10 (a) The ^{13}C NMR spectrum of virginae butanolide A, with peaks numbered according to the formula in Figure 3.9 (a). Un-numbered lines are solvent or impurities. (b) The spectrum after a labelling experiment with [1,3-^{13}C$_2$]glycerol. Labelling of atoms 2 and 4 (less than that of 4 and 5) was caused by degradation of glycerol to acetate and its incorporation. (c) Part of the spectrum showing coupling between adjacent ^{13}C atoms that shows that the β-keto-ester of Figure 3.9 (e) was incorporated intact. Adapted from Sakuda and Yamada (1999).

coupling can be a neat way of showing that a precursor is incorporated intact in a more elaborate molecule, and that the label has not been scrambled. The example of virginae butanolide A (VBA, Section 3.3.4) illustrates this. In Figure 3.9(e) a β-keto-thioester was synthesized with C-2 and C-3 labelled with ^{13}C and used in biosynthetic experiments and the VBA isolated. The NMR spectrum of the resulting VBA (Figure 3.10(c)) showed by the coupled carbon atoms at 47 ppm and 73 ppm that these two labels were still together in many of the molecules of VBA. If the intermediate thioester had been degraded to acetate and then re-incorporated into VBA, there could have been ^{13}C atoms in VBA but the probability of the two ^{13}C atoms still being in the same molecule and still next to each other would be very small and no ^{13}C-^{13}C coupling would have been visible. Doubly labelled acetic acid, ^{13}CH$_3$^{13}COOH, also gives ^{13}C-^{13}C coupling, so whenever it is used and an acetate unit is incorporated intact, this coupling will be seen. It is a useful technique to detect bond-breaking and re-arrangement of the carbon chain.

3.4 ANALYTICAL ASPECTS

The techniques of gas chromatography used to separate compounds, when linked to mass spectrometry to obtain molecular masses and fragmentation patterns for the determination of structures of compounds, are important tools in the identification of volatile insect substances and for the study of their biosynthesis. Gas chromatography linked to Fourier transform infrared spectroscopy can also be very useful in some circumstances for identification (Leal, 1998). Preparative gas chromatography is probably the best method for obtaining microquantities of pure compound for bioassay, further reaction or identification (Heath and Dueben, 1998).

The positions of double bonds in alkenes can be learned through reaction with dimethyl disulfide, through chemical ionization mass spectrometry or through ozonolysis. These, and other micro-chemical techniques, are discussed by Attygalle (1998). The position of methyl branches in monomethylalkanes can be deduced from mass spectra. As one progresses to more methyl groups per molecule, interpretation becomes more difficult. For a detailed consideration of methylalkane identification see Carlson et al. (1998).

3.4.1 Chirality

The importance of chirality in biosynthesis has already been emphasized in Section 2.5. Many, but not all, insect pheromones and other compounds are chiral. The response of the insect to different chiral forms is important, particularly for pheromones. Mori (2007) has surveyed pheromones according to response and has found ten different possibilities. 1. Only one enantiomer is bioactive and its opposite enantiomer has no effect upon the response to the natural enantiomer. 2. Only one enantiomer is active and the unnatural enantiomer inhibits the response to the active enantiomer. 3. Where there are more than one chiral centre in the molecule, only one of these is active, and its diastereomer (*i.e.* just one chiral centre is altered) inhibits the response of the active enantiomer. 4. The natural pheromone consists of a single enantiomer, but the opposite enantiomer or diastereomer is also active. 5. The natural pheromone is a mixture of enantiomers or diastereomers, and all are active. 6. Different enantiomers or diastereomers are used by different species. 7. Both enantiomers are necessary for activity. 8. One enantiomer is more active than the other, but a mixture is more active than one enantiomer alone. 9. One enantiomer is active to males, and the other is active to females. 10. Only the *meso*-isomer is active. This is a special case where there are at least two chiral centres but internal symmetry makes the compound achiral. (For example, 2,4-pentandiol has two chiral and one *meso*-forms.)

For any chiral pheromone compound, it is impossible to predict which enantiomer will be active or whether one or both or a blend of enantiomers will be the naturally occurring substance. It is therefore important to determine chirality of the natural compound or mixture. To do that, the compound has to be isolated chemically pure, then the proportions of enantiomers found, *i.e.* the enantiomeric purity. To identify enantiomers, at least one of the enantiomer of known chirality must be available for comparison. Optical rotation, optical rotatory dispersion (ORD) and circular dichroism (CD) are of limited use in determining chirality because insect substances are rarely available in sufficient quantity for these techniques. Gas chromatography on a chiral stationary phase is one convenient way to separate and quantify small amounts of enantiomers (Mori, 1998). With some luck a chiral phase, usually based on a derivative of cyclodextrin can be found that will separate the enantiomers, especially if the compound is moderately volatile, since the chiral phases do not have very high temperature stability. An alternative method is to convert the unknown compound or mixture into a derivative with a chiral reagent, creating two diastereomeric derivatives, which are then separable by normal gas chromatography. Other methods include liquid chromatography on a chiral phase and the use of a chiral NMR shift reagent, or the conversion with a chiral reagent to diastereomers for NMR spectroscopy (Mori, 1998).

3.5 MOLECULAR BIOLOGY IN BIOSYNTHETIC STUDIES

The availability of molecular biology techniques has revolutionized both our understanding of biosynthetic processes and our approaches to study them. The language and techniques of molecular biology are quite different from those of conventional biochemistry and chemistry. It is not possible here to explain those methods, nor to give more than a few examples. Where brief definition of a concept can be given, it will be found in the glossary at the end of the book. An introduction to molecular biology can be found in texts like that of Miller and Tanner (2008) or Brown (2006).

The search for genes coding for particular biochemical reactions or products can be conducted in two different ways; either through conventional molecular biology or through functional genomics. The type of approach chosen will depend upon the goal of the study and how much we already know about the organism to be studied. The discovery of the polymerase chain reaction (PCR) has allowed almost any organism to be studied by molecular biology, so long as homologous genes are known from other species. This can be called a reductionist approach concentrating on the isolation of a single gene, or a few genes. The alternative may be described as a global or holistic approach, to look at many genes simultaneously through functional genomics. This approach concentrates upon dynamic aspects of gene expression, like which genes are turned on in a particular situation. For those accustomed to concentrating on a few enzymes or a single gene, functional genetics may appear unfocused. Most of the genes discovered and annotated during genomic studies may appear peripheral to the task in hand, but some may turn out to be of great interest.

Both the conventional approach of molecular biology and functional genomics can be used with non-model insects, but to apply the full power of modern approaches, one needs to work with model species with completely sequenced genomes, like *Drosophila* among the insects. The purpose of this section is to outline briefly molecular approaches to biosynthetic studies of insects by concentrating on a few examples.

3.5.1 Using the Polymerase Chain Reaction

To identify a gene, one needs to know something about the nucleotide sequence in that gene. The older molecular cloning techniques used fragments of nucleic acid or antibodies to identify possible genes from genome libraries or from libraries of complementary DNA (cDNA). That method has largely been replaced by cloning using the polymerase chain reaction (PCR). As with screening through a gene library, PCR-based cloning relies on comparing sequences of nucleic acids from known genes in related species with sequences in the studied species to identify a gene useful to the study. This avoids the expensive and time-consuming steps of constructing a gene library, and screening it for a sequence similar to the known one.

For example, the Δ11-desaturases of lepidopteran pheromone glands (Section 5.1.3.1), which control key steps in the biosynthesis of a wide range of pheromones derived from fatty acids, were isolated by PCR using cDNA made from pheromone gland mRNA. The primers constructed for PCR were designed to amplify a central region of the desaturase enzyme genes bounded by two patterns of nucleotides described as histidine boxes, that is conserved patterns of nucleotides bounded by ones that code for histidine residues. The histidine boxes are implicated in iron binding and catalytic functions (Shanklin et al., 1994). In this way a Δ11-desaturase was identified from the ni moth or cabbage looper *Trichoplusia ni* that is 55% identical and 72% similar to a rat Δ9-desaturase (Knipple et al., 1998). A number of other Δ11-desaturases were isolated by similar techniques, expressed in bacteria, and assayed for function (Rosenfield et al., 2001; Liu et al., 2002).

In another example, the first cDNA fragments of 3-hydroxy-3-methylglutaryl-CoA reductase (a key enzyme of terpene synthesis (Section 7.2.1)) were isolated from the bark beetle *Ips parconfusus* by the 3′RACE PCR procedure (rapid amplification of cDNA 3′-ends) (Tittiger et al., 1999). Alignment of the nucleotides in genes for animal hydroxymethylglutaryl-CoA reductase (HMG-R) enzymes in GenBank revealed numerous highly conserved regions, particularly in the catalytic (carboxyl terminus) portion of the enzyme, which allowed the designing of primers to obtain the full-length sequence of the HMG-R gene of *Ips paraconfusus* (Tittiger et al., 1999; 2003), and the related bark beetles *Ips pini* (Hall et al., 2002) and *Dendroctonus jeffreyi* (Tittiger et al., 1999).

The first animal geranyl diphosphate synthase (Section 7.2.1) gene was isolated from *Ips pini* by PCR using primers designed from conserved regions of isoprenyl transferases, enzymes required for producing terpenes (Gilg et al., 2005). The gene was expressed in Sf9 tissue cells and the protein produced was able to condense isopentenyl diphosphate and dimethallyl diphosphate to make geranyl diphosphate. Then subsequent work showed that this protein also converted some geranyl diphosphate to myrcene (Gilg et al., 2009) (Figure 3.11). The enzyme then has a dual function, giving two products. The isolation of the gene and its expression in Sf9 cells was central to recognition of its dual functions.

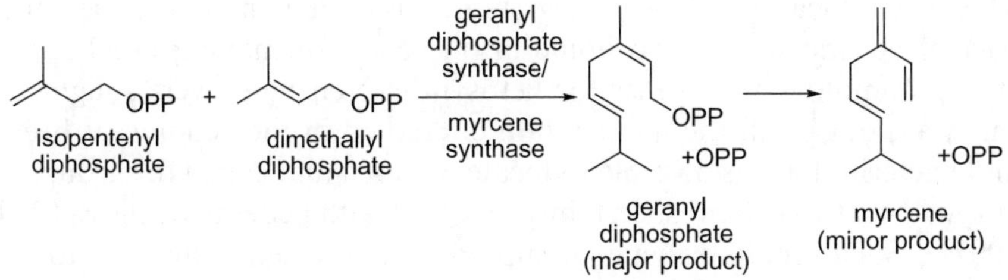

Figure 3.11 Geranyl diphosphate synthase-myrcene synthase condenses isopentenyl diphosphate and dimethallyl diphosphate to a mixture of geranyl diphosphate and myrcene. Cloning the enzyme and expressing it made it possible to show that one enzyme was producing both compounds.

The PCR approach can be used if a partial sequence of the protein is known, or if suitable homologous genes are available. When mRNA from the target insect, tissue or gland is isolated and reverse transcribed into cDNA, a large number of cDNA portions would be present. The one of interest is isolated and copied by use of the appropriate primers and PCR. Once the required cDNA is isolated, it can be inserted into an expression system (such as bacterial cells) and the protein isolated. In the examples above, the $\Delta 11$-desaturase was expressed in a mutant yeast cell lacking desaturases, and the unsaturated fatty-acid product of the mutant yeast cells with the $\Delta 11$-desaturase inserted were isolated and characterized by GC-MS. In the case of the geranyl diphosphate synthase-myrcene synthase, the enzyme was expressed in a bacterial system and assayed by the use of radio-labelled substrates. Beyond providing a method of producing desired proteins, the clones can be used to gain other insight. For example, Hall et al. (2002) used riboprobes for HMG-R in *Ips pini* to locate the site of production of monoterpenoid aggregation pheromone in male midgut tissue. Sequence comparisons may also be used for molecular evolution studies.

PCR can be used to clone and map the DNA from micro-organisms that have not yet been isolated and cultured. This is what has been done in the study of biosynthesis of pederin (Section 6.9). The total DNA of the *Paederus* beetle and the unidentified bacterium suspected to be its symbiont was isolated, broken into segments by restriction enzymes and amplified by PCR. A cosmid library was prepared, essentially a library of large gene fragments. Sufficient DNA was thus produced that large sections of it could be sequenced (the order of the nucleotides in the genome determined, a process largely automated), and compared with DNA libraries. By this comparison, sections of DNA could be recognized as belonging to a bacterium and, further, a bacterium close to *Pseudomonas aeruginosa*, the genome of which was already known.

These identifications depend heavily on computer methods, the other part of genetic studies. To home in on genes for making polyketides, strong homology with genes for ketosynthase enzymes were sought. In this way, out of all the information contained in the genome of beetle and bacterium it was possible to locate sections that coded for groups of enzymes that were almost certainly involved with pederin synthesis (Piel, 2002). Such multi-enzyme complexes are often referred to as metabolons.

3.5.2 Functional Genomics

Functional genomics is a process that is more powerful than the older cloning methods and greatly speeds up gene discovery. It concentrates on data from sequencing and genome mapping to study function and interaction; dynamic aspects of transcription, translation, protein formation and enzyme action. It makes use of high-throughput screening methods, which in turn rely heavily on control software, robotics, data processing, liquid handling devices and an essential laboratory tool, the microlitre plate. High-throughput refers to rapid, simultaneous screening of hundreds or thousands of samples, and is most used in genomics and drug discovery. Because of the repetitive nature of the work, it makes use of robots, which means high capital cost, so much of the work is contracted out to companies or central facilities. High-throughput screening with functional genomics is very helpful with non-model organisms to learn about biosynthetic processes. It can reveal physiological interactions that might otherwise be missed using conventional biochemical or molecular biological approaches. The use of functional genomics in chemical ecology to study the honeybee *Apis mellifera*, the silkworm *Bombyx mori* and bark beetles has been reviewed by Tittiger (2004).

These techniques were used to learn more about the hormonal regulation of attractant pheromone biosynthesis in the pine engraver beetle *Ips pini* (Section 7.3.1), and to identify genes that respond to juvenile hormone (JH) (Section 7.6.1) in this insect, where pheromone production appeared to be controlled by the hormone. Expressed sequence tags (EST) were first identified. Expressed sequence tags are short (200–500 nucleotides) but unique sections of DNA within a gene that is useful for identifying the whole gene, derived from cDNA. A small-scale EST project recovered 574 tentatively unique genes expressed in the anterior midgut tissue of *Ips pini* males that had been fed, and treated with juvenile hormone (Eigenheer *et al.*, 2003). Microarrays were prepared and hybridized to cDNA from midguts of fed and JH-treated beetles and

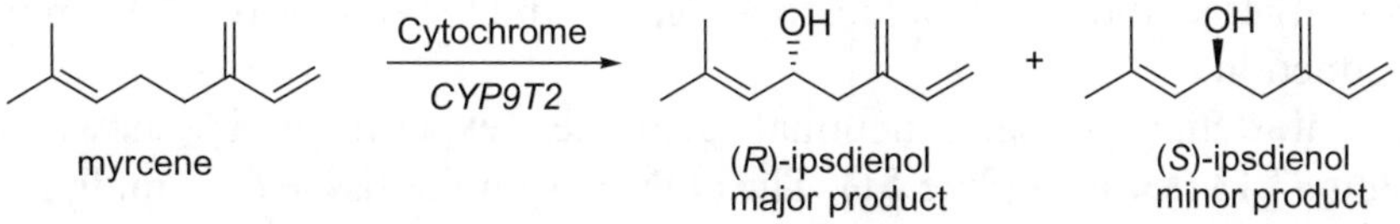

Figure 3.12 The oxidation of myrcene to a mixture of (R)- and (S)-ipsdienol, not necessarily the enantiomer mixture found in a pheromone. It was shown that the reaction took place in the midgut of male *Ips* beetles.

from control beetles (Keeling *et al.*, 2006). The results from the microarray showed that feeding and treating with JH work together co-ordinate to stimulate genes of the mevalonate pathway and many other genes implicated in pheromone biosynthesis in this beetle (Keeling *et al.*, 2006). The "pheromone biosynthetic" cluster of genes includes those for the mevalonate pathway, for the geranyl diphosphate synthase-myrcene synthase and for a specific cytochrome P450 that hydroxylates myrcene to ipsdienol (Figure 3.12) (Sandstrom *et al.*, 2006; 2008), and a dehydrogenase that inter-converts ipsdienol and ipsdienone (Blomquist and Tittiger, personal communication). A number of other potentially important genes for pheromone production were identified by this method, and surprisingly, although the pheromone is male-produced, expression of many of the genes was stimulated in females also.

Microarrays can lead to ambiguous results and tend to under-represent changes in gene expression, so it is important to confirm microarray data with other ways of measuring expression. Quantitative real-time PCR (RT-PCR or qPCR) is used for this purpose. In the case of *Ips pini* studies, qPCR confirmed the effect of feeding on the upward regulation of mevalonate pathway genes, and provided additional data that were crucial to the identification of pheromone biosynthesis genes (Keeling *et al.*, 2006). For example, qPCR revealed sex differences in basal levels of mevalonate pathway genes that were consistent with their role in male-specific pheromone biosynthesis. By combining data from qPCR and microarrays, the preliminary identification of the cytochrome P450 and ipsdienol-ipsdienone genes was possible. A comparison now of the basal levels of expression of the mevalonate pathway genes made clearer why males alone produce pheromone, since males have a six to forty times higher level of mevalonate genes than females. A key gene in monoterpene production, the geranyl diphosphate synthase-myrcene synthase gene is at low level of activity in females and its activity is depressed by feeding. This was the first study of development and hormonal regulation of insect pheromone biosynthesis based on microarrays. It answered a number of questions about juvenile hormone regulation of pheromone biosynthesis, and contributed to a clearer understanding of pheromone production.

A flowchart of a functional genomics experiment is shown in Figure 3.13 (see also plate 51). The cDNA from the tissue (*e.g.* midgut of bark beetle) is used to generate ESTs (left side of figure). Modern "next-generation" sequencing techniques can avoid the requirement to produce and array a gene library. Each clone is sequenced once. On the right, microarrays are produced, based on the sequence date, and hybridized simultaneously with labelled cDNA from control and experimental tissue. The control cDNA is labelled with the fluorescent dye cyanine 3, which has a green fluorescence, and then treated with red-fluorescing cyanine 5. They are mixed and hybridized on the microarray plates. The fluorescence is detected by a scanner. cDNA from the experimental tissue fluoresces red, that common to both controls and experiment fluoresces yellow. The lower right panel shows how results of experiments at different time points are clustered using bioinformatics support. Each column represents a time, and each row a different gene. Colours, usually blue and yellow, indicate relative expression of that gene, and colour intensity the degree of difference (plate 51). Black means no difference in expression. For further explanation see websites for microarrays, for example www.bio.davidson.edu/courses/genomics/chip/chip.html.

3.5.3 Model Organisms

The use of PCR and functional genomics have relieved researchers from their dependence on genetically familiar model species. *Drosophila melanogaster* (plate 4) was the undisputed model for many insect studies because of its extensive genetic history, and the relative ease with which transgenic strains could be created and studied. Now, however, complete genomes are available from several insect genera representing several orders, including *Aedes aegypti* and *Anopheles gambiae* for Diptera, *Apis mellifera* for Hymenoptera, *Tribolium castaneum* for Coleoptera, *Bombyx mori* for Lepidoptera and *Acyrthosiphon pisum* for Hemiptera. Extensive EST data are available for scores of others. As sequencing costs fall, more and more data will be available, making full genomes or comprehensive transcriptome databases for any insect increasingly available, and removing the necessity for model species.

One area where non-model species continue to lag is transgenic insects. They are especially useful because they provide a means to study the effect of adding or removing (silencing) a gene. Transgenic techniques typically require extensive investment to develop suitable vectors and markers. Advances in interference RNA techniques (RNAi) offer

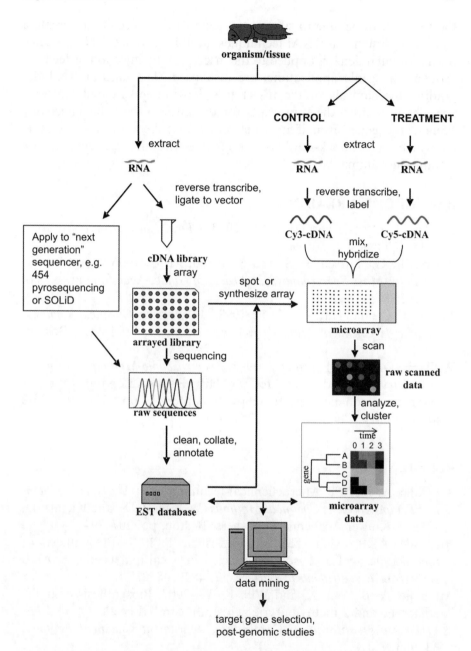

Figure 3.13 Scheme of a study using functional genomics (See Plate 51 for a colour version). Figure modified from C. Tittiger, Functional genomics and insect chemical ecology, *J. Chem. Ecol.*, 2004, **30**, 2342. Reproduced with permission of Springer.

the prospect to be able to silence any gene without needing to create a transgenic system. In RNAi techniques, double-stranded RNA, specific for a particular gene, is applied to the insect, *e.g.* by injection or feeding, though transgenic applications are also used. The natural RNA-degrading mechanisms of the insect use the double-stranded RNA to specifically remove the transcripts for the target gene. This effectively silences the gene, even if it is actively transcribed. In this way, the equivalent of a "knock-out" insect is created without requiring to create a transgenic animal.

BACKGROUND READING

Beesley, T. E. and Scott, R. P. W. 1998. *Chiral Chromatography*, John Wiley, Chichester, 506 pp.

Kitson, F. G., Larsen, B. S. and McEwen, M. C. 1996. *Gas Chromatography-Mass Spectrometry, A Practical Guide*, Academic Press, San Diego, 381 pp.

Millar, J. C. and Haynes, K. F. 1998. *Methods in Chemical Ecology, Volume 1, Chemical Methods*, Kluwer Academic Publishers, Boston, 390 pp.

Williams, D. H. and Fleming, I. 1995. *Spectroscopic Methods in Organic Chemistry*, 5th edition, McGraw-Hill Books, London (Nuclear magnetic resonance spectroscopy, Chapter 3, pp. 63–169; Mass spectrometry, Chapter 4, pp. 170–225).

REFERENCES

Attygalle, A. B. 1998. Microchemical techniques. In: *Methods in Chemical Ecology, Vol. 1, Chemical Methods* (Webster, F. X. and Kiemle, D. J., ed.), Kluwer Academic Publishers, Boston, pp. 207–294.

Attygalle, A. B., Meinwald, J. and Eisner, T. 1991. Biosynthesis of methacrylic acid and isobutyric acids in a carabid beetle, *Scarites subterraneus*. *Tetrahedron Letters*, **32**, 4849–4852.

Attygalle, A. B., Wu, X. and Will, K. W. 2007. Biosynthesis of tiglic, ethacrylic and 2-methylbutyric acids in a carabid beetle *Pterostichus (Hypherpes)californicus* (Carabidae). *Journal of Chemical Ecology*, **33**, 963–970.

Brown, T. A. 2006. *Gene Cloning and DNA Analysis*, 5th edition, Blackwell Publishing, Oxford, 407 pp.

Campbell, I. M. 1974. Incorporation and dilution values – their calculation in mass spectrally assayed stable isotope labelling experiments. *Bioorganic Chemistry*, **3**, 386–397.

Carlson, D. A., Bernier, U. R. and Sutton, B. D. 1998. Elution patterns from capillary GC for methyl-branched alkanes. *Journal of Chemical Ecology*, **24**, 1845–1865.

Crewe, R. M. and Ross, F. P. 1975. Pheromone biosynthesis – formation of sulfides by ant *Paltothyreus tarsatus*. *Insect Biochemistry*, **5**, 839–843.

Eigenheer, A. L., Keeling, C. I., Young, S. and Tittiger, C. 2003. Comparison of gene representation in midguts from two phytophagous insects, *Bombyx mori* and *Ips pini*, using expressed sequence tags. *Gene*, **316**, 127–136.

Fletcher, M. T., Wood, B. J., Schwartz, B. D., Rahm, F., Lambert, L. K., Brereton, I. M., Moore, C. J., de Voss, J. J. and Kitching, W. 2004. A precision apparatus, with solid phase micro-extraction monitoring capability, for incorporation studies of gaseous precursors into insect-derived metabolites. *ARKIVOC*, **2004**, 109–117.

Gilg, A. B., Bearfield, J. C., Tittiger, C., Welch, W. H. and Blomquist, G. J. 2005. Isolation and functional expression of an animal geranyl diphosphate synthase and its role in bark beetle pheromone biosynthesis. *Proceedings of the National Academy of Sciences, USA*, **102**, 9760–9765.

Gilg, A. B., Tittiger, C. and Blomquist, G. J. 2009. Unique animal prenyltransferase with monoterpene synthesis activity. *Naturwissenschaften*, **96**, 731–735.

Hall, G. M., Tittiger, C., Andrews, G. L., Mastick, G. S., Kuenzli, M., Luo, X., Seybold, S. J. and Blomquist, G. J. 2002. Midgut tissue of male pine engraver, *Ips pini*, synthesizes monoterpenoid pheromone component ipsdienol *de novo*. *Naturwissenschaften*, **89**, 79–83.

Heath, R. R. and Dueben, B. D. 1998. Analytical and preparative gas chromatography. In: *Methods in Chemical Ecology, Vol. 1, Chemical Methods* (Webster, F. X. and Kiemle, D. J., ed.), Kluwer Academic Publishers, Boston, pp. 85–126.

Keeling, C. I., Bearfield, J.C., Young, S., Blomquist, G. J. and Tittiger, C. 2006. Effects of juvenile hormone on gene expression in the pheromone-producing midgut of the pine engraver beetle, *Ips pini*. *Insect Molecular Biology*, **15**, 207–216.

Knipple, D. C., Rosenfield, C.-L., Miller, S. J., Liu, W., Tang, J., Ma, P. K. W. and Roelofs, W. L. 1998. Cloning and functional expression of a cDNA encoding a pheromone gland-specific acyl-CoA Δ11-desaturase of the cabbage looper moth, *Trichoplusia ni*. *Proceedings of the National Academy of Sciences, USA*, **95**, 15287–15292.

Leal, W. S. 1998. Infrared and ultraviolet spectroscopy techniques. In: *Methods in Chemical Ecology, Vol. 1, Chemical Methods* (Webster,

F. X. and Kiemle, D. J., ed.), Kluwer Academic Publishers, Boston, pp. 295–338.

Liu, W. T., Jiao, H.M., Murray, N. C., O'Conner, M. and Roelofs, W. L. 2002. Gene characterized for membrane desaturase that produces (E)-11 isomers of mono- and di-unsaturated fatty acids. *Proceedings of the National Academy of Sciences, USA*, **99**, 620–624.

Miller, A. and Tanner, J. 2008. *Essentials of Chemical Biology*, John Wiley & Sons, Chichester, Chapter 3, Molecular biology as a toolset for chemical biology, pp. 139–174.

Mori. K. 1998. Separation of enantiomers and determination of absolute configuration. In: *Methods in Chemical Ecology, Vol. 1, Chemical Methods* (Webster, F. X. and Kiemle, D. J., ed.), Kluwer Academic Publishers, Boston, pp. 185–206.

Mori, K. 2007. Significance of chirality in pheromone science. *Bioorganic and Medicinal Chemistry*, **15**, 7505–7523.

Nolis, P., Gil, S., Espinosa, J. F. and Parellaa, T. 2009. Improved NMR methods for the direct C-13-satellite-selective excitation in overlapped H-1-NMR spectra. *Magnetic Resonance in Chemistry*, **47**, 121–132.

Piel, J. 2002. A polyketide synthase-peptide synthetase gene cluster from an uncultured bacterial symbiont of *Paederus* beetles. *Proceedings of the National Academy of Sciences, USA*, **99**, 14002–14007.

Rosenfield, C.-L., You, K. M., Marsella-Herrick, P., Roelofs, W. L. and Knipple, D. C. 2001. Structural and functional conservation and divergenc among acyl-CoA desaturases of two noctuid species, the corn earworm, *Helicverpa zea*, and the cabbage looper, *Trichoplusia ni*. *Insect Biochemistry and Molecular Biology*, **31**, 949–964.

Sakuda, S. and Yamada, Y. 1999. Biosynthesis of butyrolactones and cyclopentanoid skeletons formed by aldol. In: *Comprehensive Natural Product Chemistry*, vol. 1 (Sankawa, U., ed.), Pergamon Press, Oxford, pp. 139–158.

Sandstrom, P., Welch, W. H., Blomquist, G. J. and Tittiger, C. 2006. Functional expression of a bark beetle cytochrome P450 that hydroxylates myrcene to ipsdienol. *Insect Biochemistry and Molecular Biology*, **36**, 835–845.

Sandstrom, P., Ginzel, M. D., Bearfield, J. C., Welch, W. H., Blomquist, G. J. and Tittiger, C. 2008. Myrcene hydroxylases do not determine enantiomeric composition of pheromonal ipsdienol in *Ips* spp. *Journal of Chemical Ecology*, **34**, 1584–1592.

Shanklin, J., Whittle, E. and Fox, B. G. 1994. Eight histidine residues are catalytically essential in a membrane associated iron enzyme, stearyl-CoA desaturase and are conserved in alkane hydroxylase and xylene monooxygenase. *Biochemistry*, **33**, 12787–12794.

Skiba, P. J. and Jackson, L. L. 1993. (Z)-10-Heptadecen-2-one and 2-tridecanone biosynthesis from [1-14C]acetate by *Drosophila buzzatii*. *Insect Biochemistry and Molecular Biology*, **23**, 375–380.

Tecle, B., Brophy, J. J. and Toia, R. F. 1986. Biosynthesis of 2-hydroxy-6-methylacetophenone in an Australian ponerine ant, *Rhytidoponera aciculata* (Smith). *Insect Biochemistry*, **16**, 333–336.

Tittiger, C. 2004. Functional genomics and insect chemical ecology. *Journal of Chemical Ecology*, **30**, 2335–2358.

Tittiger, C., Blomquist, G. J., Ivarsson, P., Borgeson, C. E. and Seybold, S. J. 1999. Juvenile hormone regulation of HMG-R gene expression in the bark beetle *Ips paraconfusus* (Coleoptera: Scolytidae): implications for male aggregation pheromone biosynthesis. *Cellular and Molecular Life Sciences*, **55**, 121–127.

Tittiger, C., Barkawi, L., Bengoa, C., Blomquist, G. J. and Seybold, S. J. 2003. Structure and juvenile hormone-mediated regulation of the HMG-CoA reductase gene from the Jeffrey pine beetle, *Dendroctonus jeffreyi*. *Molecular and Cellular Endocrinology*, **199**, 11–21.

CHAPTER 4
Fatty Acids and Cuticular Hydrocarbons

4.1 INTRODUCTION

The principal source of stored energy in the body is fat, which in insects is 90% in the form of triglycerides (or triacylglycerol). Triglycerides are esters of one molecule of glycerol and three of various fatty acids. Triglycerides and fatty acids can be regarded as primary metabolites. The triglycerides are stored as lipid droplets in the cells of the fat body. The glycerides are released from the fat body in the form of diglycerides (or 1,2-diacylglycerol), and transported by lipophorin (Arrese *et al.*, 2001), a lipoprotein that also carries other lipids, such as hydrocarbons, sterols and fatty acids. The synthesis, desaturation and chain elongation of fatty acids should be understood as integral parts of the one operation that leads on to several groups of insect compounds. Together with terpenoids, fatty acids are the source of the greatest number of insect pheromones, and if one includes cuticular hydrocarbons, the fatty acids account for the greatest number of derived compounds in insects known at present.

4.2 FATTY-ACID BIOSYNTHESIS

J. N. Collie (1907) noticed that the acids of natural fats always contained even numbers of carbon atoms (Figure 4.1), and suggested that they, and many other compounds, are made by head-to-tail linking of acetic

---

Biosynthesis in Insects, Advanced Edition
By E. David Morgan
© E. David Morgan 2010
Published by the Royal Society of Chemistry, www.rsc.org

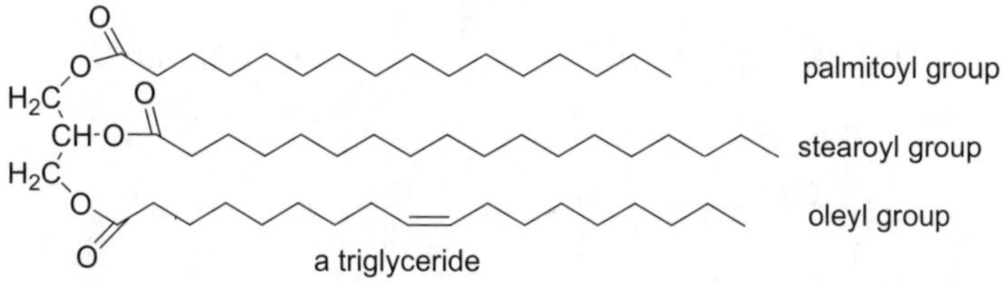

Fatty Acids and Cuticular Hydrocarbons

Figure 4.1 The greater part of fatty acids in living tissues are bound up as triglycerides or glycerol triesters (the structure shown is 1-palmitoyl-2-stearoyl-3-olein). Shown are the formulae of the most common fatty acids, their common names and abbreviations.

CH$_3$(CH$_2$)$_{12}$COOH — tetradecanoic or myristic acid — C$_{14:0}$
CH$_3$(CH$_2$)$_{14}$COOH — hexadecanoic or palmitic acid — C$_{16:0}$
CH$_3$(CH$_2$)$_{16}$COOH — octadecanoic or stearic acid — C$_{18:0}$
CH$_3$(CH$_2$)$_7$CH=CH(CH$_2$)$_7$COOH — (Z)-9-octadecenoic or oleic acid — C$_{18:1}$
CH$_3$(CH$_2$)$_4$CH=CHCH$_2$CH=CH(CH$_2$)$_7$COOH — (Z,Z)-9,12-octadecadienoic or linoleic acid — C$_{18:2}$

acid units. Later, in the 1950s, when the radioactive isotope ^{14}C was available, Arthur Birch at Manchester used labelled acetic acid to show that Collie's proposal was indeed correct.

Fatty acids are synthesized by head-to-tail condensation of two-carbon units, using the *Claisen condensation* to make new carbon-to-carbon bonds. While in the laboratory a strong base (ethoxide ions) and anhydrous conditions are required for this reaction between two molecules of ethyl acetate, essentially the same condensation occurs in cells in an aqueous system and at neutral pH, but in order to make an acetate group sufficiently reactive, two activating effects are applied. First, the acetate is converted to a thioester (Section 2.2.2) and, secondly, the CH$_3$ group reactivity is increased by conversion of acetate to malonate (Figure 4.2). The acidity value for the CH$_2$ group in a malonic thioester is not readily available, but it is evidently very high. In this biosynthetic reaction the combined effect of thioester, an additional carboxylate to activate the CH$_3$ portion of acetic acid and the driving force of decarboxylation are sufficient for an enzyme-catalysed condensation.

4.2.1 The Synthetic Enzyme Complex

In higher organisms, the whole process of fatty-acid synthesis takes place on a complex of enzymes, a megasynthase (Asturias *et al.*, 2005), or multifunctional enzyme, located in the cytosol of cells. The growing molecule remains attached to the complex through a thioester of acyl carrier protein (ACP) (Section 2.2.2), itself a part of the megasynthase,

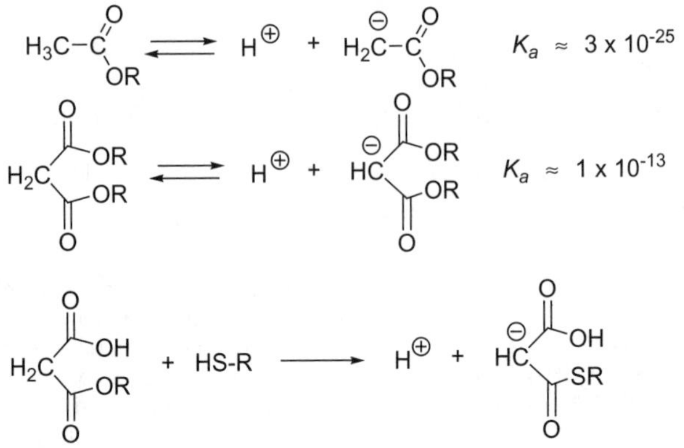

Figure 4.2 The removal of a proton from the CH_3 of an acetate derivative requires a very strong base, that is, for this reaction the CH_3 group of acetate is a very weak acid. Conversion to a malonate derivative increases the acidity almost 10^{12} times, use of a thiomalonate increases it still more.

and the fatty acid is only released from the enzyme complex when the molecule has reached its full length. The complex consists of a single protein chain, with seven catalytic domains along its length, but not arranged in the order of the reaction steps. Two molecules of the complex associate to form a dimer, and both parts of the dimer are used in producing a single molecule of fatty acid, with two molecules of fatty acid being produced at the same time (Figure 4.3). This complex of enzymes is called a type I fatty-acid synthase. In bacteria there are discrete, separable enzymes, collectively called a type II fatty-acid synthase, which carry out the same biosynthesis.

The seven domains are the ACP, a β-keto-synthase (KS), which condenses a malonyl group with the growing acyl group to give a β-keto-acyl group; a β-keto-reductase (KR), which reduces C=O to CHOH; dehydratase (DH), which produces an α,β-unsaturated acyl derivative; enoyl reductase (ER), which reduces the double bond; malonyl-acetyl transferase (MAT), which supplies malonyl-SCoA molecules for the condensations; and thioesterase (TE), which cuts the link to acyl carrier protein when the fatty-acid molecule has reached its full length (Figure 4.4). The acyl carrier protein part, by which the growing chain is attached to the synthase, is a single peptide chain of 77 amino acids, ending in serine, to which is attached phosphate, pantothenic acid and finally mercaptoethylamine, so the terminus of acyl carrier protein resembles co-enzyme A.

The co-enzyme biotin is the carrier of CO_2 to convert acetyl CoA into malonyl CoA. The biotin is attached to acetyl CoA carboxylase through

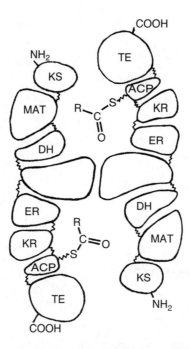

Figure 4.3 A diagrammatic representation of the fatty-acid synthase complex dimer. The two units of the dimer sit head-to-tail, and form two holes inside which the fatty-acid molecules, two at a time, are built up. Parts of both units are used to create a single fatty-acid molecule. The symbols for the domains of the complex are KS β-keto-synthase; MAT malonyl-acetyl transferase; DH dehydratase; ER enoyl reductase; KR β-keto-reductase; ACP acyl carrier protein; TE thioesterase. These symbols are also used besides the arrows in Figure 4.4.

the ε-amino-group of lysine. The source of the new carboxyl group is bicarbonate. The details of the biotin reaction are still not firmly understood, but are best represented as in Figure 4.5. The acetyl group that is the starter unit for the chain is first transferred from co-enzyme A to the acyl carrier protein, then moved again to the β-keto-acyl synthase portion, while the malonyl CoA replaces it on the acyl carrier protein.

Condensation between the two attached groups then takes place as shown in Figure 4.6(a). The decarboxylation of malonate provides a driving force for the condensation. The remaining steps of reduction, dehydration and hydrogenation operate by conventional chemical mechanisms. For reduction of the carbonyl group to C-OH, the *pro-R* hydrogen atom on NADPH is added to the β-keto-group. Again in the reduction of the double bond, the *pro-R* hydrogen atom on NADPH is added to C-3, and H^+ is added to C-2. There is a *syn* elimination of water to give the double bond (Figure 4.6(b)). The resulting butyryl

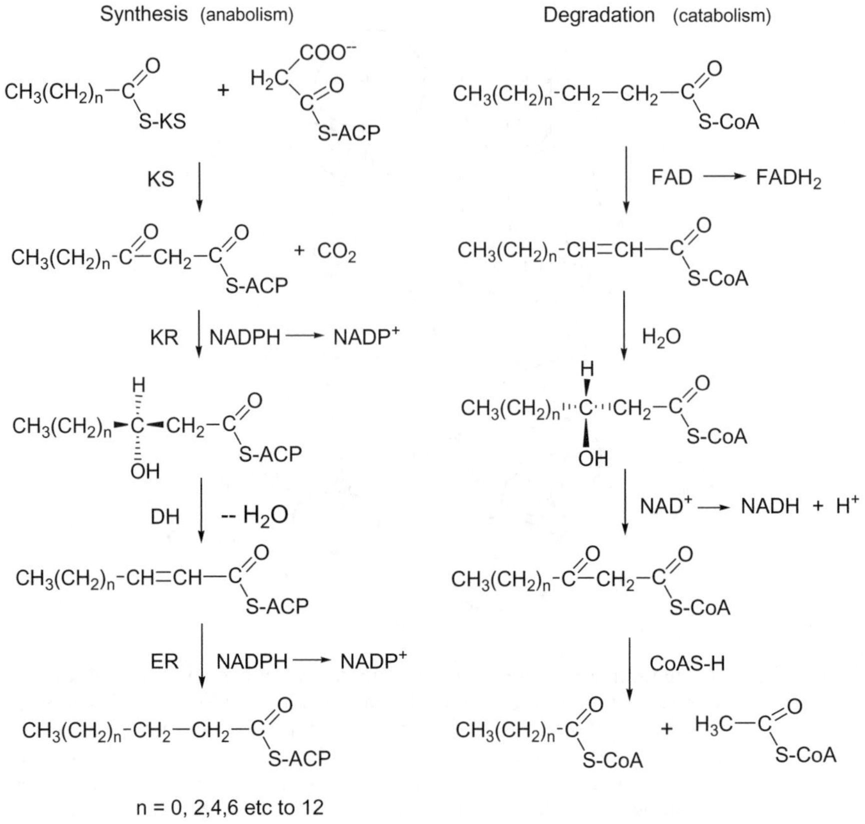

Figure 4.4 The essential steps of fatty-acid synthesis and degradation. The growing chain in synthesis is attached to acyl carrier protein (ACP) through a thioester. In degradation the shrinking molecule is attached to co-enzyme A, also through a thioester. Note that the β-hydroxyl is the (R)-enantiomer in the synthesis process, it is the (S)-enantiomer in degradation; and while NADPH is used to reduce the double bond, FAD is used to produce it. KS is β-keto-synthase, KR is β-keto-reductase, DH is dehydratase and ER is enoyl reductase, as in Figure 4.3.

group moves to the thiol group on the β-keto-acyl transferase, permitting another malonyl group to be attached and the process is repeated. When the chain grows to sixteen carbon atoms it is released from the acyl carrier protein to give palmitic acid. The overall stoichiometric equation is given in Figure 4.6(c).

If there is a pool of propionic acid, it can serve as the starter unit, so the final fatty acid has an odd number of carbon atoms. These odd-numbered acids are frequently found in small quantities in insects, accompanying the even-chain-length acids.

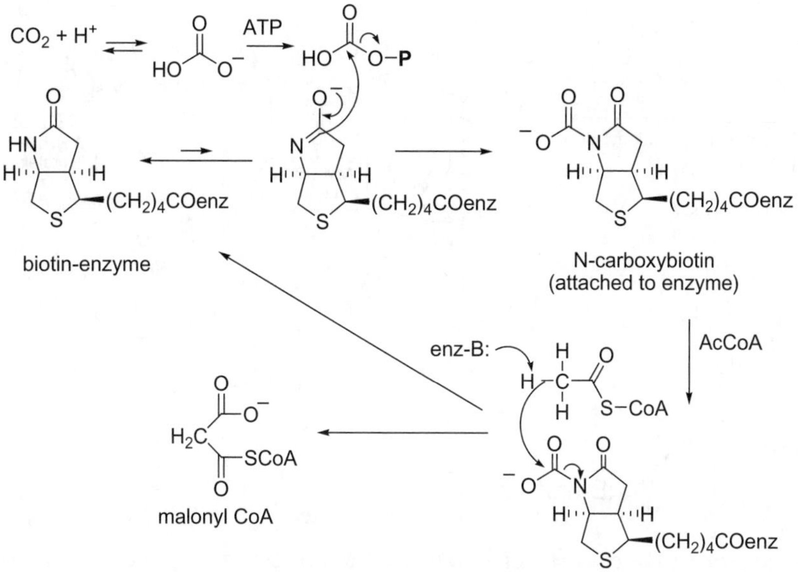

Figure 4.5 Carbon dioxide, as HCO_3^- is attached to biotin, and then transferred to acetyl co-enzyme A forming malonyl co-enzyme A for fatty-acid synthesis. :B$^-$ is a basic group on the enzyme. The acidity produced by dissolution of carbon dioxide is absorbed by the negative charges on protein and phosphate in the haemolymph and cell.

Further lengthening of the chain or desaturation take place elsewhere. Chain lengthening takes place by addition of acetyl groups catalysed by enzymes on the endoplasmic reticulum membrane, requiring malonate and NADPH and following the same group of reactions.

The degradation and synthesis of fatty acids are very similar, so the degradation of fatty acids is summarized in Figure 4.4 also, for comparison. Though the process is similar, there are differences that nature maintains so that independent regulatory mechanisms can control them. While synthesis takes place in the cytosol, the degradation occurs in the mitochondria of the cells, and individual enzymes, not the complex, are used. Synthesis uses NADPH for reduction of a double bond, while in degradation the double bond is formed by the use of FAD. NADPH is also used for the reduction of C=O to CHOH in synthesis, while in degradation NAD$^+$ is used in the oxidation of CHOH to C=O. The stereochemistry of the hydroxyl intermediate is also different. In synthesis an (*R*)-OH group is formed by reduction, while in degradation an (*S*)-OH is formed by hydration of the double bond (Figure 4.4). Finally, in synthesis, the growing chain remains attached to ACP, while in degradation, the fatty acid is attached to CoA. These are further

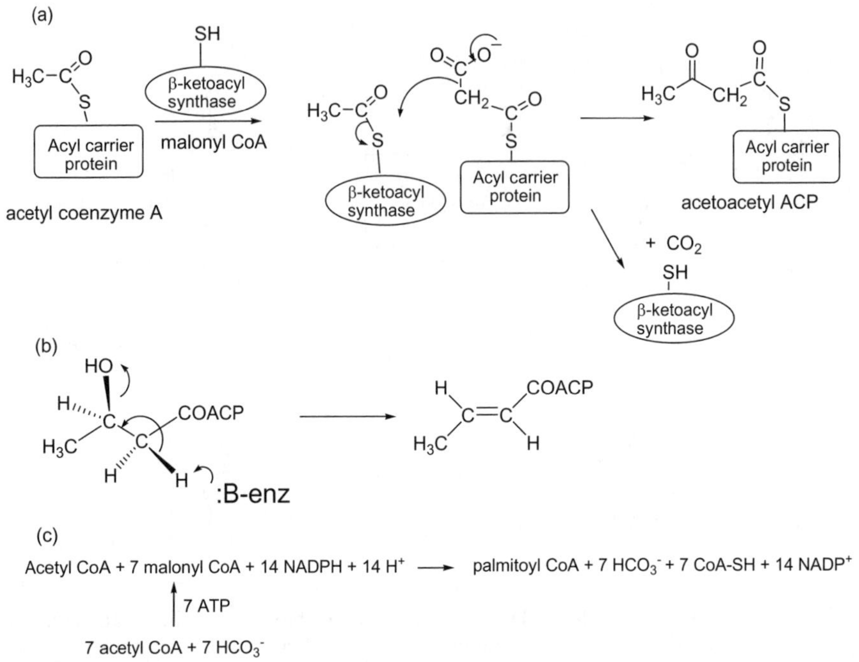

Figure 4.6 (a) The central reaction of formation of a new carbon–carbon bond in the synthesis of a fatty acid in animals. (b) The elimination of water on the dehydratase enzyme is a *syn* elimination. (c) The overall stoichiometric equation to produce a molecule of palmitic acid.

examples of the way that degradation and synthesis are separated as chemical processes.

4.2.2 Unsaturated Acids and Desaturase Enzymes

Double bonds, as in oleic acid (Figure 4.1), can be introduced in two ways: either a double bond is left in the growing chain (in anaerobic bacteria) or by removal of two hydrogen atoms from the complete molecule (requiring oxygen, in plants and animals, including insects). This latter is a most remarkable reaction, still unachievable in the laboratory, and hence has received a great deal of investigation. Double bonds are introduced in fatty acids by desaturase enzymes. They are remarkable because they can remove hydrogen atoms from a non-activated alkyl chain, with great precision of position and stereochemistry, something that chemists have not yet learned to do. They are also of special interest because of their importance in synthesis of lepidopteran sex pheromones (Section 5.1). The C–H bond is very stable;

it has a bond enthalpy of $413\,kJ\,mol^{-1}$ or $98.7\,kcal\,mole^{-1}$. It requires the oxidative power of O_2 to break this bond. The by-product is water.

There are two types of desaturase; the first are soluble enzymes, found only in plants, and located in the plastids. Their substrate is a fatty acid attached to acyl carrier protein (acyl ACP). The second type, found in animals, including insects, is integrally bound to the membrane of the endoplasmic reticulum. Their substrate is the fatty acid bound to co-enzyme A (acyl CoA).

Soluble enzymes are much easier to study, so more is known of the first type, but from many studies with a variety of spectroscopic, X-ray and molecular biological techniques, it seems the mechanism of reaction is the same in both types. The general appearance is shown for a soluble Δ9-desaturase from the castor oil plant in Figure 4.7. Although the full story of these enzymes is not yet known, the description here summarizes our present knowledge of membrane-bound fatty-acid desaturases found in insects. The description is of a Δ9-desaturase, the most common type, which converts stearic acid to oleic acid (Z)-9-octadecenoic acid.

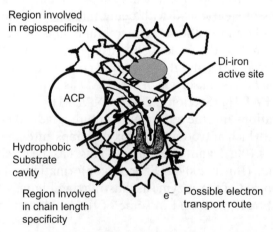

Figure 4.7   A drawing of the Δ9-desaturase enzyme from the castor oil plant showing the monomer with the long, narrow pocket into which the fatty acid, shown as a curved line, fits. While the fatty acid is attached to acyl carrier protein in plants, as here, it is attached to CoA in animals. Note the di-iron cluster, and the region at the bottom of the cavity, which determines how deep the fatty acid can fit, and consequently where the double bond will be inserted. When stearic acid fills the pocket, C-9 is about 0.55 nm from one of the iron atoms, sufficient space to accommodate O_2. The direction from which the electrons come from cytochrome, which re-oxidize the iron cluster, is also shown. The protein chain is represented by a jagged black line. Figure provided by J. Shanklin and E. Cahoon. See also Plate 5.

The location of the double bond is measured from the carboxylate end of the molecule. Palmitic acid with this enzyme gives palmitoleic acid (Z)-9-hexadecenoic acid. If an unnatural C_{17} or C_{19} acid is supplied to the desaturase enzyme, a Δ-9 acid is always formed. The double bonds introduced in this way in fatty acids always have a Z or cis configuration. The desaturases of insect fat body tend to produce more oleic acid than palmitoleic acid, *i.e.* they are partially selective for stearic acid.

The protein chain consists essentially of four α-helix coils (Plate 5), embedded in the membrane, with a long narrow pocket into which the alkyl end of the fatty acid fits. Note the bent configuration that the molecule assumes at the C-9 atom (regardless of chain length), where the new double bond will be formed, and the two iron atoms within catalytic distance of this position. The iron atoms are buried deep inside the molecule, held by several histidine residues. When the substrate fits into the pocket, oxygen becomes attached to the di-iron cluster (Figure 4.8), which then attacks the hydrocarbon chain (Buist and Behrouzian, 1996). The subsequent steps are still not known in great detail.

It is known that the *pro-R* hydrogen atoms are removed from C-9 and C-10 (Figure 4.9) from specific labelling of *pro-R* and *pro-S* hydrogens at these positions of stearic acid with deuterium and tritium. Two further enzymes, both bound in the same membrane, and two co-enzymes are required to complete the cycle. Cytochrome b_5 restores the desaturase iron to its reduced state, and it in turn is re-converted to its reduced state by cytochrome b_5 reductase, which uses FAD as co-enzyme. The FAD is converted to $FADH_2$ (Section 2.2.4) by NADPH.

Small variations in the enzyme structure give other desaturases, hydroxylases (which introduce hydroxyl groups into the fatty acid at C-9) (Figure 4.10(a)) and epoxidases. This has been clearly demonstrated in plants (Buist, 2007). Different conformations of the acid alkyl chain in the pocket of the enzyme as illustrated in Figure 4.10(b) can give *trans* double bonds, as found in some of the lepidopteran pheromones

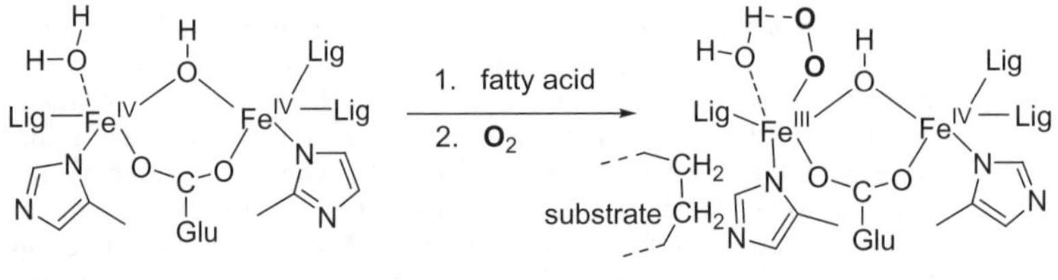

Figure 4.8 The proposed arrangement of ligands (including histidine and glutamic acid residues) at the di-iron cluster in a desaturase enzyme.

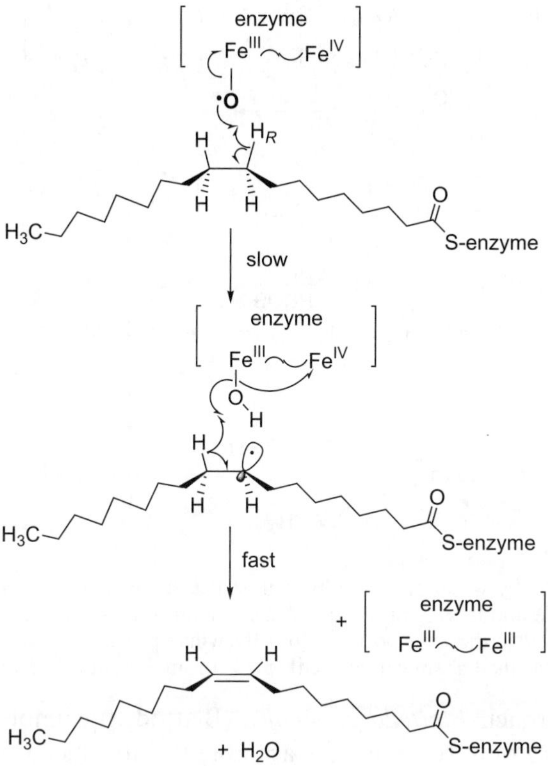

Fatty Acids and Cuticular Hydrocarbons 75

Figure 4.9 The removal of the *pro*-H_R from C-9 and the formation of a double bond in a fatty acid, illustrated by the conversion of stearic acid (as a CoA thioester) to oleic acid by a Δ9-desaturase enzyme. It is a free radical reaction with the removal of two *syn*-oriented vicinyl hydrogen atoms.

(see later). Changes in protein folding produce Δ-7 or Δ-11 desaturases (Sperling *et al.*, 2003). Acetylenases, methyl oxidases (converting methane to methanol) and other enzymes have similar di-iron active sites (Shanklin and Cahoon, 1998).

Linoleic ((9Z,12Z)-9,12-octadecadienoic) and linolenic ((9Z,12Z, 15Z)-9,12,15-octadecatrienoic) acids (Figure 4.11) are made by plants by further desaturation of oleic acid. Usually, animals cannot make linoleic acid, thus it is essential in their diets, but they are able to introduce double bonds between C-9 and the carboxyl group (usually at C-6). Radio-labelling experiments with insects indicated that some of them are capable of synthesizing linoleic acid. Thirty-eight species were screened and, of them, twelve were found capable of making linoleic acid (Cripps *et al.*, 1986). The cabbage looper *Tricoplusia ni* (Lepidoptera: Noctuidae), the cecropia silk moth *Hyalophora cecropia* (Lepidoptera: Saturniidae) and the

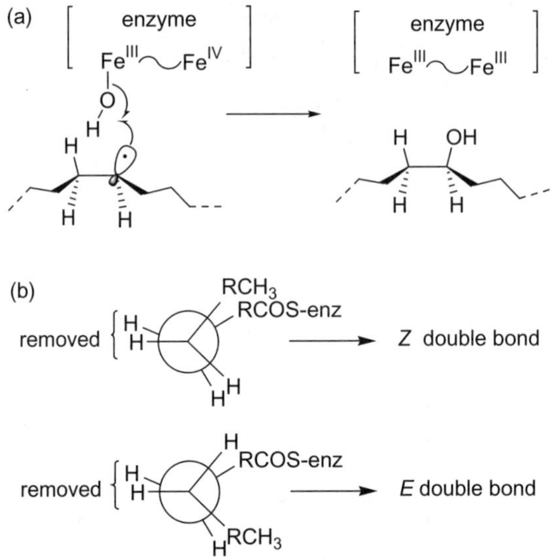

Figure 4.10 (a) By variation of the protein structure, hydroxylase enzymes similar to desaturases are produced, which can introduce hydroxyl groups into a fatty acid at the same position. (b) A Newman projection to show how a twist of the alkyl chain can give either a Z or an E double bond with desaturases.

German cockroach *Blattella germanica* (Blattodea) cannot make linoleic acid. Therefore only some insects are able to introduce unsaturation on either side of the double bond of oleic acid. Experiments have shown that the desaturases are in the endoplasmic reticulum, and the use of axenic insects have eliminated the possibility that symbiotic bacteria are responsible. Recently the first two genes for insect $\Delta 12$-desaturases have been isolated and characterized, one from the house cricket, *Achaeta domestica* (Orthoptera: Gryllidae), and the other from the red flour beetle *Tribolium castaneum* (Coleoptera: Tenebrionidae) (Zhou et al., 2008). These genes have evolved independently from all those previously known for $\Delta 12$-desaturases and are only distantly related to each other. A slug and a garden snail can also make linoleic acid (Weinert et al., 1993). There is no indication that linolenic acid can be made by higher animals.

Arachidonic acid (5,8,11,14-eicosatetraenoic acid) is made only by animals, by chain extension and desaturation from linoleic acid in the endoplasmic reticulum (Figure 4.11).

4.2.3 Eicosanoids

A series of compounds called collectively eicosanoids, because they are derived from C_{20} unsaturated acids, have now been found in virtually

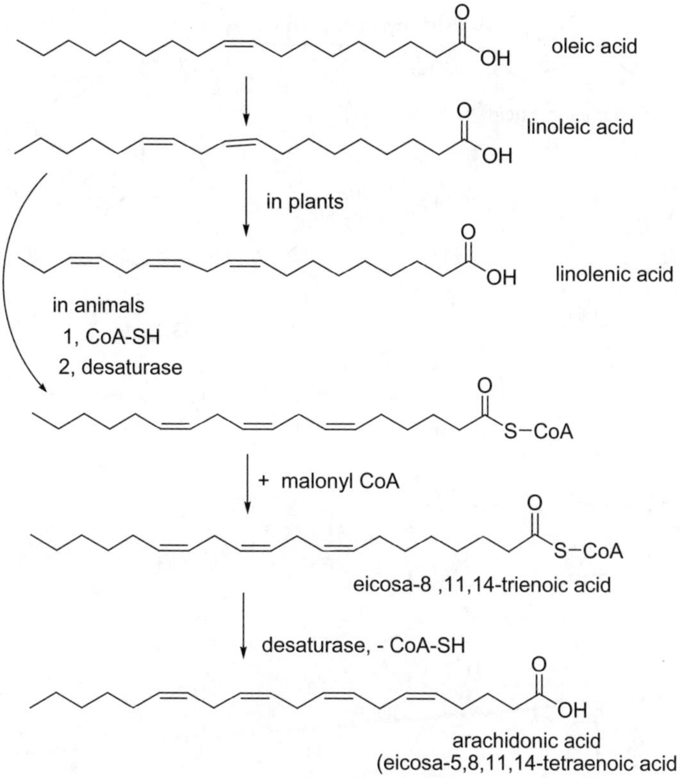

Fatty Acids and Cuticular Hydrocarbons 77

Figure 4.11 The biosynthetic relationship between oleic, linoleic, linolenic and arachidonic acids.

every part of the animal kingdom that has been scrutinized (Howard and Stanley, 1999). They were first discovered in man and are important in human physiology, and have been extensively studied. They include prostaglandins, prostacyclins, thromboxanes, leucotrienes and lipoxins. Prostaglandins, prostacyclins, and thromboxanes are made from arachidonic acid through the catalysis of cyclo-oxygenase enzymes (Figure 4.12). Little or nothing is known at present of thromboxanes and lipoxins in insects. Prostaglandins are widely distributed in various tissues and exert hormone-like effects, and are effective at very low concentrations. Prostaglandins are named with a letter to indicate the state of oxygenation and a subscript number which tells the number of double bonds in the molecule. They are biosynthesized chiefly by cytosolic, but also by microsomal, preparations, from the midgut of the tobacco hornworm *Manduca sexta* (Lepidoptera: Sphingidae) (Plate 6), the American cockroach *Periplaneta americana* (Blattoidea),

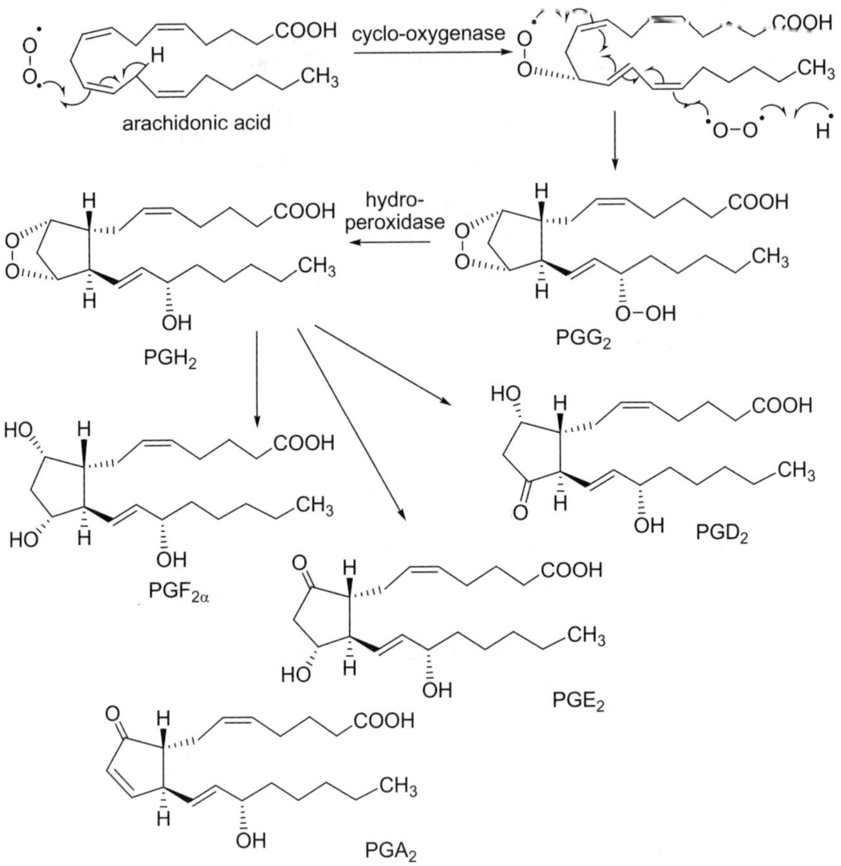

Figure 4.12 The probable formation of some prostaglandins found in insects, based upon the known reactions in vertebrates. The first step in the series is inhibited by aspirin. The prostaglandins PGG_2 and $PGH_{2\alpha}$ are intermediates in the sequence. PGA_2 in insects is probably formed by a separate pathway.

Helicoverpa zea (Lepidoptera: Noctuidae) and larvae of the beetle *Zophobas atratus* (Coleoptera: Tenebrionidae).

Arachidonic acid is converted by a cyclo-oxygenase to the unstable endoperoxide prostaglandin G_2 (PGG_2). A hydroperoxidase reduces PGG_2 to the more stable PGH_2. The two enzymes required here are associated with the same holoprotein, collectively called COX (cyclo-oxygenase). The conversion of PGH_2 is catalysed by various cell-specific enzymes; by PGF reductase to $PGF_{2\alpha}$, by PGD isomerase to PGD_2 and by another isomerase to PGE_2. PGA_2 is probably formed by another pathway. The study with *Manduca* midgut was made with arachidonic acid labelled with tritium on all the double bonds, *i.e.* positions 5, 6, 8, 9,

11, 12, 14 and 15. The major product identified was PGA_2, but PGD_2, PGE_2 and $PGF_{2\alpha}$ were all present (Büyükgüzel et al., 2002).

Prostaglandins are sometimes described as local hormones. In insects they release egg-laying behaviour, salt and water transport, fever response and cellular immune defences (Stanley, 2006). In *Manduca sexta* they have been shown to induce biosynthesis of antimicrobial peptides and proteins, haemocyte migration to the source of a wound or infection and cellular reaction or encapsulation (Merchant et al., 2008). Prostaglandin GA_2 (PGA_2) (Figure 4.12) seems to be important in insect immune response.

The salivary glands of the lone star tick *Amblyomma americanum* (Ixodida) have been shown to synthesize prostaglandins. The tick cannot synthesize arachidonic acid, but sequesters it from its host so that the glands contain high levels of arachidonic acid.

The defensive secretion of the red flour beetle *Tribolium castaneum* (Plate 7) contains more than 0.25% of two aromatic hydroxyketones that are potent inhibitors of prostaglandin H synthesis (about 100 times more potent than aspirin) (Howard et al., 1986). Insects in several orders contain substantial inhibitors of prostaglandin synthesis.

4.2.3.1 Insect Immunity. Insects have three sources of protection from infection and injury: their cuticle presents a physical barrier, when infection reaches inside their bodies, there is induced formation of antimicrobial peptides, lysozyme (Section 2.1.1) and polyphenol oxidases, and cellular immune reactions of phagocytosis, nodulation and encapsulation. Phagocytosis (the engulfing of invading organisms by specialized cells in the haemolymph) begins immediately after infection. Larger invading organisms are encapsulated in a chitinous barrier (Section 9.7), which eventually forms a sclerotized nodule, surrounded with an impenetrable layer of cuticle. More slowly, there is formation of antibacterial peptides. Eicosanoids have an important but not yet fully understood place in insect immune response (Stanley et al., 2009). Prostaglandins have been demonstrated to react to invasions of fungi, protozoa, viruses and parasitoids.

4.2.4 Branched Fatty Acids

Branched fatty acids, known as iso-acids (n-2 methyl) and anteiso-acids (n-3 methyl), occur normally in small quantities in fats. The synthesis of iso-acids begins with the amino acids valine and leucine (Figure 4.13). This has been demonstrated with radio-labelled isotopes, by radioactive monitoring and with stable isotopes by ^{13}C nuclear magnetic resonance

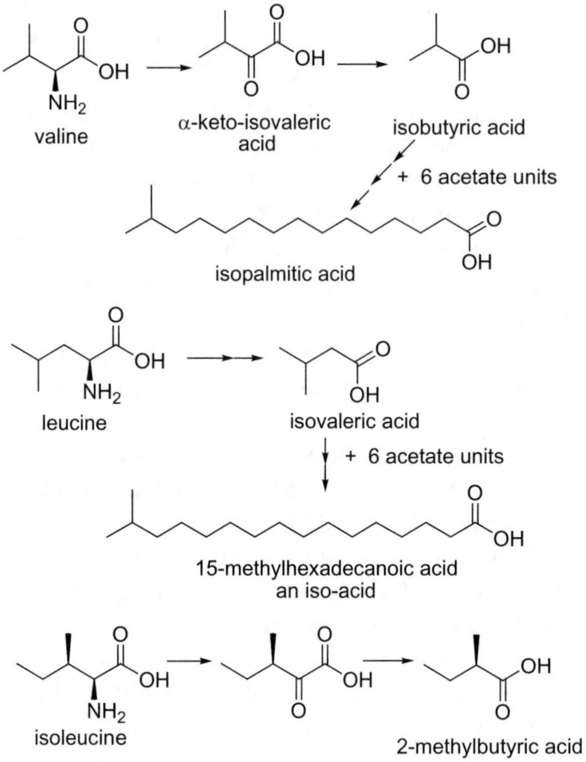

Figure 4.13 The origin of branched iso- and anteiso-fatty acids illustrated for the C_{16} and C_{17} acids, respectively. Iso-acids have an even number of carbon atoms, while anteiso-acids have an odd number. In insects, 3-methyl fatty acids are not made from isobuytyric acid but from acetate and malonate.

spectroscopy or mass spectrometry. Isobutyric acid, from valine, and isovaleric acid, from leucine, by chain extension give iso-fatty acids. Isobutyric acid gives fatty acids with an even total number of carbon atoms, while isovaleric acid gives fatty acids with an odd total number of carbons. 2-Methylbutyric acid is derived from isoleucine, and has a chiral centre. Both isobutyric acid and 2-methylbutyric acid are common defensive compounds among insects. Larvae of the European leaf beetle *Chrysomela lapponica* (Coleoptera: Chrysomelidae) feeding on birch produce a deterrent complex mixture of esters of isobutyric and (S)-2-methylbutyric acids from nine pairs of abdominal glands. Application of deuterated valine gave esters with deuterated isobutyric acid (Schulz et al., 1997). The alcohol portions mostly come from the plant. Synthesis of these branched acids from valine and isoleucine has been demonstrated in at least one case each in Chapter 3 (Section 3.3.3).

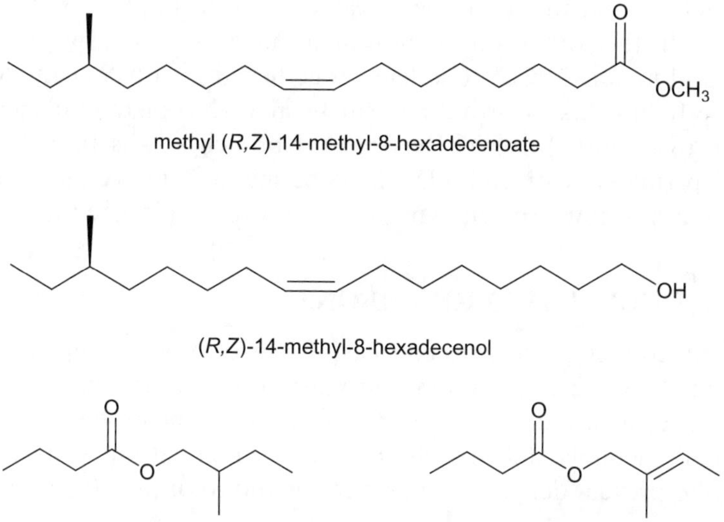

Figure 4.14 Two unsaturated anteiso-compounds that are part of the sex pheromone of some *Trogoderma* species, and two small-molecule esters that form the sex pheromone of the bug *Alydus eurinus*. They are probably butyric esters of alcohols derived from valine *via* isobutyric acid.

Plants use isobutyric acid to gives anteiso-fatty acids with an odd total number of carbon atoms. In insects, anteiso-acids are made with an acetate starter unit followed by a propionate. Females of the broad-headed bug *Alydus eurinus* (Hemiptera) release 2-methylbutyl butyrate and (*E*)-2-methylbutenyl butyrate as an attractant pheromone (Figure 4.14). Adults, when disturbed, also produce a defensive secretion of butyric and hexanoic acids from metathoracic glands.

In both the housefly *Musca domestica* (Diptera: Muscidae) and the German cockroach *Blatella germanica* it has been shown that the branched-chain fatty acids that are the precursors of hydrocarbons are synthesized by a microsomal synthase in the integument (Juarez *et al.*, 1992). The cuticular lipids of the cabbage seed weevil *Ceutorhynchus assimilis* (Coleoptera: Curculionidae) contain small amounts of iso-acids and anteiso-acids esterified with long-chain alcohols, iso-alcohols and anteiso-alcohols derived from the corresponding acids. The kapra beetle *Trogoderma granarium* (Coleoptera: Dermestidae) and other species of *Trogoderma* use derivatives of the anteiso-unsaturated acid (*R*,*Z*)-14-methyl-8-hexadecenoic acid (Figure 4.14), including the methyl ester, the corresponding alcohol, the aldehyde and derivatives with the *E*-double bond, as part of the female-produced sex attractant. The formation of a whole range of methyl-branched fatty acids has been demonstrated in the housefly *Musca domestica* (Blomquist *et al.*, 1994).

In bacteria and fungi another means of producing methyl branches operates. In the formation of the fungal metabolite lovastatin, acetyl-SCoA and malonyl-SCoA condense together and with S-adenosyl methionine by an unknown mechanism to give $(2R)$-methylbutyric acid. Pederin (Section 6.9), which has methyl branches, was thought from early experiments with radio-labels to be made from acetate and malonate, but has now been shown not to involve propionate units.

4.3 CUTICULAR HYDROCARBONS

The outer covering of insects consists of a layer of water-repellent lipid compounds, which can contain a mixture of hydrocarbons, long-chain alcohols, acetates, aldehydes, ketones, fatty acids, their ester, or esters of long-chain alcohols, and triacyl glycerides. The lipid outer layer is important to prevent dehydration or wetting and to protect from bacterial infection.

The amount of triglycerides in cuticular extracts is sometimes uncertain because of the possibility of contamination from internal lipids. Part of the lipid is extractable only by polar solvents, and probably occurs as a lipid-sclerotin mixture. Lockey (1988) gives a table of composition of cuticular lipids for 40 species across 10 classes in the Exopterygota and Endopterygota. All but one, the Colorado potato beetle, *Leptinotarsa decemlineata* (Coleoptera: Chrysomelidae), are listed as containing hydrocarbons, and later work has shown that it too has methyl-branched alkanes. Hydrocarbons appear to be universally present on the cuticle, and can comprise anything from 100% of the lipids, as in the field cricket *Nemobius fasciatus* (Orthoptera: Gryllidae) to as little as 0.5% in the cuticle of tobacco budworm pupae *Heliothis virescens* (Lepidoptera: Noctuidae). The hydrocarbons, in turn, may be made up of mixtures of alkanes, methylalkanes and alkenes. In recent years attention has focused on the hydrocarbons, since they also serve a communication function to recognize species, sex, fertility, dominance, mimicry and, in social insects (bees, wasps, ants and termites), the mixture is characteristic of the group (Nelson and Blomquist, 1995, Howard and Blomquist, 2005). The available evidence suggests the mixture helps individuals to distinguish between nest mates, individuals from another colony and even patrilines and the tasks that workers are performing. What is not yet clear is whether all the hydrocarbons are used in this way. Linear hydrocarbons have rather featureless molecules, and there is evidence that they are not used for recognition, and only alkenes and methyl-branched alkanes, which have less symmetry, or more characteristic shape, are used (Dani *et al.*, 2005). Early work showed that honeybee olfactory neurons did not respond to C_{23} to C_{27} alkanes, and in the ant *Formica exsecta* (Hymenoptera: Formicidae) only (Z)-9-alkenes elicited aggression in inter-colonial tests (Martin et al., 2008).

4.3.1 Hydrocarbon Biosynthesis

It was known for some time that hydrocarbons were synthesized by cells close to the epidermis, probably the oenocytes, while fat body synthesized triglycerides and polar lipids but not hydrocarbons. Ferveur et al. (1997) by manipulation of genes were able to induce adult male fruit flies *Drosophila melanogaster* to produce feminine pheromone hydrocarbons, and showed that these compounds were made in the oenocytes. Fan et al. (2003) were able to separate oenocytes clearly from other tissues and used radio-labelled propionate with *Blatella germanica* to prove that oenocytes synthesized hydrocarbons, not only in insects where oenocytes are localized in the haemocoel but also in those where oenocytes are in the abdominal integument. The hydrocarbons are transported from the oenocytes by lipophorin (Section 4.1), the general transporter of lipids through the haemolymph. How they reach the outer surface of the cuticle is not yet clear.

The amount of hydrocarbons detectable internally is much greater early in an instar in several studies. The main component of the female sex pheromone found on the cuticle of *Musca domestica*, (Z)-9-tricosene, is detectable internally 24 hours before it can be detected on the cuticle, and in six-day-old adults there is three times more of it internally than on the cuticle. In *Blatella germanica*, using radio-labelled precursors, about 50% of labelled maternal hydrocarbons can be detected in oocytes and the maternal hydrocarbons can still be found in embryos and the cuticle of first, second and even later instars of progeny (Fan et al., 2008). In the bug *Triatoma infestans* (Hemiptera: Reduviidae) hydrocarbons can be synthesized by the eggs (Juarez and Fernandez, 2007).

Cuticular hydrocarbons are derived from fatty acids through chain lengthening and decarboxylation. Chain lengthening in *Musca domestica* takes place in the microsomal fraction (Blomquist, 2003). The carbon chain is extended with more acetate groups (converted to malonate for the synthesis) to make longer-chain acyl CoA, which is decarboxylated to hydrocarbons (Figure 4.15). In *Periplaneta americana* stearic acid is chain-lengthened only to hexacosanoic acid and decarboxylated to pentacosane, and linoleic acid is lengthened to a C_{28}-dienoic acid to give heptacosadiene. These two compounds, with 3-methylpentacosane, are the principal cuticle hydrocarbons of the

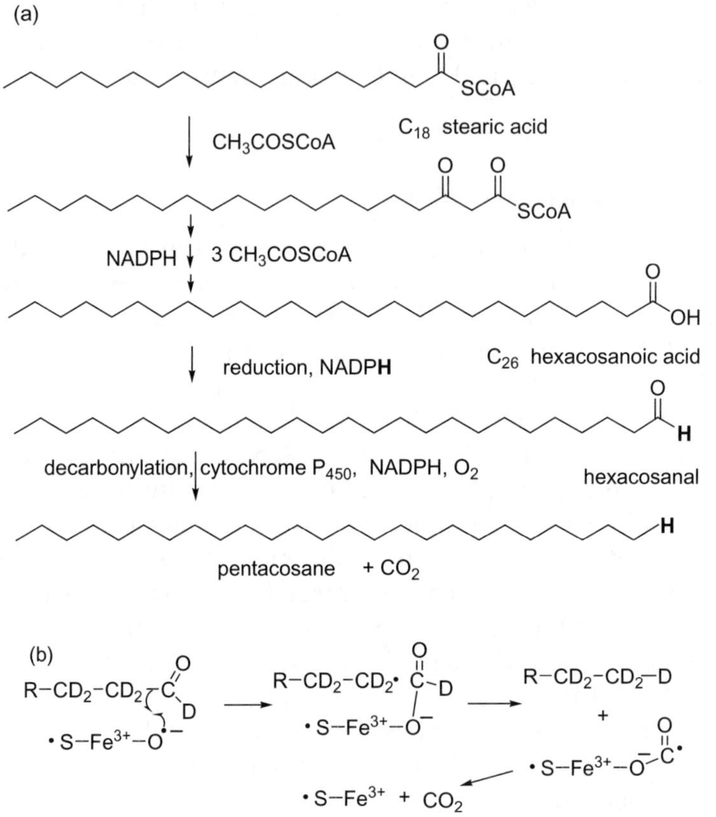

Figure 4.15 (a) Outline of the synthesis of a C_{25} hydrocarbon by chain-lengthening from stearic acid and decarboxylation. The bold **H** atom indicates that a labelled hydrogen atom from NADPH is retained on the terminal carbon atom of the alkane. (b) A simplified version of the proposed mechanism for decarboxylation, involving a diradical perferryl group on the enzyme. Deuterium on carbon atoms 2 and 3 are unaffected in the reaction, showing that only the aldehyde group is attacked.

species. The greatest contributions to the study of hydrocarbon biosynthesis have come from the group of Blomquist. Major and Blomquist (1978) showed that in *Periplaneta* long-chain fatty acids are converted to hydrocarbons. Later he showed that the carboxyl group is reduced to aldehyde with NADPH and, with the aid of oxygen and a cytochrome P450, the carbonyl group is lost as CO_2 (Reed *et al.*, 1994). This route has been confirmed by others, with other insects, in contradiction to Dennis and Kolattakudy (1992) who claimed that, in bacteria, plants and animals, a microsomal cobalt-porphyrin-containing enzyme converted the intermediate aldehyde to carbon monoxide and hydrocarbon.

The enzymes involved have not yet been cloned to settle the uncertainty. It has been shown, by labelling, that the hydrogen of aldehyde becomes attached to the end of the hydrocarbon chain. The decarbonylation occurs in the microsomal fraction. Chain length is controlled at the microsomal elongation of fatty acyl-CoA stage, rather than in the reduction to hydrocarbon. Hydrocarbons, like fatty acids, are then transported in the haemolymph by lipophorin.

As the common fatty acids have an even number of carbon atoms, most hydrocarbons have an odd number of carbon atoms. The low proportion of fatty acids in lipids with an odd number of carbon atoms is matched by a low proportion of even-numbered alkanes.

Until recently, it was thought that cuticular hydrocarbons had up to 40 carbon atoms in the chain, which conveniently matched the limit of detection by gas chromatography, but the arrival of new high-temperature stationary phases and new mass spectrometric techniques has enabled longer chains to be detected. In at least one case, the ant *Formica truncorum* (Hymenoptera: Formicidae), although the series of cuticular hydrocarbons formerly recognized runs from about C_{21} to C_{37}, another series exists between C_{41} and C_{48}, chiefly multiply methyl-branched, which accounts for 56% of the total. The technique of matrix-assisted laser desorption-ionization mass spectrometry (MALDI) coupled with time-of-flight mass analysis (TOF) has revealed hydrocarbons with 70 carbon atoms, saturated and unsaturated, in an examination of 12 species of diverse insects (Cvacka *et al.*, 2006). In the region of molecular weights where they overlapped, GC-MS and MALDI-TOF techniques gave good congruency.

4.3.2 Branched-chain Hydrocarbons

Methyl-branched hydrocarbons, other than those with a 2- or (in some cases) 3-methyl branch, are produced through the substitution of propionic acid (in the form of methylmalonyl CoA) for acetic acid, in building up the carbon chain. The intermediate methyl-branched fatty acids and the microsomal synthase for making them have been isolated (Blomquist *et al.*, 1994). Studying the synthesis of 3-methylpentacosane by *Periplaneta americana* (Plate 8) with labelled propionate and ^{13}C NMR spectroscopy showed that the methyl group was introduced early in the synthesis, not near the end. Sodium 2-[methyl–^{13}C]methylmalonate was incorporated chiefly into the branch methyl group, and sodium [1-^{13}C]propionate was incorporated only into the 4-position (Figure 4.16) (Dwyer *et al.*, 1981a). For methyl- and dimethyl-alkanes in the Pacific dampwood termite *Zootermopsis angusticollis* (Isoptera: Termopsidae),

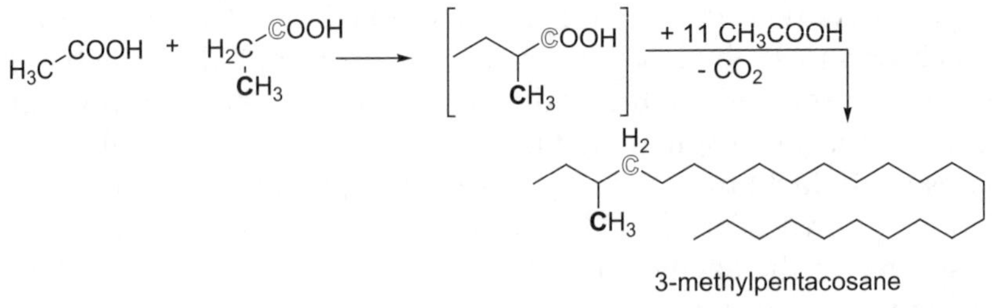

Figure 4.16 The synthesis of 3-methylpentacosane from acetate and propionate. The labelling of carbon in methyl and carboxylate, and their appearance in the final hydrocarbon are indicated by the bold and outlined C, respectively. The intermediate CoA thioesters and methylmalonate are not shown.

the smoky brown cockroach *Periplaneta fuliginosa* (Blattodea) and *Musca domestica*, propionate was the preferred intermediate and in *Z. angusticollis* and *M. domestica* [methyl-^{14}C]methionine was not incorporated. The positions in the chain methylated (usually 5, 7, 9, 11, 13, 15) favour the methylmalonate hypothesis. Dimethylalkanes have methyl groups separated by 3, 5 or 7, and so on, CH_2 groups, because the methylmalonate units are separated by a finite number of malonate units, plus one CH_2 from the methylmalonate giving rise to the branch methyl. The enzyme system is capable of selecting between malonyl CoA and methylmalonyl CoA and adding the correct intermediate at each step in the chain lengthening, but it is not known how this selection is achieved.

In *Z. angusticollis*, succinate is converted to methylmalonate, apparently by gut micro-organisms, but mitochondrial extracts of *Blatella germanica* and the southern armyworm *Spodoptera eridania* (Lepidoptera: Noctuidae) can also convert succinate to methylmalonate. This is an interesting reaction using vitamin B_{12} as co-enzyme with methylmalonyl CoA mutase. Vitamin B_{12} has a complex structure with a central cobalt atom. One of the ligands attached to the cobalt atom is a 5'-deoxyadenosyl group. The Co–CH_2 bond of this ligand is broken homolytically to give a carbon radical which attacks an α-CH_2 group of succinyl CoA. Rearrangement of the radical intermediate and re-capture of hydrogen from the co-enzyme ligand gives methylmalonyl CoA (Figure 4.17). Examining the reaction in Figure 4.17 explains why [2,3-^{13}C]succinic acid is found in the alkanes at higher level than [1-^{13}C]succinic acid, the carboxyl carbon being more likely to be lost in decarboxylation. Though the reaction has not been investigated in insects, it has been studied in other organisms, and the X-ray crystal structure of the enzyme methylmalonyl CoA mutase has been solved.

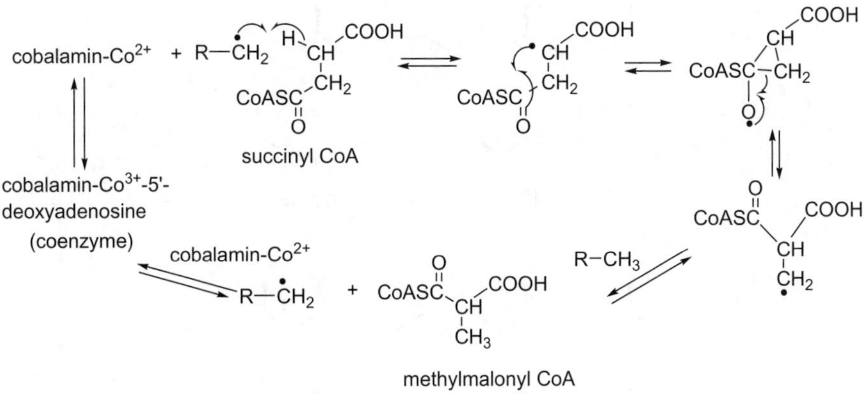

Figure 4.17 The rearrangement of succinic acid (as its CoA half ester) to malonic acid is a free-radical rearrangement initiated by the co-enzyme 5'-deoxyadenosylcobalamin (a form of vitamin B_{12}).

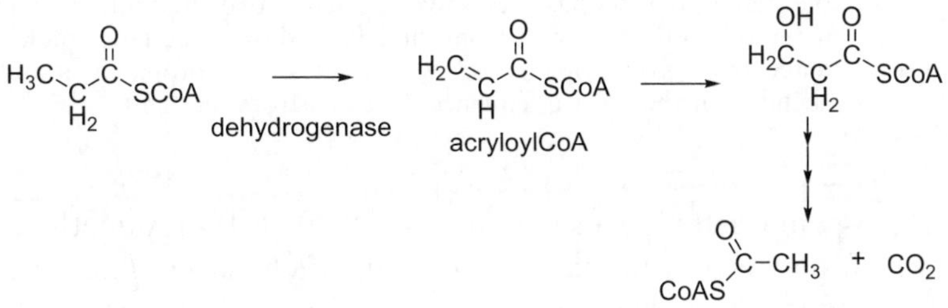

Figure 4.18 The metabolism of propionate by insects is through dehydrogenase-catalysed formation of acryloyl CoA, hydratase-catalysed hydration of the double bond and oxidation of the hydroxyl to carboxyl. The carboxyl group of propionate is lost as CO_2 in steps not yet explored.

Termites have unusually high levels of vitamin B_{12} for insects, but not all insects have vitamin B_{12} (Wakayama et al., 1984). This led to the discovery that insects metabolize propionate differently from vertebrates. Labelled propionic acid, both [2-^{13}C]- and [3-^{13}C]propionic acid can be incorporated into straight-chain fatty acids. All insects so far examined metabolize propionate to acetate as shown in Figure 4.18 (Halarnkar and Blomquist, 1989).

We know almost nothing about the chirality of the methyl branches in cuticular hydrocarbons. One exception is (3S,11S)-3,11-dimethylnonacosane, an intermediate in the formation of (3S,11S)-3,11-dimethylnonacosan-2-one, the female sex pheromone of *Blatella germanica* (Chase et al., 1990). The Australian cane beetle *Antitrogus parvulus* (Coleoptera: Scarabaeidae) produces two major hydrocarbons,

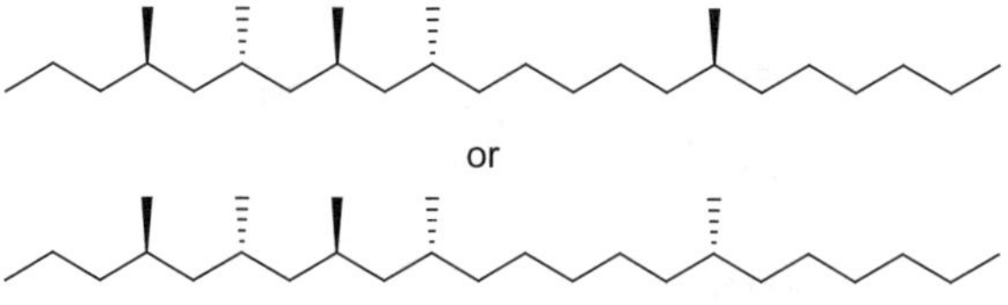

4,6,8,10,16-pentamethyldocosane

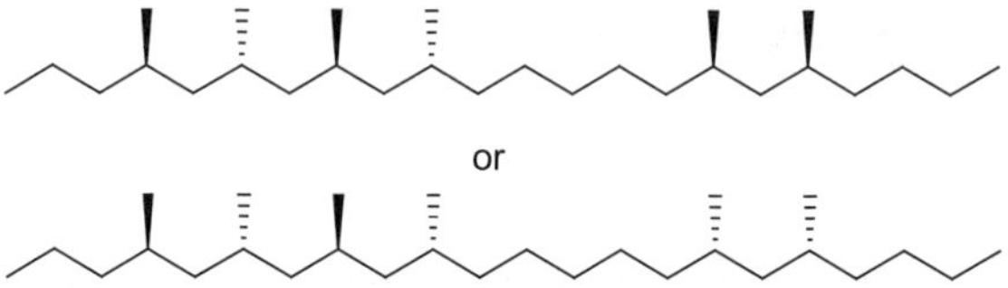

4,6,8,10,16,18-hexamethyldocosane

Figure 4.19 Two hydrocarbons from the cane beetle *Antitrogus parvulus*. In each case the stereochemistry has been solved to one of two alternatives. Note that if the biosynthesis begins from the left as shown, the latter molecule is made from five propionates, two acetates, two propionates and two acetates, or the reverse sequence if started from the right.

4,6,8,10,16-pentamethyldocosane and 4,6,8,10,16,18-hexamethyldocosane, which play a part in mate recognition (Fletcher *et al.*, 2008). The partial stereochemistry has been deduced as shown in Figure 4.19.

The cricket *Nemobius fasciatus* (Orthoptera: Gryllidae) was used to show that fully deuterated L-valine and [1-^{14}C]isobutyric acid were incorporated into 2-methylalkanes (see Figure 4.13). It seems that 2-methylalkanes with an even number of carbon atoms are made in this way. Another cricket, *Gryllus pennsylvanicus* (Orthoptera: Gryllidae), has 2-methylalkanes with both odd and even numbers of carbon atoms. It was suggested that odd-numbered 2-methylalkanes are made from leucine, metabolized to isovaleric acid (Blailock *et al.*, 1976) (Figure 2.20). The arctiid moths *Holomelina aurantiaca*, *H. immaculata* and *H. lamae*, all of which use 2-methylheptadecane as principal component of their sex pheromone, synthesize it in oenocytes, from where it is carried by lipophorin (with other hydrocarbons) and selectively taken up by the pheromone glands. Deuterium-labelled leucine and isovaleric acid, and radio-labelled acetic, malonic, 13-methyl tetradecanoic and 15-methylpalmitic acids were all incorporated into the hydrocarbon, but labelled valine and isobutyric acid were not (Figure 4.20). Similarly, in the gypsy moth *Lymantria dispar* (Lepidoptera: Lymantriidae), (*Z*)-2-methyl-7-octadecene is made, most probably in the oenocytes, beginning with

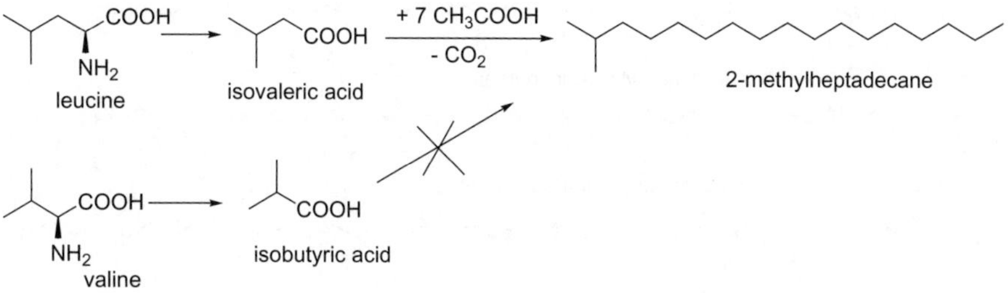

Figure 4.20 The origin of the branch methyl of 2-methylheptadecane in arctiid moths.

valine, transported to the pheromone gland and there epoxidized to the pheromone (Jurenka et al., 2003).

Several authors have reported very-long-chain hydrocarbons, alcohols and esters of these alcohols in the *internal* lipids of pupae of Lepidoptera, Coleoptera and Diptera. The unusual 2,18,20-trimethyltetratriacontane, 2,18,20-trimethylhexatriacontane and 2,24,26-trimethyldotetracontane were found in the internal lipids of the tobacco budworm *Heliothis virescens* and the corn earworm *Helicoverpa zea* (Nelson, 2001). Pupae of a number of other lepidopterans (*e.g.* the banded sunflower moth *Cochylis hospes* (Lepidoptera: Tortricidae), the southwestern corn borer *Diatraea grandiosella* (Lepidoptera: Pyralidae) and the sunflower moth *Homoeosoma electellum* (Lepidoptera: Pyralidae) make very-long-chain methyl-branched alcohols and acetates, like those shown for the southern armyworm *Spodoptera eridania* in Figure 4.21. The methyl groups tend to be at positions 5, 11, 13, 15, 17 and 21, numbering from the alkyl end of the chain. From studies of incorporation of [^{3}H]acetate and [1-^{14}C]propionate it is concluded that synthesis of these compounds begins at the alkyl end and terminates at the alcohol end so, again, the methyl branches are added early in the synthesis. The pupae of the tobacco hornworm *Manduca sexta* have very-long-chain methyl-branched alcohols, acetates and propionates. The New World screwworm *Cochliomyia hominivorax* (Diptera: Calliphoridae) and the common housefly *Musca domestica* have even longer methyl-branched alcohols (C_{48} to C_{56}) and their acetates than those in Lepidoptera; pupae of both species contain 26,36-dimethyloctatetracontanol (dimeC$_{48}$-OH) (Nelson et al., 1999). Among the Coleoptera, the boll weevil *Anthonomus grandis* (Coleoptera: Curculionidae) (Plate 9) has 26,38-dimethylpentacontanol (dimeC$_{50}$-OH) and the sunflower stem weevil *Cylindrocopturus adspersus* (Coleoptera: Curculionidae) has 28,38-dimethylpentacontanol. The fate and function of these

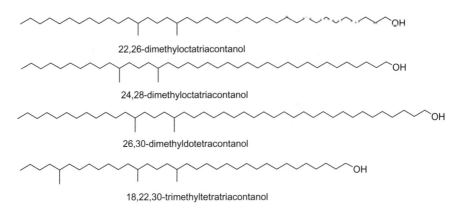

Figure 4.21 Very-long-chain alcohols synthesized by the pupae of some Lepidoptera. The first three are from *Spodoptera eridania*, the last from *Manduca sexta*. Such compounds appear only internally in the pupae and are of unknown function.

compounds is unknown. They do not appear in larvae or adults, and do not form part of the cuticular lipids.

4.3.3 Alkenes

The available evidence indicates that long-chain alkenes are formed by chain lengthening of oleic, palmitoleic and linoleic acids. *Drosophila pseudoobscura* incorporates [1-^{14}C]acetate into alkenes, dienes and trienes. Double bonds are most frequently found in monoenes at the Δ-9 position, which corresponds to oleic acid as the starting point for chain elongation, and next in frequency is the Δ-7 position, corresponding to palmitoleic acid, followed by Δ-11, and at Δ-6,9 in dienes, suggesting linoleic acid as the starter. For example, *Periplaneta americana* (Plate 8), which is known to be able to make linoleic acid (Section 4.1.2), converts it into (6Z,9Z)-6,9-heptacosadiene by elongation and decarboxylation. When labelled acetate was injected into *P. americana* it was incorporated into both saturated and unsaturated hydrocarbons, but [1-^{14}C]linoleic acid was incorporated only into the dienes, and after ozonolysis it could be shown that the label was not in the first six carbon atoms from the alkyl end (Dwyer *et al.*, 1981b). Dienes are generally less common in many insects, and the double bond positions are much more variable.

Methyl-branched alkenes are rare. The first internal alkenes with methyl branches were found in the primitive Australian ant *Nothomyrmecia macrops* (Hymenoptera: Formicidae) (Plate 10), which has

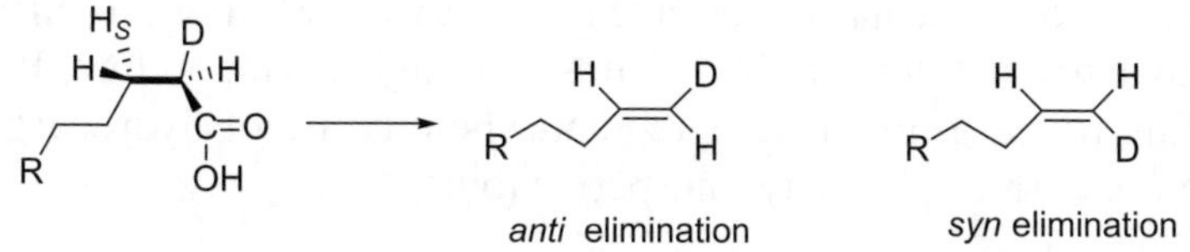

Figure 4.22 Formation of the double bond in terminal alkenes. The 2-deuterio-fatty acid must be held on the enzyme in the configuration shown during the elimination of the carboxyl group to give a 1-alkene by *anti* elimination. Had a *syn* elimination occurred, then the deuterium would have been *cis* to the alkyl group as in the third formula.

almost no alkanes but a number of C_{33} to C_{37} monoenes with one or two methyl branches.

Terminal double bonds are not usually found in cuticle alkenes but do occur in *Tribolium confusum* secretions. W. Boland's group have shown, by deuterium labelling in different positions near the carboxyl group, that they are formed by an *anti* elimination of the carboxyl group and the *pro-S* hydrogen on the second carbon atom of a fatty acid, which is held firmly in one configuration on the enzyme (Figure 4.22). The same mechanism applies to this reaction in plants (Görgen et al., 1990).

4.3.4 Physical State of Hydrocarbons

Are the cuticular hydrocarbons in a solid or a liquid state? There have been studies on the permeability of cuticle to water loss correlated to "melting temperature" of cuticular lipids. Melting temperature can be estimated from changes in infrared C–H absorptions by Fourier transform infrared spectroscopy with a microscope on tiny samples or individual cells. Mixtures of hydrocarbons have very broad melting ranges.

Beyond hexadecane (m.p. 18 °C), all straight-chain alkanes are wax-like solids. Pentacosane melts at 50 °C, triacontane at 60 °C and tetracontane ($C_{40}H_{82}$) at 80 °C. Introducing an internal double bond reduces the melting point by 20 to 50 °C compared with the alkane, and a methyl branch (depending upon where in the chain) reduces melting point by ~30 °C. To keep the cuticular-hydrocarbon surface soft and flexible, as longer-chain alkanes are present in the cuticle, so more long-chain alkenes or methyl-branched alkanes are found in the mixture. Some insects have only straight-chain alkanes and alkenes, others have alkanes and methyl-branched alkanes only, others have mixtures of all three types. The "melting temperature" of the complex mixture can vary from less than 25 °C to more than 80 °C and on most insects cuticles will be low and very broad. Experiments to examine the effect of temperature and humidity of environment on cuticular lipids have tended to show minor modifications, but a move towards longer alkyl chains with higher temperature. It will be interesting to learn how insects regulate the mixture of hydrocarbons to achieve the correct plasticity and permeability.

BACKGROUND READING

Mann, J. 1994. *Chemical Aspects of Biosynthesis*, Oxford University Press, Oxford (Secondary metabolites derived from acetate: fatty acids and polyketides, Chapter 2, pp. 25–93).

Rees, H. H. 1977. *Insect Biochemistry*, Chapman and Hall, London, 64 pp.

Tillman, J. A., Seybold, S. J., Jurenka, R. A. and Blomquist, G. J. 1999. Insect pheromones – an overview of biosynthesis and endocrine regulation. *Insect Biochemistry and Molecular Biology*, **29**, 481–514.

REFERENCES

Arrese, E. L., Canavoso, L. E., Jouni, Z. E., Pennington, J. E., Tsuchida, K. and Wells, M. A. 2001. Lipid storage and mobilization in insects: current status and future directions. *Insect Biochemistry and Molecular Biology*, **31**, 7–17.

Asturias, F. J., Chadick, J. Z., Cheung, I. K. Stark, H., Witkowski, A., Joshi, A. K. and Smith, S. 2005. Structure and molecular organization of mammalian fatty acid synthase. *Nature Structural and Molecular Biology*, **12**, 225–232.

Blailock, T. T., Blomquist, G. J. and Jackson, L. L. 1976. Biosynthesis of 2-methylalkanes in the crickets *Nemobius fasciatus* and *Gryllus pennsylvanicus*. *Biochemical and Biophysical Research Communications*, **68**, 841–849.

Blomquist, G. J. 2003. Biosynthesis and ecdysteroid regulation of housefly sex pheromone production. In: *Insect Pheromone Biochemistry and Molecular Biology* (Blomquist, G. J. and Vogt, R. G., ed.) Elsevier, London and San Diego, pp. 231–252.

Blomquist, G. J., Guo, L., Gu, P. D., Blomquist, C., Reitz, R. C. and Reid, J. R. 1994. Methyl-branched fatty acids and their biosynthesis in the housefly *Musca domestica* L. (Diptera, Muscidae). *Insect Biochemistry and Molecular Biology*, **24**, 803–820.

Buist, P. H. 2007. Exotic biomodifications of fatty acids. *Natural Products Reports*, **24**, 1110–1127.

Buist, P. H. and Behrouzian, B. 1996. Use of deuterium isotope effects to probe the crypto-regio-chemistry of $\Delta 9$ desaturation. *Journal of the American Chemical Society*, **118**, 6295–6296.

Büyükgüzel, K., Tunaz, H., Putnam, S. M. and Stanley, D. W. 2002. Prostaglandin biosynthesis by midgut tissue isolated from the tobacco hornworm, *Manduca sexta*. *Insect Biochemistry And Molecular Biology*, **32**, 435–443.

Chase, J., Jurenka, R. A., Schal, C., Halarnkar, P. P. and Blomquist, G. J. 1990. Biosynthesis of methyl branched hydrocarbons of the German cockroach *Blattella germanica* (L) (Orthoptera, Blattellidae). *Insect Biochemistry*, **20**, 149–156.

Collie, J. N. 1907. Derivatives of the multiple keten group. *Journal of the Chemical Society*, **91**, 1806–1813.

Cripps, C., Blomquist, C. J. and de Renobales, M. 1986. De novo biosynthesis of linoleic acid in insects. *Biochimica et Biophysica Acta*, **876**, 572–580.

Cvacka, J., Jiros, P., Sobotnik, J., Hanus, R. and Svatos, A. 2006 Analysis of insect cuticular hydrocarbons using matrix-assisted laser desorption/ionization mass spectrometry. *Journal of Chemical Ecology*, **32**, 409–433.

Dani, F. R., Jones, G. R., Corsi, S., Beard, R., Pradella, D. and Turillazzi, S. 2005. Nestmate recognition clues in the honeybee: differential importance of cuticular alkanes and alkenes. *Chemical Senses*, **30**, 477–489.

Dennis, M. and Kolattukudy, P. E. 1992. A cobalt-porphyrin enzyme converts a fatty aldehyde to a hydrocarbon and CO. *Proceedings of the National Academy of Science, USA*, **89**, 5306–5310.

Dwyer, L. A., Blomquist, G. J., Nelson, J. H. and Pomonis, J. G. 1981a. A ^{13}C-NMR study of the biosynthesis of 3-methylpentacosane in the American cockroach. *Biochimica et Biophysica Acta*, **663**, 536–544.

Dwyer, L. A., de Renobales, M. and Blomquist, G. J. 1981b. Biosynthesis of (Z,Z)-6,9-heptacosadiene in the American cockroach. *Lipids*, **16**, 810–814.

Fan, Y. L., Zurek, L., Dijkstra, M. J. and Schal C. 2003. Hydrocarbon synthesis by enzymatically dissociated oenocytes of the abdominal integument of the German cockroach, *Blattella germanica*. *Naturwissenschaften*, **90**, 121–126.

Fan, Y., Eliyahu, D. and Schal, C. 2008. Cuticular hydrocarbons as maternal provisions in embryos and nymphs of the cockroach *Blattella germanica*. *Journal of Experimental Biology*, **211**, 548–554.

Ferveur, J.-F., Savarit, F., O'Kane, C. J., Sureau, G., Greenspan, R. J. and Jallon, J.-M. 1997. Genetic feminization of pheromones and its behavioral consequences in *Drosophila* males. *Science*, **276**, 1555–1558.

Fletcher, M. T., Allsopp, P. G., McGrath, M. J., Chow, S., Gallagher, O. P., Hull, C., Cribb, B. W., Moore, C. J. and Kitching, W. 2008. Diverse cuticular hydrocarbons from Australian canebeetles (Coleoptera; Scarabaeidae). *Australian Journal of Entomology*, **47**, 153–159.

Görgen, G., Fross, C., Boland, W. and Dettner, K. 1990. Biosynthesis of 1-alkenes in the defensive secretion of *Tribolium confusum* (Tenebionidae) – stereochemical implications. *Experientia*, **46**, 700–704.

Halarnkar, P. P. and Blomquist, G. J. 1989. Partial characterization of the pathway from propionate to acetate in the cabbage looper, *Trichoplusia ni*. *Insect Biochemistry*, **19**, 7–13.

Howard, R. W. and Blomquist, G. J. 2005. Ecological, behavioral and biochemical aspects of insect hydrocarbons. *Annual Review of Entomology*, **50**, 371–393.

Howard, R. W. and Stanley, D. W. 1999. The tie that binds: Eicosanoids in invertebrate biology. *Annals of the Entomological Society of America*, **92**, 880–890.

Howard, R. W., Jurenka, R. A. and Blomquist, G. J. 1986. Prostaglandin synthetase inhibitors in the defensive secretion of the red flour beetle *Tribolium castaneum* (Herbst) (Coleoptera, Tenebrionidae). *Insect Biochemistry*, **16**, 757–760.

Juarez, M. P. and Fernandez, G. C. 2007. Cuticular hydrocarbons of triatomines. *Comparative Biochemistry and Physiology A*, **147**, 711–730.

Juarez, P., Chase, J. and Blomquist, G. J. 1992. A microsomal fatty acid synthetase from the integument of *Blatella germanica* synthesises methyl-branched fatty acids, precursors to hydrocarbons and contact sex pheromones. *Archives of Biochemistry and Biophysics*, **293**, 333–341.

Jurenka, R. A., Subchev, M., Abad, J. L., Choi, M. Y. and Fabrias, G. 2003. Sex pheromone biosynthetic pathway for disparlure in the gypsy moth, *Lymantria dispar*. *Proceedings of the National Academy of Science, USA*, **100**, 809–814.

Lockey, K. H. 1988. Lipids of the insect cuticle: origin, composition and function, *Comparative Biochemistry and Physiology Part B*, **89**, 595–645.

Major, M. A. and Blomquist, G. J. 1978. Biosynthesis of hydrocarbons in insects – decarboxylation of long-chain acids to normal alkanes in *Periplaneta*. *Lipids*, **13**, 323–328.

Martin, S. J., Vitikainen, E., Helanterä, H. and Drijfhout, F. P. 2008. Chemical basis of nest-mate discrimination in the ant *Formica exsecta*. *Proceedings of the Royal Society B*, **275**, 1271–1278.

Merchant, D., Ertl, R. L., Rennard, S. I., Stanley, D. W. and Miller, J. S. 2008. Eicosanoids mediate insect hemocyte migration. *Journal of Insect Physiology*, **54**, 215–221.

Nelson, D. R. 2001. Discovery of novel trimethylalkanes in the internal hydrocarbons of developing pupae of *Heliothis virescens* and *Helicoverpa zea*. *Comparative Biochemistry and Physiology B*, **128**, 647–659.

Nelson, D. R. and Blomquist, G. J. 1995. Insect waxes. In: *Waxes: Chemistry, Molecular Biology and Functions* (Hamilton, R. J., ed.), The Oily Press, Dundee, pp. 1–90. [This review contains 651 relevant references.].

Nelson, D. R., Fatland, C. L. and Adams, T. S. 1999. Very long-chain methyl-branched alcohols and their acetate esters in pupal internal lipids of developing Coleoptera, *Anthonomus grandis* and *Cylindrocopturus adspersus*, and Diptera, *Cochliomyia hominivorax* and *Musca domestica*. *Comparative Biochemistry and Physiology B*, **122**, 223–233.

Reed, J. R., Vanderwel, S., Choi, S., Pomonis, J. G., Reitz, R. C. and Blomquist, G. J. 1994. Unusual mechanism of hydrocarbon formation in the housefly: cytochrome P450 converts aldehyde to the sex pheromone component (Z)-9-tricosene and CO_2. *Proceedings of the National Academy of Science, USA*, **91**, 10000–10004.

Schulz, S., Gross, J. and Hilker, M. 1997. Origin of the defensive secretion of the leaf beetle *Chrysomela lapponica*. *Tetrahedron*, **53**, 9203–9212.

Shanklin, J. and Cahoon, E. B. 1998. Desaturation and related modifications of fatty acids. *Annual Review of Plant Physiology and Plant Molecular Biology*, **49**, 611–641.

Sperling, P., Ternes, P., Zank, T. K. and Heinz, E. 2003. The evolution of desaturases. *Prostaglandins, Leukotrienes and Essential Fatty Acids*, **68**, 73–95.

Stanley, D. 2006. Prostaglandins and other eicosanoids in insects: biological significance. *Annual Review of Entomology*, **51**, 25–44.

Stanley, D., Miller, J. and Tunaz, H. 2009. Eicosanoid actions in insect immunity. *Journal of Innate Immunity*, **1**, 282–290.

Wakayama, E. J., Dillwith, J. W., Howard, R. W. and Blomquist, G. J. 1984. Vitamin B_{12} levels in selected insects. *Insect Biochemistry*, **14**, 175–179.

Weinert, J., Blomquist, G. J. and Borgeson, C. E. 1993. De-novo biosynthesis of linoleic acid in two non-insect invertebrates – the land slug and the garden snail. *Experientia*, **49**, 919–921.

Zhou, X.-R., Horne, I., Damcevski, K., Haritos, V. Green, A. and Singh, S. 2008. Isolation and functional characterization of two independently-evolved fatty acid Δ12-desaturase genes from insects. *Insect Molecular Biology*, **17**, 667–676.

CHAPTER 5
Aliphatic Compounds from Fatty Acids

The chemical definition of aliphatic compounds as those which contain chains of carbon atoms allows us to consider together a large number of insect compounds, chiefly pheromone and defensive secretions, that can be seen as derived from fatty acids, but are metabolized further than the simple changes that give the cuticular hydrocarbons. It is here more useful to consider them by insect order rather than by structure. One large group is the lepidopteran sex pheromones, which range across several chemical types, but have common biochemical origins.

5.1 LEPIDOPTERAN SEX PHEROMONES

The Lepidoptera evolved with flowering plants (see Figure 1.1), and today are the second largest group of insects, representing 16% of all species. They have been particularly successful as predators on plants and, since plants form a large part of our food, Lepidoptera are, to us, important pests. Adult females of Lepidoptera have abdominal glands from which they release volatile chemicals, sex pheromones, which attract mature males, when the females are ready for copulation. Following the isolation and structure determination of the very first insect pheromone compound, bombykol, a component of the sex pheromone of the silkworm *Bombyx mori*, by Butenandt *et al.* (1959), lepidopteran pheromones have been intensively studied for five decades, and we now know much about their biosynthesis (Jurenka, 2003). Additionally, many of these species are easy to rear and pheromone synthesis was relatively easy to relate to fatty acids.

Females of very many species of Lepidoptera release a peptide from the brain or suboesophageal ganglion called pheromone biosynthesis activating neuropeptide (PBAN) (Rafaeli and Jurenka, 2003). PBAN is not a single substance but a family of peptides, varying with species, but all ending at the carboxylate end in the sequence: phenylalanine-x-proline-arginine-leucine amide (x indicates one of several amino acids). PBAN in turn induces release of volatile chemicals from the pheromone gland in the tip of the abdomen to attract males for mating. Just how PBAN works to release pheromone varies with species and is not fully understood. It may act by stimulating enzyme activity in production of fatty acids, or the supply of intermediates for those acids, or converting the fatty acids to final products, but not, apparently, the desaturation stage (Section 5.1.3.1).

Most of these pheromone compounds have the following characteristics: 1) they are made of straight chains of 10, 12, 14, 16 or 18 carbon atoms; 2) they fall into the classes of hydrocarbons or epoxides or have a primary alcohol, ester, aldehyde or acetate at one end of a carbon chain; and 3) they have up to three double bonds with an *E* or *Z* configuration, and sometimes a triple bond. Some of the questions about their biosynthesis were: 1) are they derived from material in their diet, made from fatty acids, or synthesized from acetic acid units in the gland; 2) are the short chains made by stopping the synthesis at an earlier stage, or are the longer chains of fatty acids shortened; and 3) at which stage are the double bonds introduced?

It was surprisingly difficult at the time to answer the first question, but experiments with radio-labelled and heavy-atom-labelled acetic acid and labelled fatty-acid molecules have answered this question. We know now that these pheromones are either made from 16- and 18-carbon fatty acids circulating in the haemolymph (insect blood) as part of triglycerides, or are made in the pheromone gland from acetate units.

5.1.1 Hydrocarbons

Moths of three families, Geometridae, Arctiidae and sometimes Noctuidae, use hydrocarbons or epoxides as sex pheromones. Like cuticular hydrocarbons, these are produced in the oenocytes, carried through the haemolymph by lipophorin (Schal *et al.*, 1998), and selectively absorbed in some unknown way from it by the pheromone gland in the abdomen. This has been demonstrated, for example, in the arctiid moths *Holomelina aurantiaca* and *H. lamae*, both of which use 2-methylheptadecane as pheromone (Schal *et al.*, 1998). The branch methyl in this case is derived from leucine (Figure 4.20). There are a few monomethyl-branched

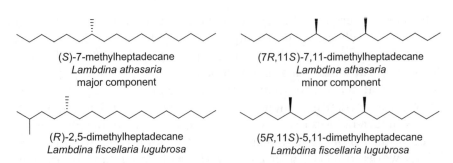

Figure 5.1 Examples of some saturated hydrocarbon sex pheromones of Lepidoptera. In each of these examples the chirality of the compounds is known. The ratio of the two compounds in the pheromone of *L. fiscellaria lugubrosa* is 1 : 1. Note that all such hydrocarbon pheromones have either a methyl branch or a double bond in the carbon chain. In these particular examples the opposite enantiomers of the compounds are inactive but do not interfere with the activity of the natural compounds.

pheromones where the chirality is known, and also a few with two methyl branches where chirality has been established (see Figure 5.1). I am not aware of any example of a linear alkane as a lepidopteran pheromone (*cf.* Section 4.2).

Alkenyl pheromones tend to have double bonds at positions 3, 6 and 9 (derived from linolenic acid) or at 6 and 9 (derived from linoleic acid). While biosynthesis of hydrocarbons with an odd number of carbon atoms in the chain can easily be shown to be derived from decarboxylation of even-numbered fatty acids, the origin of straight-chain hydrocarbons with an even number of carbon atoms was less well understood. The pheromone of the winter moth *Erannis* (*Agriopis*) *bajaria* (Geometridae) consists of C_{18} and C_{19} trienes (Figure 5.2). By use of suitable deuterium-labelled unsaturated fatty acids, applied topically in dimethyl sulfoxide (the usual method for applying such precursors), to the abdominal tip, labelled C_{18} hydrocarbon was obtained. Eicosatrienoic acid labelled with deuterium on C-3 and C-4 (Figure 5.2a) was converted to octadecatriene with the deuterium intact, showing that two carbons were being removed from the carboxylate end. Nonadecatrienoic acid labelled on C-2 and C-3 was converted to the same compound with deuterium intact, showing that a nonadecatrienoic acid was an intermediate. The result indicated that linolenic acid must be the precursor, which is chain-lengthened to the C_{20} acid, and either decarboxylated to the C_{19} triene (Figure 5.2b), or shortened by α-oxidation and decarboxylation to the C_{18} triene (Goller *et al.*, 2007).

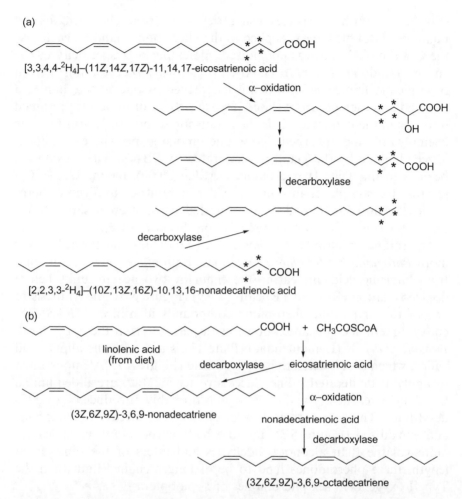

Figure 5.2 The biosynthesis of the pheromone of the winter moth *Erannis* (*Agriopis*) *bajaria* from linolenic acid. The asterisks show the position of labelling with deuterium in model compounds which were converted by the pheromone gland into the C_{18} component without loss of deuterium, showing that a C_{20} acid is degraded by one carbon atom, and then decarboxylated to the pheromone compound. Details of the α-oxidation have not been investigated.

5.1.2 Epoxides

Epoxides can be produced in the pheromone gland by oxidation of a hydrocarbon double bond. For example, a deuterium-labelled precursor [3-^{2}H]-3,6,9-nonadecatriene was applied topically to the pheromone gland of the Japanese giant looper *Ascotis selenaria cetacea* (Geometridae), which converted it into the deuterated epoxide pheromone

(Figure 5.3). With this species, using trienes of various chain lengths and numbers of double bonds, applied to the pheromone gland, in each case the C-3 double bond was epoxidized, indicating regiospecificity of the enzyme (Ando et al., 2008). The triene in *A. selenaria cetacea* is derived from dietary linoleic acid, which is lengthened by one acetate unit to a C_{20} acid and then decarboxylated. Hydrocarbon formation is presumed to occur in the oenocytes; it is then transported by lipophorin into the haemolymph and absorbed by the pheromone gland where it is epoxidized. Uptake by the gland is accelerated by PBAN. These examples belongs to the Type II pheromones (Millar, 2000), having C_{17} to C_{23} unsaturated straight chains and epoxides, in contrast to Type I pheromones (Section 5.1.3), made from saturated acids, where desaturation is one of the steps in their final preparation (Section 5.1.3).

The saltmarsh moth *Estigmene acrea* (Arctiidae) and the ruby tiger moth *Phragmatobia fuliginosa* (Arctiidae) both make their pheromones from linolenic acid, increasing the chain by two acetate units before decarboxylation (Rule and Roelofs, 1989) (Figure 5.3). Biosynthesis of one of the first moth pheromone compounds identified, (7*R*,8*S*)-7,8-epoxy-2-methyloctadecane, known as disparlure, from the gypsy moth *Lymantria dispar* (Lymantriidae) (Plate 11), starts with the amino acid valine, which is converted to isobutyric acid (Figure 4.13), demonstrated by applying deuterated valine. Successive acetate units are added until a C_{18} chain is reached. The double bond is probably introduced by a $\Delta 12$-desaturase. The acid is then decarboxylated to 2-methyl-7-octadecene, and epoxidized (Figure 5.3). Because hydrocarbons and epoxides are more volatile than alcohols, aldehydes and esters of the same chain length, these pheromones tend to be of longer chain length than the Type I oxygenated sex pheromones of Lepidoptera.

5.1.3 Other Oxygenated Compounds

The largest group of lepidopteran sex pheromones are those oxygenated at the end of the chain as in the example of bombykol (Figure 5.4). These Type I compounds are synthesized from acetyl-CoA in the pheromone gland in the abdomen. The combination of chain length, number, position and geometry of double bonds and oxidation state of the oxygen function make a large number of structures possible (Figure 5.4), and blends of three or four compounds in different proportions make possible still more individual species pheromones.

5.1.3.1 Desaturase Enzymes. Acyl-CoA desaturases are key enzymes in the biosynthesis of type I pheromones. A number of desaturases have

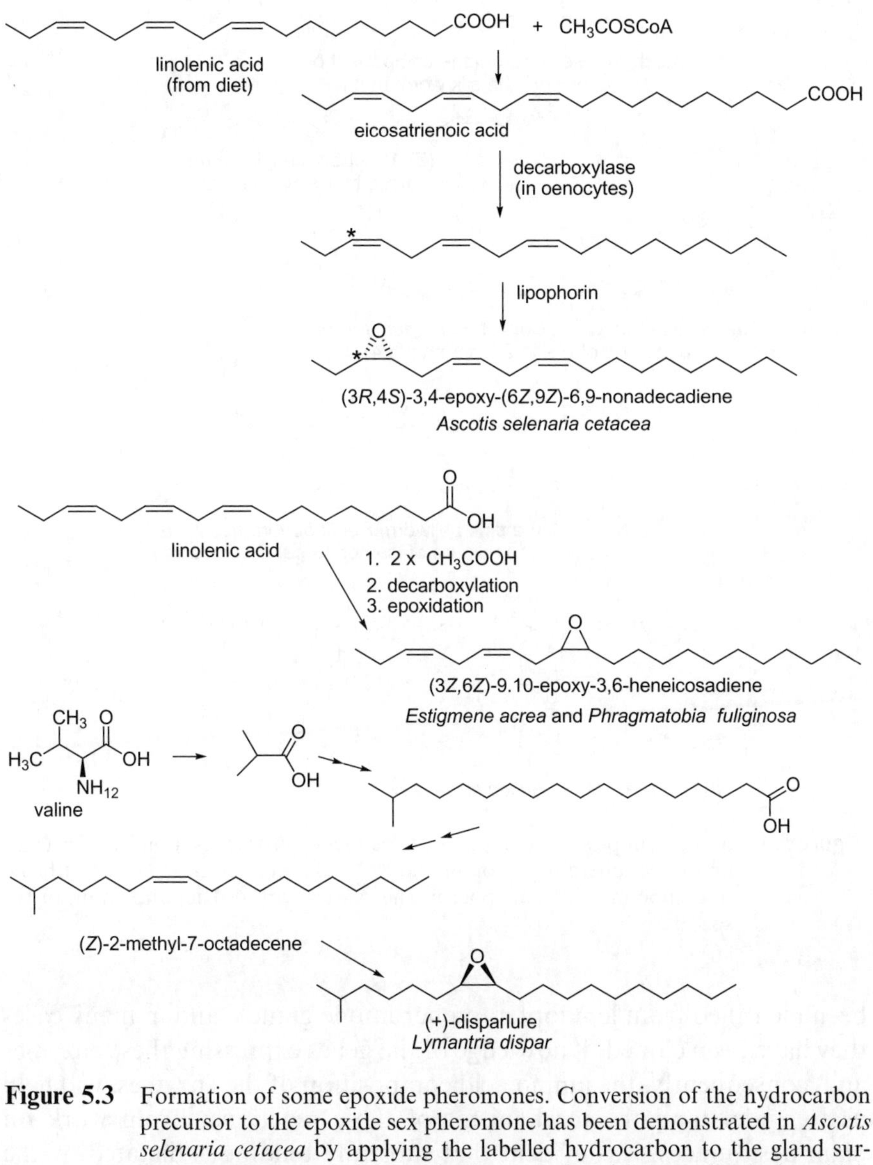

Figure 5.3 Formation of some epoxide pheromones. Conversion of the hydrocarbon precursor to the epoxide sex pheromone has been demonstrated in *Ascotis selenaria cetacea* by applying the labelled hydrocarbon to the gland surface. In *Estigmene acrea* and *Phragmatobia fuliginosa* linolenic acid is lengthened by two acetate units before decarboxylation and epoxidation. The chiralities of the pheromone compounds of these two species are unknown. The total pheromone blend is differentiated by different minor components. Synthesis of disparlure begins with isobutyric acid to give the branched chain.

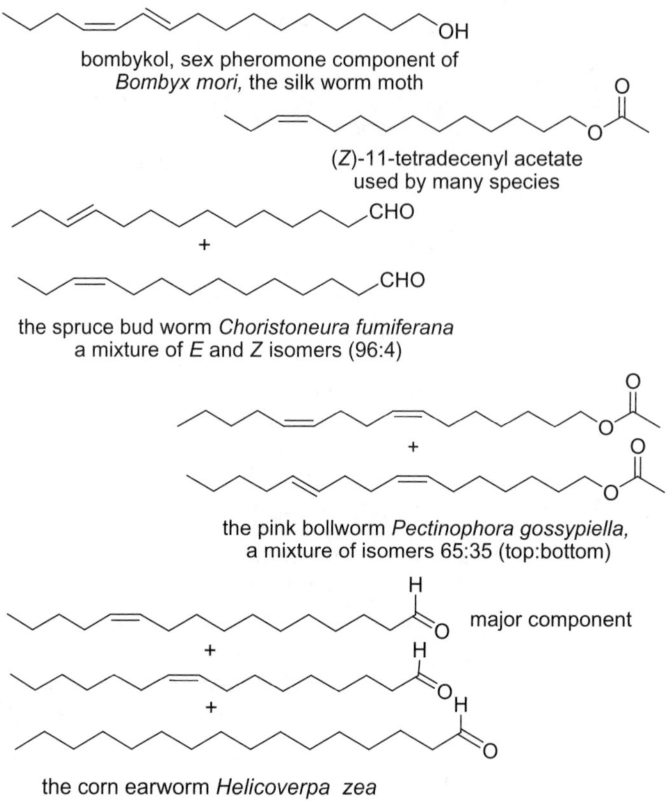

Figure 5.4 Some examples of lepidopteran Type I sex pheromones. Bombykol was the first insect pheromone compound isolated and identified. The exact blend of compounds in some pheromones varies with district and strain of the species.

been identified from lepidopteran pheromone glands, and in many cases they have been cloned. Knowledge of the genes expressing these enzymes and, consequently, the amino acid composition of the enzymes, will help us to understand the mechanisms of these reactions. From work on plant desaturases, it is evident that substrate binding is dictated by the curvature of the substrate pocket at the active site (Whittle et al., 2008). The Δ9-desaturase enzymes for making unsaturated acids are common to plants and animals. They are non-haem, iron-containing membrane proteins located in the endoplasmic reticulum. They require molecular oxygen. The desaturases used by the pheromone glands are unique to insects. Some work only on saturated carbon chains, others are specific to one position on the chain, others are more accommodating to chain length and unsaturation. Double bonds can be inserted at positions 5, 9,

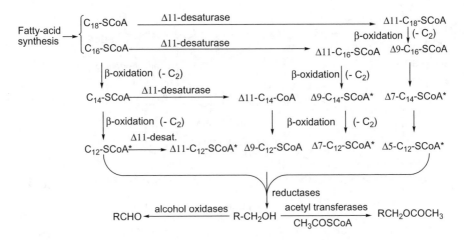

Figure 5.5 A scheme showing how many of the lepidopteran sex pheromone compounds are made by a combination of chain shortening by one acetate unit at a time and the action of a Δ11-desaturase enzyme. Note that a compound can be reached in more than one way. For example, 9-dodecanoyl-SCoA (Δ-9-C_{12}-SCoA) can be made from 11-tetradecanoyl-SCoA by β-oxidation or from dodecanoic acid catalysed by a Δ-9 desaturase. The final stages of formation of pheromone compounds are summarized at the bottom. Reduced derivatives of all the compounds indicated with an asterisk are found in the pheromone gland of the cabbage looper moth *Trichoplusia ni*.

10, 11 or 14 carbon atoms from the carboxyl group. The scheme in Figure 5.5 shows how a number of pheromone compounds can be created by a combination of one desaturase and chain shortening.

Some unusual desaturases have been identified in the study of female lepidopteran sex pheromones. Δ11-Desaturases appear to be unique to insects. Generally *trans* double bonds are uncommon in nature, but they are less so in lepidopteran pheromones. These are made not by isomerization from *cis* double bonds but directly from the saturated chain. Several genes encoding such enzymes have been cloned. The pheromone of the red-banded leafroller *Argyrotaenia velutinana* (Tortricidae), an important agricultural pest, contains a mixture of *cis* and *trans* isomers, and was one of the first studied. Sodium [^{14}C]acetate was applied in dimethylsulfoxide to the surface of the pheromone gland. The biosynthetic reaction (Bjostad and Roelofs, 1981) is outlined in Figure 5.6. The ratio of (Z)- to (E)-11-tetradecenoic acids produced by the one desaturase is 40:60, but the ratio of the two final products which comprise the pheromone, (Z)- and (E)-11-tetradecen-1-yl acetate is 92:8. There was no evidence of an E/Z isomerase present in the gland. The precise ratio in the pheromone therefore depends on the selectivity of the enzymes in the

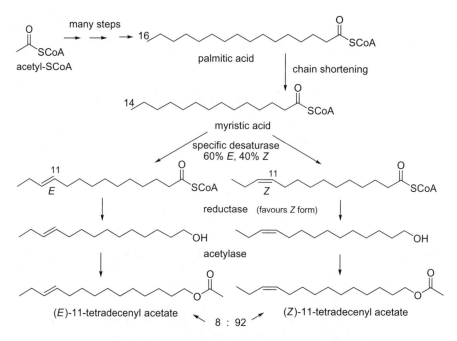

Figure 5.6 Biosynthesis of the sex pheromone of *A. velutinana*, catalysed by a desaturase that gives a mixture of *E* and *Z* double bonds at C-11. The sex pheromone is a mixture of (*Z*)- and (*E*)-11-tetradecenyl acetate in an isomer ratio of 92 : 8.

sequence. (4*E*)-Tetradecenyl acetate provides the pheromone of the tomato pinworm *Keiferia lycopersicella* (Gelechiidae) and its double bond is both *trans* and at the unusual 4-position. Into what compound, between a saturated fatty acid and the final pheromone, the double bond is introduced has not yet been investigated.

The light brown apple moth *Epiphyas postvittana* (Tortricidae) provides an example of a desaturase that can desaturate both saturated and unsaturated C_{14} acids to give *trans* isomers (Liu et al., 2002) (Figure 5.7a). Stereochemical studies have shown that Δ11-desaturases catalyse the removal of the *pro-R* C-11 hydrogen atom and the *pro-R* C-12 hydrogen selectively to give a *cis* double bond, as happens with the more common Δ9-desaturases. When a *trans* double bond is formed, the C-11 *pro-R* and C-12 *pro-S* hydrogens are selectively removed (Figure 5.7b).

The formation of the two double bonds of bombykol (Figure 5.4) is catalysed by a single desaturase, which has been cloned and expressed in insect cells. First a *Z*-11-double bond is introduced into palmitoyl-SCoA, and then two allylic hydrogen atoms are removed to give both (10*E*,12*Z*)-conjugated diene and (10*E*,12*E*)-diene in the proportions of 5:1

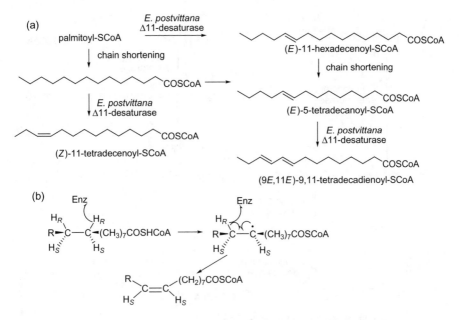

Figure 5.7 (a) The desaturase of *Epiphyas postvittana* catalyses the desaturation of both saturated and unsaturated acids. (b) The stereochemistry of Δ11 double-bond formation requires the loss of *pro-R* hydrogens from C-11 and C-12.

(Figure 5.8), the same ratio as is found in the final alcohols of the pheromone (Moto *et al.*, 2004).

The pine processionary moth *Thaumetopoea pityocampa* (Thaumetopoeidae) provides another example of an unusual desaturase that catalyses introduction of a *cis*-11 double bond, then the desaturation of this to a triple bond and then acting on the product to add an additional *cis*-13 double bond (Figure 5.8). The cDNA was isolated from the female *T. pityocampa* glands and expressed in yeast and shown to have the expected activity (Serra *et al.*, 2007). The ability to remove hydrogen atoms from the double bond can be more easily understood from models which show that 11-hexadecynoic acid is almost superimposable on palmitic acid. Unlike most lepidopteran species, which produce a blend of compounds as pheromone, the final compound (*Z*)-13-hexadecen-11-ynyl acetate, alone provides the sex pheromone of this species.

The pheromone of the Egyptian armyworm *Spodoptera littoralis* (Noctuidae) has been studied in great detail. The pheromone blend can consist of up to seven compounds, depending upon the region in which the insect is collected, but the most abundant component and the one most active in electroantennography, wind tunnel and field tests is

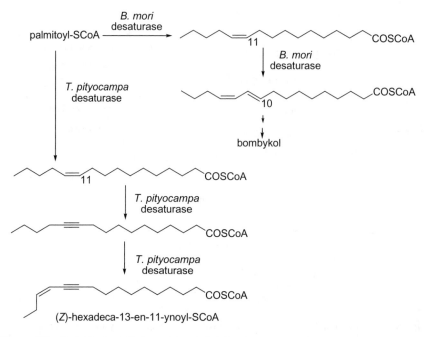

Figure 5.8 Products of the desaturases of *Bombyx mori* and *Thaumetopoea pityocampa*.

(9Z,11E)-9,11-tetradecadienyl acetate (Muñoz *et al.*, 2008). The Δ-11 desaturase of *S. littoralis*, besides catalyzing the formation of (Z)-11-hexadecenoic acid from palmitic acid, also catalyses the production of both (E)-11- and (Z)-11-tetradecenoic acids from myristic acid, and converts (Z)-11-tetradecenoic acid into (10E,12E)-tetradecadienoic acid and (Z)-11-hexadecenoic acid into (10E,12Z)-hexadecadienoic acid (Serra *et al.*, 2006). Studies with deuterium labelling on carbon atoms 11 and 12 of myristic acid showed that the mixture of (Z)-11- and (E)-11-tetradecenoic acid is catalysed by the single enzyme. The possibility of an isomerase in the gland could be ruled out because treating with either deuterated (Z)-11- or (E)-11-tetradecenoic acid alone did not produce any of the other geometric isomer. Unlike the case of bombykol, here a *second* enzyme catalysed the further unsaturation of the *trans* isomer of 11-tetradecenoic acid only (Figure 5.9).

The tobacco hornworm *Manduca sexta* produces a particularly complex pheromone consisting of about 12 aldehydes. Like *B. mori*, it uses one desaturase to catalyse the formation of a monoene and then from that a diene (Matousková *et al.*, 2007). Whether it produces *cis* or *trans* double bonds depends on the rotational conformation the molecule adopts inside the enzyme pocket (Figure 5.10). There are trienes in the pheromone blend too, but it is not known how they are made.

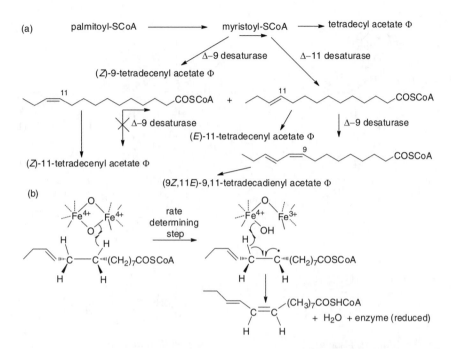

Figure 5.9 The formation of the components of the sex pheromone of *Spodoptera litoralis* each marked Φ. The mechanism for the insertion of the second double bond is shown in (b). The minor components (10E,12E)-10,12-tetradecadienyl and (9Z, 12E)-9,12-tetradecadienyl acetate are through further enzyme catalysis.

The only generalization one can make about lepidopteran pheromone biosynthesis appears to be: one cannot generalize.

5.1.3.2 *Chain Shortening.* Insects do not make C_{10}, C_{12} or C_{14} acids directly by synthesis, rather they are made by shortening the chains of C_{16} or C_{18} acids by β-oxidation (see Figure 4.4). This was first demonstrated by applying radio-labelled palmitic acid to the glands of the orange tortrix moth *Argyrotaenia citrana* (Tortricidae), where it was shortened to myristic acid. Later it was shown that the Egyptian armyworm *Spodoptera littoralis* can shorten both palmitic acid to myristic acid and palmitoleic acid ($C_{16:1}$) to myristoleic acid ($C_{14:1}$). Some enzymes shorten by four carbon atoms, others by only two atoms. The enzymes responsible for chain-shortening of pheromone intermediates are located in the peroxisomes. They have not been intensively studied like the desaturases. The reaction sequence is carried out by an enzyme complex consisting of an acyl-SCoA oxidase, an enoyl-SCoA hydratase, a 3-hydroxyacyl-SCoA dehydrogenase and a 3-oxoacyl-SCoA thiolase (Figure 4.4). Whereas the metabolic chain-shortening

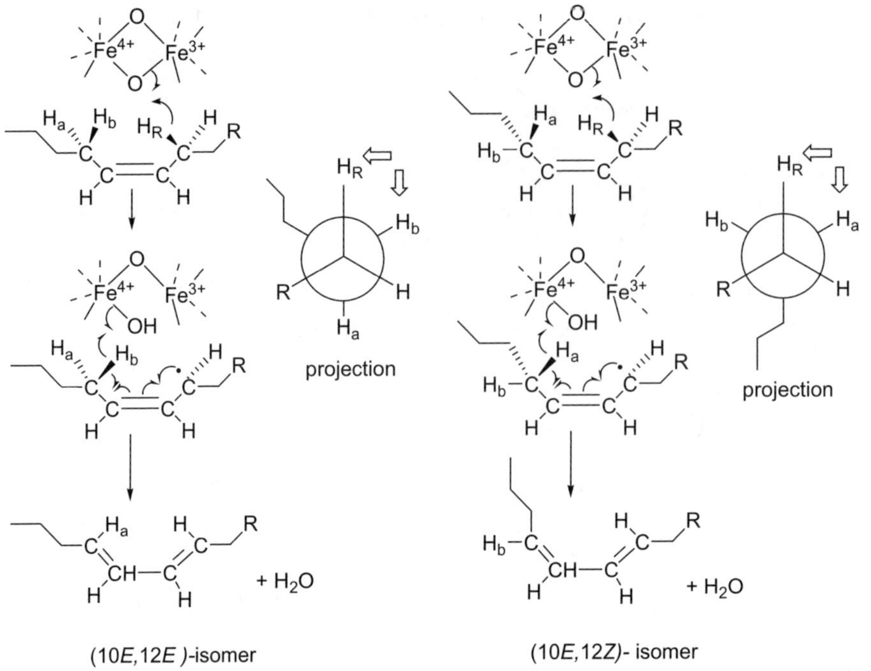

Figure 5.10 The probable mechanisms for the formation of the E,E- and E,Z-dienes in *Manduca sexta*. A Newman projection for the rotational conformation required is shown to the right of both sequences, with arrows indicating the hydrogen atoms removed. The CH=CH is omitted from the projections for simplicity.

enzymes completely degrade the fatty-acid chain, the pheromone gland enzymes have ability limited to removing one or two acetate units.

The optimum chain lengths are apparently 12 or 14 carbon atoms, less than this is too volatile. There are fewer identified pheromone compounds with 16 carbons and still fewer with 18. With still larger molecules volatility is perhaps too low (except for hydrocarbons) for efficient detection by the males.

5.1.3.3 Reductase. The key enzyme for producing the final form of the oxygen function of the Type I pheromones is a fatty-acid reductase, which reduces the carboxyl group directly to alcohol, without the formation of an isolatable aldehyde. Alcohols can be converted to acetate by an acetyl-SCoA:fatty alcohol acetyltransferase, or oxidized to the corresponding aldehyde by an alcohol oxidase. In one unusual case, in the cotton leaf perforator *Bucculatrix thurberiella* (Lyonetiidae)

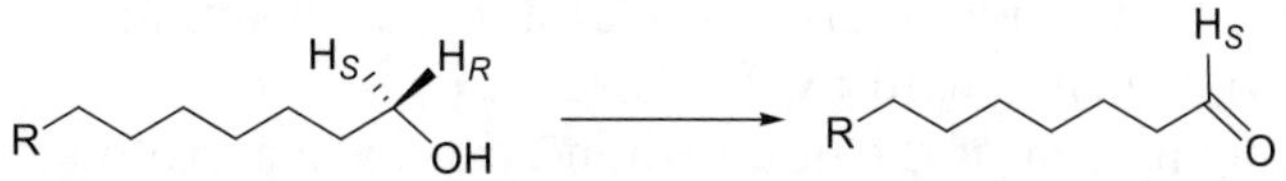

Figure 5.11 Experiments with hexadecanol, chirally labelled with deuterium on C-1, have shown that the *pro-R* hydrogen (or deuterium) is removed in the oxidation of long-chain alcohols to aldehydes.

the alcohol is esterified with nitric acid to give a pheromone mixture of (8Z)- and (9Z)-tetradecenyl nitrates.

The reductase of the silkmoth *Bombyx mori* (Bombycidae) (Plate 12) has been cloned and expressed in yeast. It converted (10E,12Z)-hexadecadienoic acid to (10E,12Z)-hexadecadienol (bombykol) (Moto *et al.*, 2003). It is upon this reductase that PBAN operates to control pheromone production in this species. The final pheromone blend for *B. mori* consists of (10E,12Z)-hexadecadienol, (10E,12E)-hexadecadienol in the ratio 84:165 and a trace of (10E,12Z)-hexadecadienal.

The stereochemistry of oxidation of alcohol to pheromone aldehyde has been studied in *Manduca sexta*. Six different hexadecanols and hexadecenols were prepared, chirally labelled with deuterium on C-1 (Figure 5.11). Using both *in vivo* and *in vitro* experiments, it was shown that the *pro-R* hydrogen (or deuterium) atom is removed, *i.e.* there is *Re*-specificity to give the aldehyde. Experiments with the chirally deuterated unnatural alcohols salicyl alcohol and 2-thiophenylmethanol gave the same result, showing that this stereospecificity of oxidation is intrinsically linked to the oxidation process (Hoskovec *et al.*, 2002).

5.1.3.4 Conclusion. There are probably 300,000 species of Lepidoptera in the world, much more than the number of suitable compounds of the types described for pheromones. Therefore species use blends. Within a blend of any species it is usually found that the components vary by one or more of five possible differences. 1. The chain length may vary by one acetate unit. 2. The functional group may be altered, *e.g.* from alcohol to aldehyde or acetate. 3. There may be one more or fewer double bonds. 4. The double bond may move two positions along the chain in unsaturated alcohols. 5. A Z double bond may be converted to E and *vice versa*. For example, the cabbage looper moth *Trichoplusia ni* (Noctuidae) uses a blend of six acetates, all of which are considered essential to the pheromone blend. They are dodecyl acetate, 11-dodecenyl acetate, (Z)-7-dodecenyl acetate, (Z)-5-dodecenyl acetate, (Z)-9-tetradecenyl acetate and (Z)-7-tetradecenyl acetate. The most frequently encountered compounds are (Z)-9-tetradecenyl acetate, which is used in the blends of 223 lepidopteran species and (Z)-11-tetradecenyl acetate (Figure 5.4), which is used by 214 species.

The configuration of the double bond can have a profound effect on reception by the male insect. Sometimes a wrong isomer can completely inhibit the response to a pheromone. For example, the moth *Eupoecilia ambiguella* (Tortricidae) produces (Z)-9-dodecenyl acetate. Males are inhibited by as little as 0.1% of the (E)-isomer. A pheromone lure containing some of this (E)-isomer can be used as a practical means to disrupt mating for this species.

5.2 LEPIDOPTERAN DEFENCE

The larvae of Lepidoptera have various forms of defence, some of which are discussed in Chapter 11. A number of species are protected with glandular hairs. The caterpillars of the garden tiger moth *Arctia caja* (Arctiidae) have long irritating hairs containing toxic proteins. Those of the European cabbage butterfly *Pieris rapae* (Pieridae) have been investigated and were found to contain a fluid consisting mainly of unsaturated lipids, called mayolenes, which are deterrent to ants. Biosynthesis of mayolenes has not been investigated, but their structures indicate that they are made from hydroxylated linolenic acid, esterified with a range of fatty acids, chiefly palmitic and stearic acids (Figure 5.12) (Smedley *et al.*, 2002). It has been noted that small amounts of hydroxy-acids have been found along with unsaturated acids in the study of lepidopteran desaturase enzymes. If the free radical first formed in desaturation (Figures 5.9 or 5.10) reacts with a water molecule, before the second hydrogen is removed, such hydroxyl groups can be introduced into a fatty acid. An esterase would complete the synthesis of mayolenes.

5.3 COLEOPTERAN COMPOUNDS

The Coleoptera or beetles form the largest insect order, with over 300,000 species identified, or 38% of all insects, with possibly 8 million species in total. They all have sclerotized fore-wings that form a protective cover of their active wings. About half of them are Curculionidae or weevils. From the remark attributed to J. B. S. Haldane that God must have an inordinate fondness for beetles, and their number, we might expect them to have diverse biosynthetic schemes. We do not find the uniformity of pheromone use as described for the Lepidoptera.

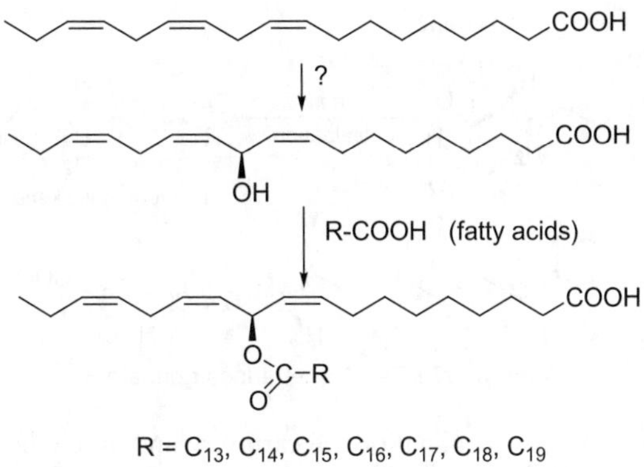

Aliphatic Compounds from Fatty Acids 111

Figure 5.12 An outline of the probable route to the defensive mayolenes from linolenic acid and other fatty acids. The most abundant mayolenes in the secretion of *Pieris rapae* are where R is 15 or 17, *i.e.* using palmitic or stearic acids.

5.3.1 Hydrocarbons, Acids and Lactones

Sometimes hydrocarbons are part of a defensive secretion, as in the confused flour beetle, *Tribolium confusum* (Tenebrionidae), which uses quinones and hydroquinones (Section 9.4 and Section 9.5) and terminal alkenes (probably serving as spreading agents for the quinones) chiefly 1-pentadecene, with lesser amounts of 1,6-pentadecadiene, 1-hexadecene and 1-heptadecene. The beetle *Eleodes beamerii* (Tenebrionidae) has 1-nonene, 1-undecene and 1-nonen-3-one, and *E. longicollis* has 1-undecene and 1-tridecene, all in their defensive secretions. Terminal double bonds are not usually found in cuticular hydrocarbons. W. Boland's group showed for *T. confusum*, by deuterium labelling in several places in the fatty-acid chain, that its terminal alkenes are formed by an *anti* elimination of the carboxyl group and the *pro-S* hydrogen of palmitic or stearic acids (Figure 5.13a), as in plants (Görgen *et al.*, 1990).

Females of the spotted longicorn beetle *Neoclytus acuminatus acuminatus* (Cerambycidae) incorporate six methylalkanes into their cuticle lipids, of which the major one, 9-methylheptacosane, is synergized by 7-methylpentacosane and 7-methylheptacosane to act together as a sex pheromone which males only detect by antennal contact with the female surface. The female rove beetle *Aleochara curtula* produces a sex attractant pheromone consisting of (Z)-9-heneicosene and (Z)-7-tricosene, and this is sometimes also produced by males to mimic females in this species. Melolonthine scarab beetles, known as cane grubs, are a serious pest of sugarcane in Australia. Their cuticles contain a range of unusual allenic hydrocarbons represented by the formula

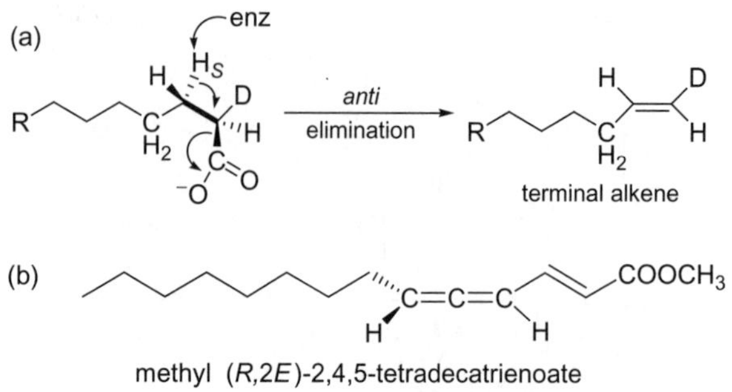

(a)

[Structure: anti elimination to terminal alkene]

(b)

[Structure of methyl (R,2E)-2,4,5-tetradecatrienoate]

methyl (R,2E)-2,4,5-tetradecatrienoate

Figure 5.13 (a) The *anti* elimination of carboxyl and *pro-S* hydrogen to give a terminal alkene in the defensive secretion of *Tribolium confusum*. (b) The allenic ester pheromone of the dried bean weevil *Acanthoscelides obtectus*.

$CH_3(CH_2)_nCH=C=CH(CH_2)_7CH_3$, where n is 11–14, 17 or 19, therefore a total chain length of 23 to 31 carbon atoms. Allenes are chiral and these are more than 85% of the (*R*)-configuration. Their biosynthesis has not yet been studied, nor have they been shown to be pheromones. A comparison with the sex pheromone of the dried bean weevil *Acanthoscelides obtectus* (below), which also has the (*R*)-configuration, is interesting. Males of the spruce engraver *Pityogenes chalcographus* (Scolitidae), emit methyl (2*E*,4*Z*)-(2,4)-decadienoate from head and thorax as part of an aggregation pheromone. *P. hopkinsi* uses male-produced ethyl dodecanoate similarly. The black carpet beetle *Attagenus megatoma* (Dermestidae) has (3*E*,5*Z*)-(3,5)-tetradecadienoic acid as part of its sexual pheromone. Females of *Melanotus* species (Elateridae) produce tetradecenals and tetradecyl acetates as sex pheromones. While their biosynthesis has not been investigated in these species, their structures suggest synthesis as in Lepidoptera.

The males of the dried bean weevil *Acanthoscelides obtectus* (Bruchidae) use an unusual and chiral allenic ester, methyl (*R*,2*E*)-2,4,5-tetradecatrienoate as sexual attractant (Figure 5.13b). Its biosynthesis has not yet been investigated. A number of species of dung and chafer beetles have female-produced lactones as sex attractants. Examples are japonilure and buibuilactone found in a number of species of *Anomala* chafer grubs (Scarabaeidae). Deuterium labelling has shown they are derived from fatty acids by hydroxylation, then chain shortening and ring closure (Figure 5.14). Full deuterium labelling also showed that the hydroxyl is introduced directly, not through unsaturation and hydration since no deuterium was lost (Leal *et al.*, 1999). The formation of the

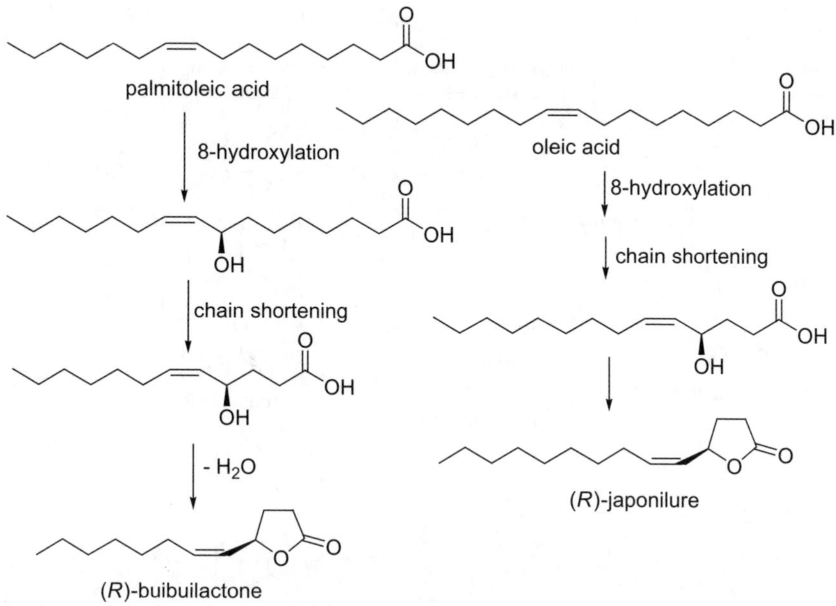

Figure 5.14 The formation of the female-produced sex pheromone lactones of *Anomala* chafer grubs.

hydroxypalmitoleic acid was found to be highly enantioselective, because when racemic hydroxypalmitoleic acid was supplied to *Anomala cuprea* (which normally produces only (*R*)-buibuilactone), it was converted into a mixture of (*R*)- and (*S*)-buibuilactones.

A group of macrocyclic lactones called cucujolides (Oehlschlager *et al.*, 1988) serve as male-produced aggregation pheromones for males and females of *Cryptolestes* (Laemophloeidae) and *Oryzaephilus* bark beetles (Silvanidae) (both formerly placed in the family Cucujiidae). One of these cucujolides is derived from farnesol (Section 7.5), but six others are from fatty acids. Unlike the above lactones, these are formed first by chain shortening and then hydroxylation by unusual hydroxylases, at or next to the alkane chain terminus, *i.e.*, the ω- or ω-1 positions (Figure 5.15). Labelling the ω-1 position with ^{18}O and ^{2}H showed that in cyclization the ^{18}O and ^{2}H were retained. When racemic (*Z*)-11-hydroxy-3-dodecenoic acid was supplied, only (*S*)-cucujolide was formed, indicating a high enantioselectivity in the cyclizing enzyme. Lauric (dodecanoic) acid and 11-hydroxydodecanoic acid were not effective precursors of cucujolide II, indicating that unsaturation is introduced at an earlier stage in the synthesis, but (*Z*)-3-dodecenoic acid and (*Z*)-11-hydroxy-3-dodecenoic acid were incorporated.

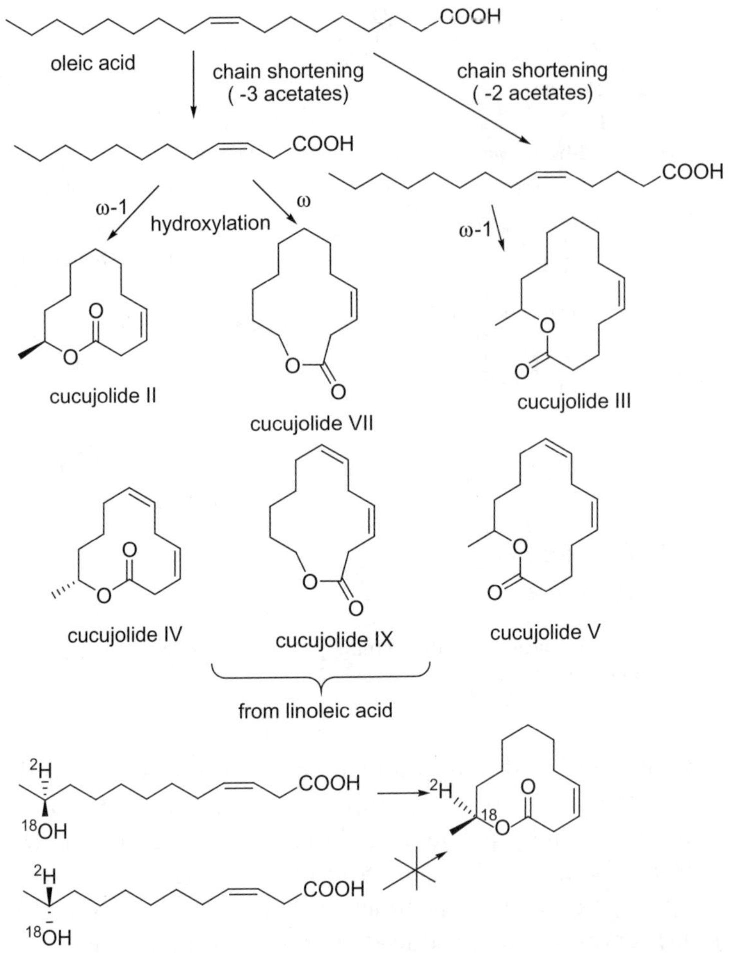

Figure 5.15 Biosynthesis of cucujolides, macrocyclic lactones of *Cryptolestes* and *Oryzaephilus* bark beetles, used as aggregation pheromones. Only the (*S*) form of the hydroxy acid is converted to cucujolide II.

5.4 COLEOPTERAN DEFENCE

Coccinellid beetles, commonly called ladybirds, many of which prey on aphids, make a range of defensive compounds. Over 50 alkaloids (see Chapter 10) of several structural types have been isolated and identified. Because of their nitrogen content all the coccinellid compounds discussed here can be classed as alkaloids, but their biosynthetic origins in fatty acids places their interest here. They are produced in fat body and stored in the haemolymph. When a ladybird is disturbed, it exudes some of this toxic and bitter-tasting haemolymph (called reflex bleeding) at its leg joints. All ladybirds display distinct aposematic colouring, by which potential predators learn to avoid their bitter and toxic haemolymph.

5.4.1 Coccinellinae

One group of such toxic compounds are the coccinellines, tricyclic amines with an azaphenylene structure, and their N-oxides, found in members of the subfamily Coccinellinae. The coccinellines were first shown to be made from acetate units by feeding experiments using radio-labelling with sodium [1-^{14}C]acetate and [2-^{14}C]acetate in 1975. More recently the experiments were repeated with *in vitro* studies using excised body parts of ladybirds that gave much higher incorporation of the radio-label (Laurent *et al.*, 2002). The synthesis was shown to occur in the fat body. The fatty-acid route was demonstrated because the formation of coccinelline was inhibited by 2-octynoic acid, a known inhibitor of fatty-acid synthesis. If the compound is derived from palmitic or stearic acid, it would require loss of one or two acetate units by β-oxidation, followed by further oxidations (Figure 5.16) at appropriate carbon atoms, requiring molecular oxygen. This proposed route was supported by experiments that showed coccinelline formation was inhibited when the experiment was performed in a nitrogen atmosphere. The origin of the nitrogen atom in coccinelline was shown to be glutamine. This was demonstrated by supplying various amino acids to a tissue culture that could synthesize the coccinelline in the presence of oxygen. An amidotransferase and pyridoxal phosphate would be required to transfer the nitrogen to the coccinelline. Precoccinelline (Figure 5.16) is basic; the final stage is oxidation to the weakly basic N-oxide. A number of isomers are possible by altering the stereochemistry at the ring junctions. The true shape and difference between these isomers is shown in Figure 5.16.

More complex dimers of hippodamine have also been found in other species of the subfamily Coccinellinae. Psylloborine A was isolated from *Psyllobora vigintiduo-punctata* (sometimes written *P. 22-punctata*) and isopsylloborine A from *Halyzia sedecimguttata* and *Vibidia duodecimguttata*. Nothing is yet known of the way in which these dimers are formed. The cream spotted ladybird *Calvia quattuordecimguttata* produces calvine, with a twelve-carbon chain with the carboxyl group still intact (Figure 5.17). Note that it also contains the N-CH$_2$-CH$_2$-O fragment of ethanolamine, probably derived from serine (Laurent *et al.*, 2003).

Two further compounds, recognized to have very similar origin to the coccinellines, are (-)-adaline and (-)-adalinine (Figure 5.18), from *Adalia*

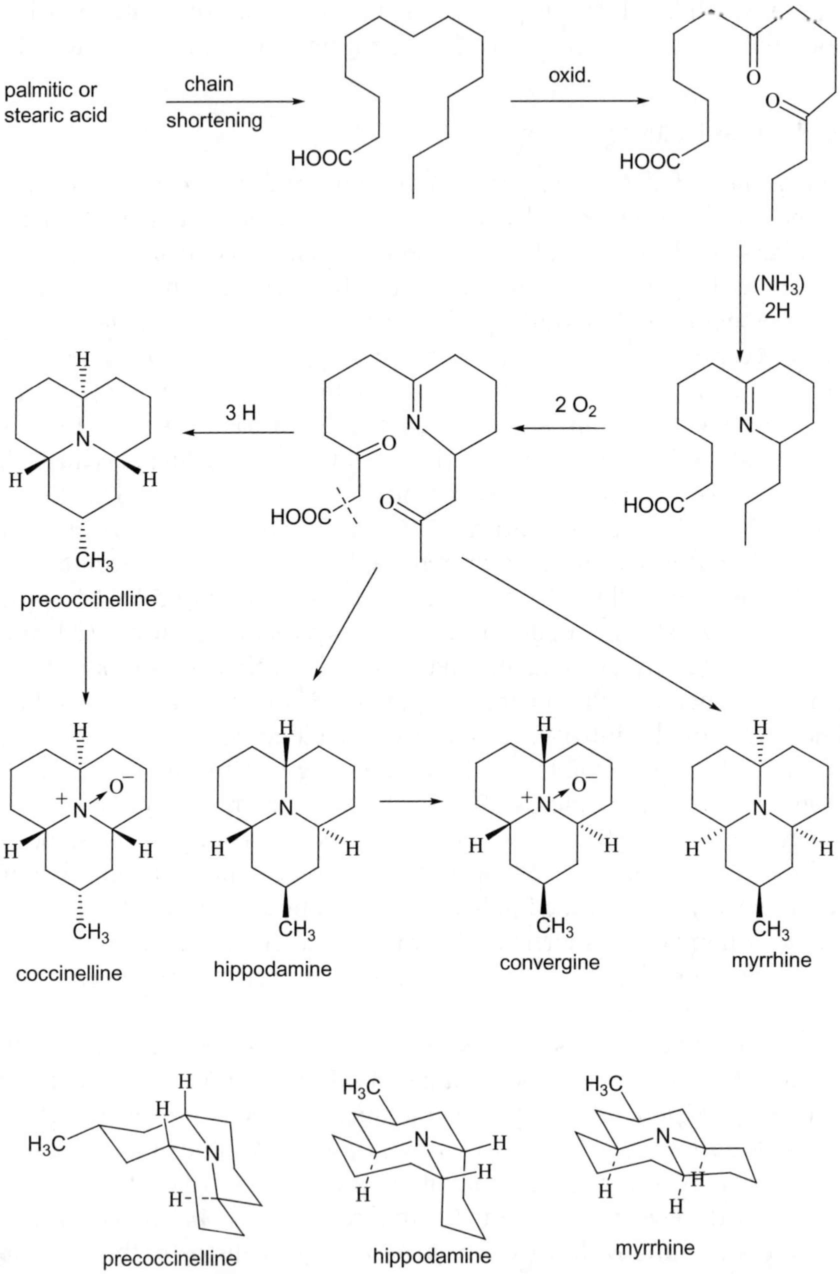

Figure 5.16 Formation of coccinelline and related tricyclic amines and amine oxides of ladybirds. Myrrhine is the thermodynamically most stable of the group, but apparently does not occur in the ladybirds as the N-oxide. Ammonia is shown in brackets because it never appears as the free compound.

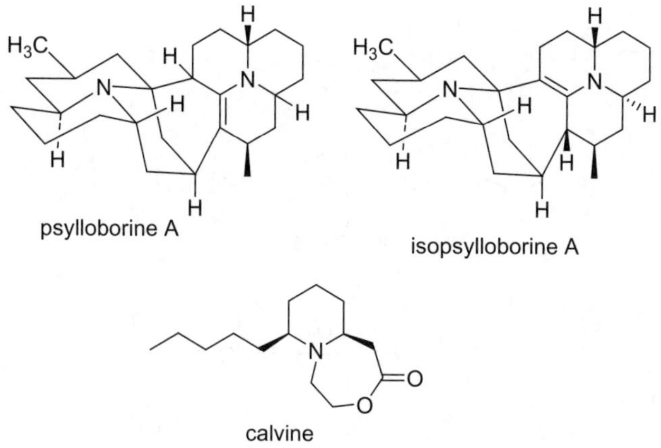

Figure 5.17 Compounds identified in more species of ladybird from the subfamily Coccinellinae.

beetles (also from the subfamily Coccinellinae). These compounds were first shown to be fatty-acid derivatives through feeding with ^{14}C-acetate. When *Adalia bipunctata* (Plate 13) was provided with (-)-adaline deuterated in the side-chain, they produced deuterated adalinine, but when fed (+)-adaline, similarly deuterated, there was no labelled adalinine produced, only unlabelled adalinine. From this it is concluded that adalinine is formed from (-)-adaline by a reversal of the Mannich reaction by which adaline is formed, but on the other side of the ring. Addition of water and a further oxidation gives adalinine. An attempt to prepare the intermediate piperideine (**A** in Figure 5.18) gave adaline with only a trace of the piperideine, so it is presumed the intermediate **A** spontaneously undergoes a Mannich reaction to give adaline (Laurent *et al.*, 2003).

5.4.2 Chilocorinae

Another group of still more complex defensive compounds from the subfamily Chilocorinae are collectively known as chilocorines (Laurent *et al.*, 2005). The first of these discovered, and the simplest, is exochomine from the pine ladybird *Exochomus quadripustulatus* (Figure 5.19). It is composed of one molecule of hippodamine linked by a single bond to a new ring system, an aza-acenaphthylene. Chilocorines A, B and C have all been isolated from the haemolymph of the cactus ladybird *Chilocorus cacti*. Chilocorine C has a five-membered ring in the hippodamine part, so its formation requires a further biosynthetic step. Chilocorine D is from *Chilocorus renipustulatus*. Nothing is known of the

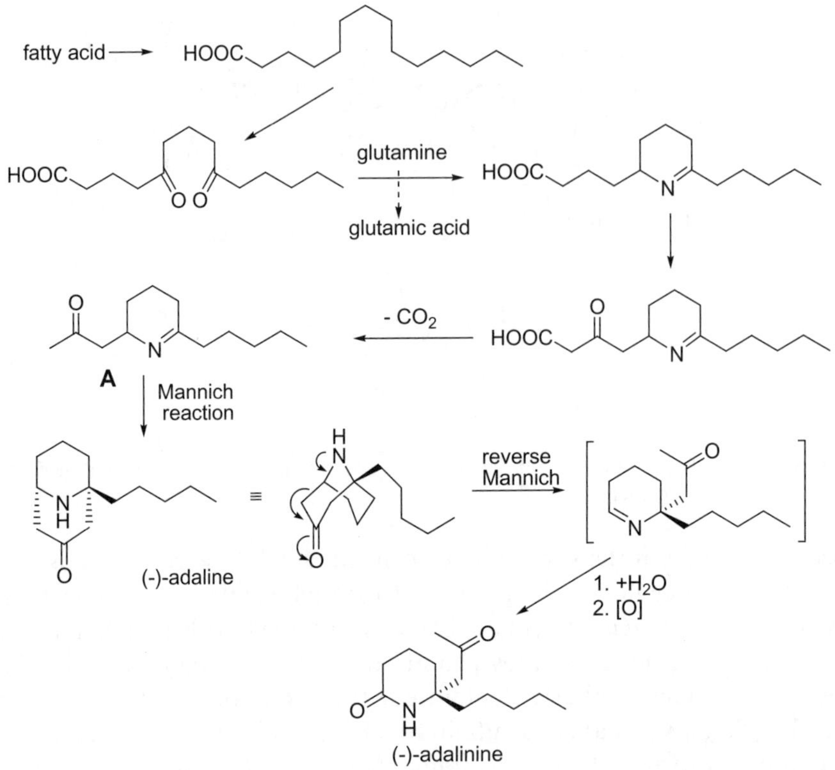

Figure 5.18 The formation of adaline and adalinine. The intermediate A appears to undergo a spontaneous Mannich reaction to give adaline, which, in turn, can undergo a reverse Mannich reaction.

biosynthesis of the aza-acenaphthylene part but as it also contains 13 carbon atoms, like hippodamine, it may also be derived from myristic acid, folded and ring-closed in a different manner.

5.4.3 Scymninae

In the subfamily Scymninae, *Hyperaspis campestris* produces (+)-hyperaspine, which also contains a chain of 13 carbon atoms. It has a partly constructed precoccinelline (Figure 5.20) with pyrrole carboxylic acid esterified to it. In the Australian ladybird *Cryptolaemus montrouzieri* (also of the Scymninae) (Plate 14) are found two compounds: 6-methylpelleterine and a second, clearly related to the first, can also be seen as a lower homologue of adaline (Figure 5.19). No biosynthetic studies have been made with this subfamily.

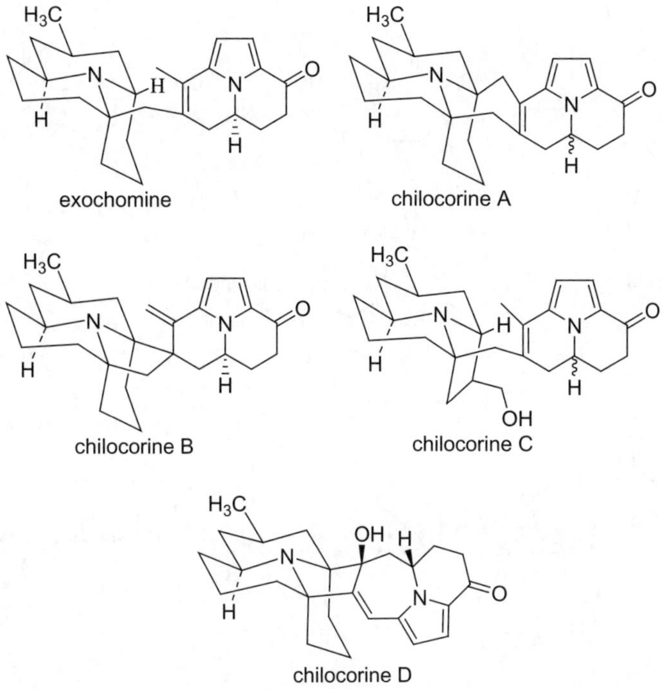

Aliphatic Compounds from Fatty Acids 119

Figure 5.19 Defensive compounds from ladybirds of the subfamily Chilicorinae.

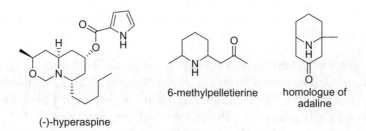

Figure 5.20 Defensive compounds from the subfamily Scymninae. The stereochemistry of 6-methylpelletierine and the homologue of adaline are undefined.

5.4.4 Epilachninae

The Mexican bean beetle *Epilachna varivestis* (subfamily Epilachninae) is a particularly serious agricultural pest. Its larvae make a quite different type of fatty-acid-derived defensive secretion. The pupae carry the toxin in droplets at the end of hairs covering the body. Epilachnine has

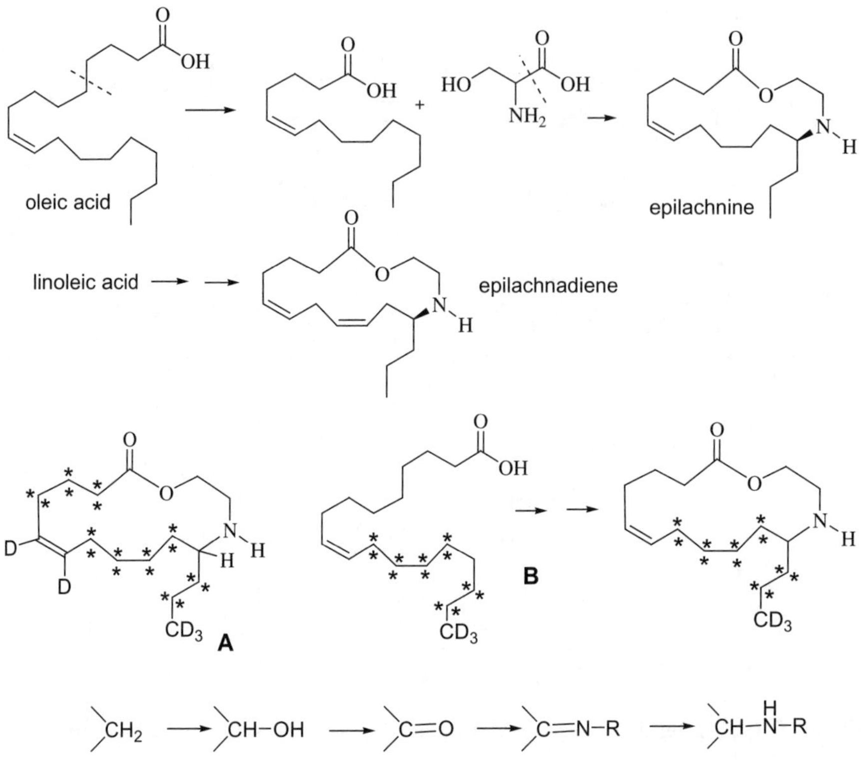

Figure 5.21 The formation of epilachnene and the similar formation of epilachnadiene from linoleic acid. Both * and D indicate deuterium labelling. Below is the sequence of reactions which are presumed to lead to ring closure.

been shown by labelling experiments to be made from oleic acid, which loses four carbon atoms from the carboxylate end, and the amino acid serine. The route to epilachnine has been studied with polydeuterated precursors (Attygalle et al., 1994). For example, fully deuterated stearic acid gave compound **A** (Figure 5.21), losing two deuterium atoms on formation of the double bond, and six more on loss of two acetate units by chain shortening. Both deuterium atoms of carbon atom 11 of epilachnene were lost in ring closure. This was confirmed by using the partially deuterated oleic acid (**B**, Figure 5.21), without deuterium at C-15, which was incorporated into epilachnine without loss of deuterium. Therefore the chain is specifically oxidized and cyclized at this carbon atom as shown at the bottom of Figure 5.21.

In another epilachnene beetle, the 24-pointed ladybird *Subcoccinella vigintiquatropunctata*, the hairs of pupae contain three dimers related to epilachnine. The starting compounds appear to be oleic and linoleic

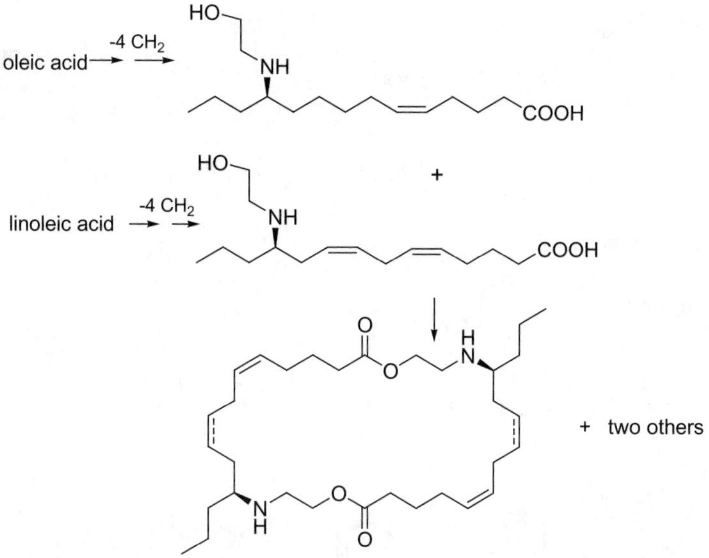

Aliphatic Compounds from Fatty Acids 121

Figure 5.22 The monomers that give rise to the macrocyclic lactones of *Subcoccinella vigintiquatropunctata* with one example of the dimers, which have a total of two, three or four double bonds. Dashed bonds show the position of additional bonds.

acids, which have been substituted at C-15 with ethanolamine (presumably derived again from serine), as in epilachnine and then dimers of these resulting acids are formed (Figure 5.22). Smaller amounts of the trimers (four possible double-bond isomers) are present in small quantity. The size of the macrocyclic rings in this species seems to be carefully controlled. The pupae of the squash beetle *Epilachna borealis* produce a wide range of extremely large macrocyclic lactones, based on shorter-chain fatty acids, substituted with ethanolamine at the ω-1 carbon atom (Figure 5.23). Mixtures of these monomers and dimers, but chiefly trimers, tetramers and pentamers, make up what the authors describe as a combinatorial library of macrocyclic polymers (Schroeder *et al.*, 2000), since there are 10 trimers, 81 tetramers, *etc.*

The fatty-acid derivative (Z)-1,17-diamino-9-octadecene (Figure 5.24) is found in several species. Its structure indicates an origin in oleic acid. Related to it is signatipennine, which has two ethanolamine substituents added to the C_{17} chain, from *Epilachna signatipennis*. Some further compounds with ring closures on this last structure are illustrated in Figure 5.24. Their biosynthesis has not received attention.

Chrysomelid or leaf beetles also arm themselves with defensive compounds. Their ability to make such compounds varies all the way from

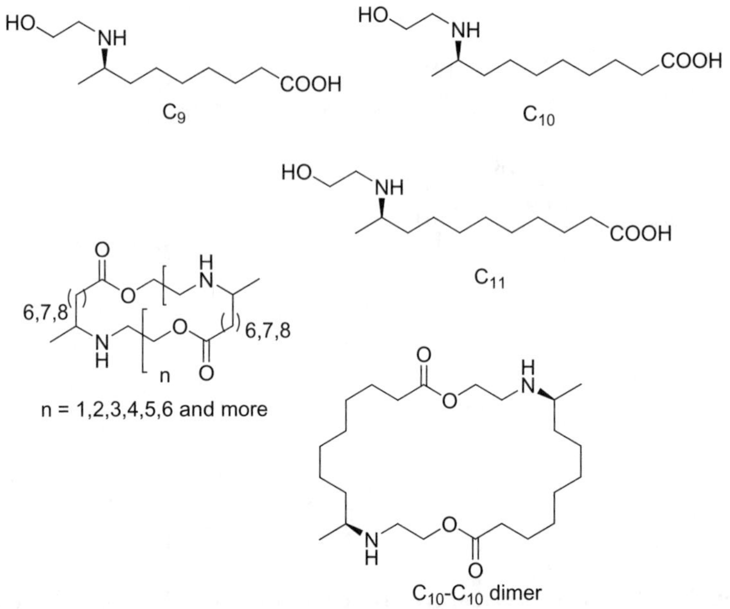

Figure 5.23 Macrocyclic lactones of the defensive secretion of the beetle *Epilachna borealis*. The three monomeric amino acids are combined in all possible ways to form dimers, trimers, tetramers and higher oligomers. The general formula and one dimer example are illustrated.

sequestering plant compounds unchanged (Chapter 11) to completely synthesizing iridoid compounds themselves (Section 7.3.2). The subject does not neatly fit into the divisions adopted here.

5.5 DIPTERAN PHEROMONES

Diptera (flies) are the fourth largest order of insects in number of species. Many species of Diptera use sex pheromones incorporated into their cuticular lipids. Their biosynthesis has been studied in detail in the housefly *Musca domestica*, the fruit fly *Drosophila melanogaster* and the tsetse fly *Glossina morsitans*.

5.5.1 *Musca*

The major sex pheromone of the female housefly *Musca domestica* (Muscidae) consists of (Z)-9-tricosene, *cis*-9,10-epoxytricosane and (Z)-14-tricosen-10-one, which mixture is sufficiently volatile to attract nearby males. Oleic acid is elongated to (Z)-15-tetracosenoic acid and decarboxylated. Incubating a mixture of (Z)-15-[1-^{14}C]- and (Z)-15-[15,16-^{3}H$_2$]tetracosenoic acid with microsomes from housefly fat body

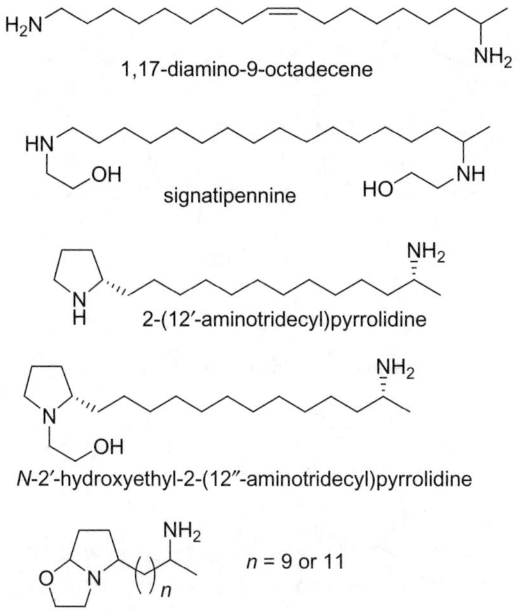

Aliphatic Compounds from Fatty Acids 123

Figure 5.24 Further compounds from epilachnine ladybirds which have not received biosynthetic attention. Some stereochemistry is undefined.

gave equal amounts of tritiated (Z)-9-tricosene and ^{14}C-labelled CO_2 (Figure 5.25) *via* the intermediate aldehyde. The aldehyde is converted to tricosene only in the presence of NADPH and oxygen. The pheromone is synthesized in the integument closely associated with oenocyte cells. The tricosene is converted by a cytochrome P450, NADPH and oxygen to the other two components of the pheromone, the epoxide (major) and the unsaturated ketone (minor) (Blomquist, 2003). For the first few days after eclosion, the female produces C_{27} hydrocarbons, then release of the hormone 20-hydroxyecdysone switches the enzymes to catalyse the production of (Z)-9-tricosene. This is an unusual case, for the only other known example, where biosynthesis of pheromone is regulated by 20-hydroxyecdysone, is in *Drosophila* (Drosophilidae). Branched-chain hydrocarbons are also involved in retaining males near the females. *Musca autumnalis* similarly uses (Z)-13-nonacosene, (Z)-14-nonacosene and (Z)-13-heptacosene.

5.5.2 *Drosophila*

The role of the pheromone compounds in the fruit fly *Drosophila melanogaster* is not as simple as in *Musca*. Females produce (7Z,11Z)-7,

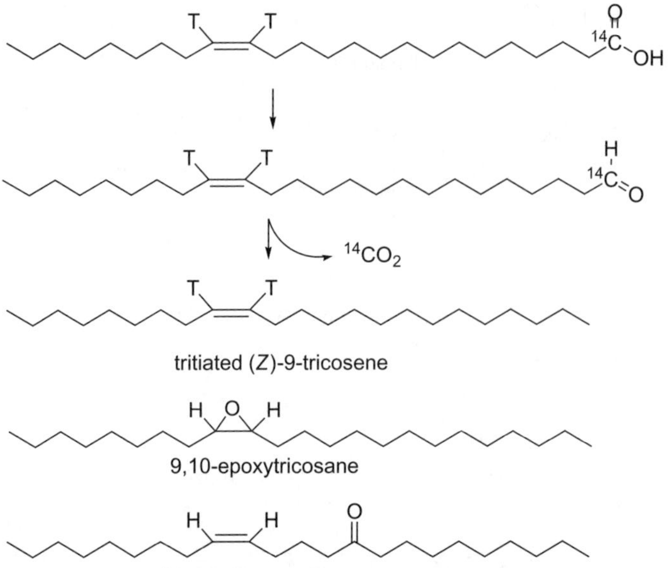

Figure 5.25 The formation of the sex pheromone of the common housefly (Z)-9-tricosene from a C_{24} fatty acid through the aldehyde. T indicates tritium. Note that using a mixture of tritiated and ^{14}C-labelled compounds for this purpose is the same as having all the labels in the same molecules. The epoxytricosene is another major component of the pheromone and the tricosenone is a minor component.

11-heptacosadiene *via* palmitic acid, which is desaturated with a Δ9-desaturase to palmitoleic acid, elongated to (Z)-11-eicosenoic acid, and then desaturated again with another Δ9-desaturase to (9Z,13Z)-9,13-eicosadienoic acid (Eigenheer *et al.*, 2002). This is again chain-lengthened to the C_{28} acid and decarboxylated to the pheromone hydrocarbon (Figure 5.26). Some strains of *D. melanogaster* produce (5Z,9Z)-5,9-heptacosadiene using a Δ11-desaturase and introducing the second double bond with another Δ11-desaturase. *D. virilis* uses (11Z)-pentacosene.

Males and females of various species of *Drosophila* also produce (Z)-7-tricosene, 7-pentacosene, (Z)-11-pentacosene, 7,11-heptacosadiene and (Z)-9-heneicosene as aggregation and anti-aphrodisiac pheromones (*i.e.* to discourage mated females from mating with another male). By manipulating genes, males of *D. melanogaster* can be feminized and induced to produce female pheromones, and in the course of these experiments, it has been proved that these hydrocarbons are made in the oenocytes (Ferveur *et al.*, 1997).

Aliphatic Compounds from Fatty Acids

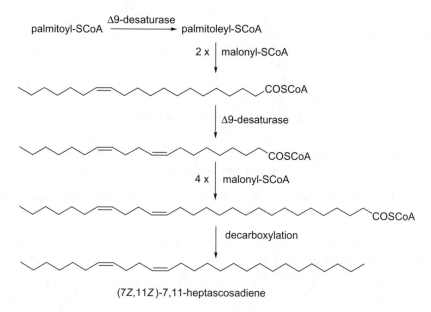

Figure 5.26 Formation of the sex pheromone hydrocarbon of *Drosophila melanogaster* from palmitic acid.

5.5.3 Glossina

Females of the tsetse fly *Glossina morsitans* (Glossinidae) produce 15,19,23-trimethyl-heptatriacontane to stimulate males to copulation. There are four possible stereoisomers and all are apparently equally active. Such involatile compounds are detected not through the air by antennae, but through contact by sensors on the legs of males. *G. pallidipes* produces 13,23-dimethylpentatriacontane, of which only the (13R,23S)-enantiomer is active, and *G. austeni* uses 15,19-dimethyltritriacontane (Figure 5.27). All these are associated with cuticle hydrocarbons. Their biosynthesis should prove interesting since it is presumed the methyl branches are added after an alkyl chain has been created.

5.6 HYMENOPTERA

As the third largest order of insects by number of recorded species, Hymenoptera has a special place because it contains the greatest number of species that have attained social status. All of the ants, and many of the bees and wasps, are social insects, and consequently have developed chemical communication to its highest level. The four orders of Coleoptera, Lepidoptera, Hymenoptera and Diptera, together with a very

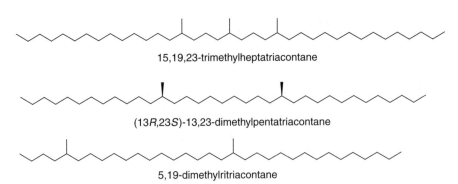

Figure 5.27 The biosynthesis of the contact sex stimulant of *Glossina morsitans*, and the pheromones of *G. pallidipes* and *G. austeni*.

small proportion of minor orders comprise the Holometabola, which accounts for 83% of all insect species.

5.6.1 Bees

Several C_{10} acids are important in honeybees *Apis mellifera* (Apidae). (*E*)-9-Oxo-2-decenoic acid (9-ODA), the major constituent of the mandibular glands, is the sex attractant of the virgin queen, synergized by the two isomers of 9-hydroxy-2-decenoic acid (Figure 5.28). After mating, the fertile queen produces a mixture of this and eight other substances, from various glands. Six of the substances are derived from fatty acid and three from aromatic compounds (Figure 5.28). Together they are known as the "queen substance" or queen retinue pheromone, which prevents queen cell construction by workers and a number of other actions which signal the presence of a dominant reproductive queen (Keeling *et al.*, 2003). It is the most complex pheromone known in insects.

Royal jelly is the food of worker larvae and of queen honeybees through their larval and adult lives. The workers produce it in their hypopharangeal and mandibular glands. It contains protein, sugars and about 5% short-chain fatty acids, chiefly 10-hydroxydec-2-enoic acid. Labelling studies have shown that stearic acid is the preferred starting material for the C_{10} acids. Hydroxylation precedes chain shortening (Figure 5.29). Both 18-hydroxylation and 17-hydroxylation are used by the mandibular glands of queens and workers, but the 17-oxidation route operates preferentially in the queen and the 18-oxidation route in the workers. The reactions are catalysed by P450 oxidases. Both enantiomers of 9-hydroxydecenoic acid have to be present for highest activity in queen pheromone, a rare example of synergism between enantiomers

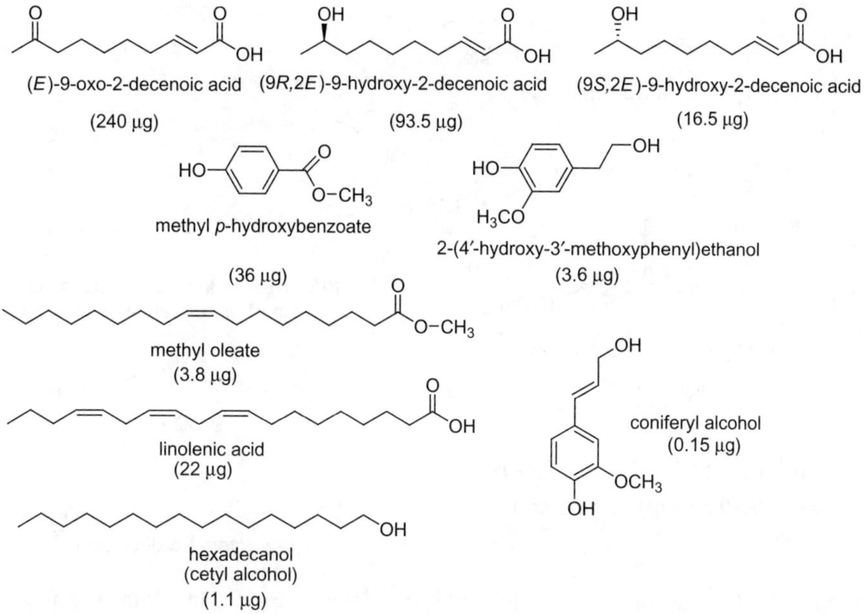

Aliphatic Compounds from Fatty Acids 127

(E)-9-oxo-2-decenoic acid (240 μg)

(9R,2E)-9-hydroxy-2-decenoic acid (93.5 μg)

(9S,2E)-9-hydroxy-2-decenoic acid (16.5 μg)

methyl p-hydroxybenzoate (36 μg)

2-(4'-hydroxy-3'-methoxyphenyl)ethanol (3.6 μg)

methyl oleate (3.8 μg)

linolenic acid (22 μg)

coniferyl alcohol (0.15 μg)

hexadecanol (cetyl alcohol) (1.1 μg)

Figure 5.28 Compounds from the queen substance or queen retinue pheromone of *Apis mellifera*. The quantity in brackets is the amount found in a single queen bee. Together they still do not account for the complete pheromone.

in insect semiochemicals. Workers make (*E*)-10-hydroxy-2-decenoic acid and (*E*)-9-carboxy-2-nonenoic acid in their mandibular glands (Plettner *et al.*, 1998). The enzymes catalysing these reactions appear not to be very specific, because 10-hydroxydecanoic acid, 12-hydroxy-2-dodecenoic acid and 12-hydroxydodecanoic acid and even some C_{14} acids are all found in mandibular glands. 10-Hydroxydec-2-enoic acid is further oxidized to (*E*)-2-decen-1,10-dioic acid and 9-oxodecenoic acid is oxidized to (*E*)-2-octen-1,8-dioic acid. These are presumably inactivation products since the authors say nothing about their activity. Recently, in an examination of royal jelly for minor components, two sugar derivatives of C_{10} acids (10-hydroxydecanoic acid 10-*O*-β-D-glucopyranoside and 10-hydroxydec-2-enoic acid 10-*O*-β-D-glucopyranoside) were isolated (Kodai *et al.*, 2007).

Foraging worker bees performing the "waggle dance" in the darkness of the hive have been shown to emit a mixture of tricosane and pentacosane with smaller amounts of (*Z*)-9-tricosene and (*Z*)-9-pentacosene onto their abdomens and into the atmosphere of the hive. These chemicals affect worker behaviour by increasing the number of foragers that leave the hive (Thom *et al.*, 2007). They are present in the normal

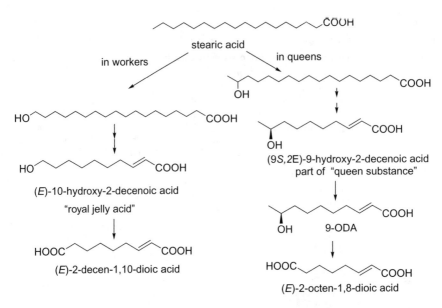

Figure 5.29 The formation of bee substances from stearic acid. Both queens and workers are capable of using both pathways but those shown actually operate preferentially. 9-Hydroxydec-2-enoic acid is also used by the queen when swarming to hold the workers in the swarm.

hydrocarbon mixture on the cuticles of bees and in the same proportions, but their sudden production and the dispersal of the odour are not yet understood.

Honeybee comb wax is produced from ventral abdominal inter-segmental glands in workers. Its composition is quite different from the predominantly hydrocarbon cuticular wax. While cuticle wax is 58% hydrocarbons, comb wax contains 13–17% hydrocarbons. According to Blomquist *et al.* (1980) the largest component of comb wax is long-chain monoesters (31–35%), followed by approximately equal proportions of diesters (10–14%) and hydrocarbons (13–17%). The wax esters are composed of palmitic and oleic acid esters of C_{20} to C_{32} primary alcohols. Comb wax production occurs between the 12th and 17th days after adult emergence, but can be strongly affected by season, the state of the queen and other factors. Wax production is not affected by application of juvenile hormone. Radio-labelled acetate, oleate and stearate were all readily incorporated into the wax (Blomquist *et al.*, 1980).

Some solitary bees line their brood cells with a waxy polymer made from hydroxy-esters. The Dufour glands of halictine and colletine bees frequently contain mixtures containing macrocyclic lactones. *Colletes* bees, for example, contain mixtures of 18-octadecanolide, 20-eicosanolide

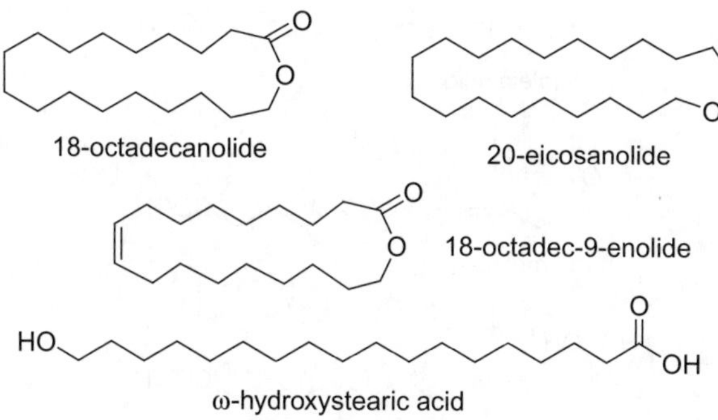

Figure 5.30 Macrocyclic lactones and a linear hydroxyacid from *Colletes* bees.

and 18-octadec-9-enolide (Figure 5.30) as well as ω-hydroxy-fatty acids (which gives the clue to the biosynthesis of the lactones) and hydrocarbons.

5.6.2 Wasps

There is only one known example of a male-produced sex pheromone from parasitic wasps. Males of the jewel wasp *Nasonia vitripennis* (Pteromalidae) attract virgin females with a mixture of (4R,5R)- and (4R,5S)-5-hydroxy-4-decanolide (Figure 5.31) (Ruther *et al.*, 2007). The site of synthesis has been traced to the rectal papillae and the secretion accumulates in the rectal vesicle. It was known that (4R,5R)-5-hydroxy-4-decanolide is a major metabolite of the *erythro* enantiomers of vernolic acid, (9Z,12R,13S)- and (9Z,12S,13R)-12,13-epoxyoctadec-9-enoic acid, in the yeast *Sporobolomyces odorus*, and *erythro*-vernolic acid is produced in plants by stereospecific oxidation of linoleic acid, catalysed by cytochrome P450. Fully labelled *cis*-[$^{13}C_{18}$]vernolic acid was synthesized and applied in solution to the abdominal tips of male *N. vitripennis*. It was possible to show that the wasps converted vernolic acid to (4R,5R)-5-hydroxy-4-decanolide. Moreover, a gene coding for an epoxide hydrolase was isolated and used to locate the site of pheromone production. It is thought this gene codes for a hydrolase that can open all four isomers of *cis*- and *trans*-vernolic acid (*i.e.* the *threo*- and *erythro*-forms of the epoxide). This is the first example of an epoxide hydrolase used in pheromone production (Abdel-Latief *et al.*, 2008). It is presumed the epoxide opening is followed by chain shortening, reduction of the double bond and lactone formation (Figure 5.31). Epoxide hydrolases degrade juvenile hormone (Section 7.6.1), and deactivate epoxide pheromones of moths (Section 5.1.2) as well as detoxify xenobiotics.

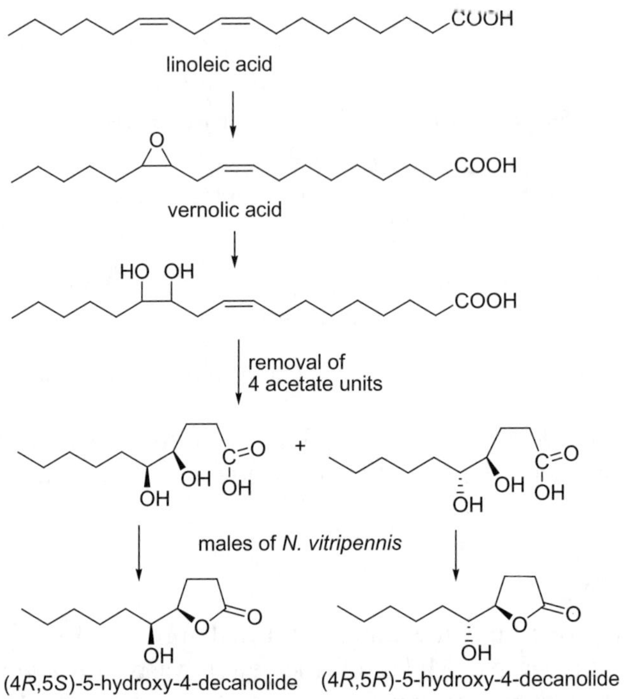

Figure 5.31 The male-produced sex pheromone of the parasitic wasp *Nasonia vitripennis*, a mixture of (4*R*,5*R*)- and (4*R*,5*S*)-5-hydroxy-4-decanolide. Its formation by epoxide opening of vernolic acid, itself derived from linoleic acid, has been demonstrated but the intermediate steps of double-bond reduction, chain shortening and lactone formation are not yet investigated.

The discovery of the female-produced sex pheromone of the Australian parasitic wasp *Neozeleboria cryptoides* (Tiphiidae) has introduced a new class of natural product, 2,5-dialkylcyclohexan-1,3-diones (Schiestl *et al.*, 2003). *N. cryptoides* and *N. monticola* both use chiloglottone A (2-ethyl-5-propylcyclohexan-1,3-dione, Figure 5.32). The discoverers presume the biosynthesis is through the coupling of a β-ketoacid derivative and an α,β-unsaturated derivative through a Claisen condensation and a Michael condensation, in unknown order (Figure 5.32) (Franke *et al.*, 2009). The male wasps are attracted as pollinators to orchids, which emit the same compound.

5.6.3 Ants

Among the many substances identified in ants as pheromones or allomones, one group of offensive compounds is worth mentioning.

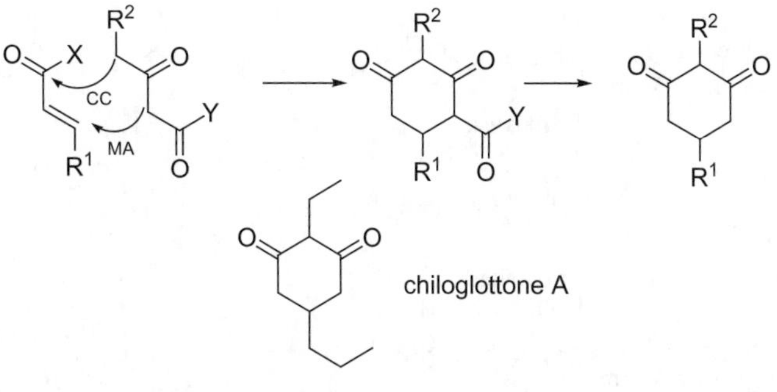

Aliphatic Compounds from Fatty Acids

Figure 5.32 The presumed biosynthetic route to the newly discovered 2,5-dialkylcycloxane-1,3-dione chilocorone A, a wasp sex pheromone, from a β-ketoacid derivative and an α,β-unsaturated acid derivative. The order of the Claisen (CC) and Michael (MA) condensation reactions necessary are unknown, the order here is merely for illustration.

Crematogaster ants are a large, aggressive, world-wide genus. They have the unusual ability to bring their abdomens forward over their heads and place a toxic secretion on other insects with their flattened sting. Some of the European species store long-chain unsaturated keto-acetates, of the form shown in Figure 5.33, in their Dufour glands (part of the poison apparatus). These are ejected with enzymes from the poison gland (an acetate esterase and an alcohol oxidase) to hydrolyse the acetate and oxidize the alcohols to aldehydes (Leclercq *et al.*, 2000). These unsaturated keto-aldehydes are strong electrophiles, reacting with OH, SH and NH_2 groups, and are strongly toxic to other ants. They are unstable and are soon oxidized to carboxylic acids or undergo rearrangement to compounds such as the lactone marked **A** in Figure 5.33. The biosynthesis of the acetates has not yet been studied but would appear to be from stearic or palmitic acids with chain extension by an unknown C_5 fragment, but the position of the double bonds indicates unusual desaturases. Other tropical *Crematogaster* ants produce similar compounds with multiple double bonds and even a triple bond (Leclercq *et al.*, 2000).

Macrocyclic lactones appear as a recurrent theme in insect defence. The Argentine ant *Linepithema humilis*, which has invaded many other countries, produces a macrocyclic lactone evidently made from a fatty acid, where synthesis begins with 2-methylbutyric acid (Figure 5.34). The African ant *Tetramorium aculeatum* produces an unsaturated lactone and its re-arrangement product (Figure 5.34) in its Dufour gland, which is the probable cause of an irritant effect of this ant. It may be noted in passing that the trail pheromone of *L. humile* is (*Z*)-9-hexadecenal (oleyl aldehyde).

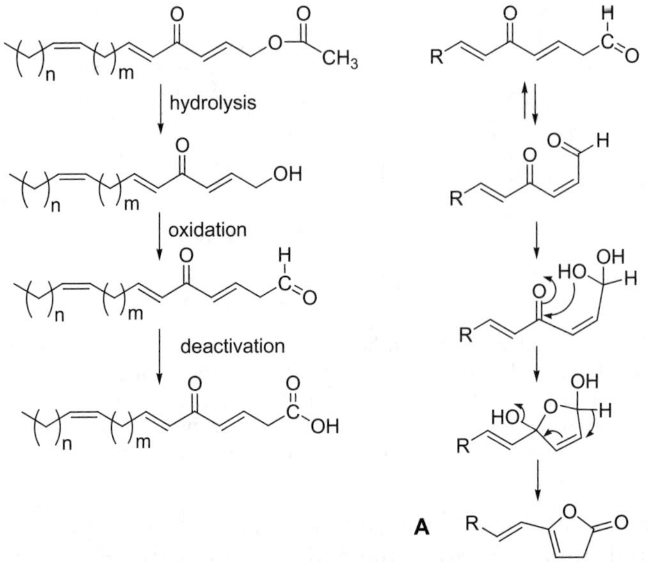

Figure 5.33 Examples of the substances stored in the Dufour gland by *Crematogaster scutellaris* ants. On ejection they are hydrolysed to alcohols catalysed by an estrase, and then oxidized to aldehydes, the true offensive compounds, by an oxygen-dependent alcohol oxidase from the poison glands. The most abundant compounds have a chain length of 21 carbon atoms, the C_{19}, C_{23} and C_{25} homologues are minor compounds, n can be 3, 5 or 7, and m varies from 3 to 13. The aldehydes are inactivated by oxidation to carboxylic acids or rearranged to lactones **A**, as shown.

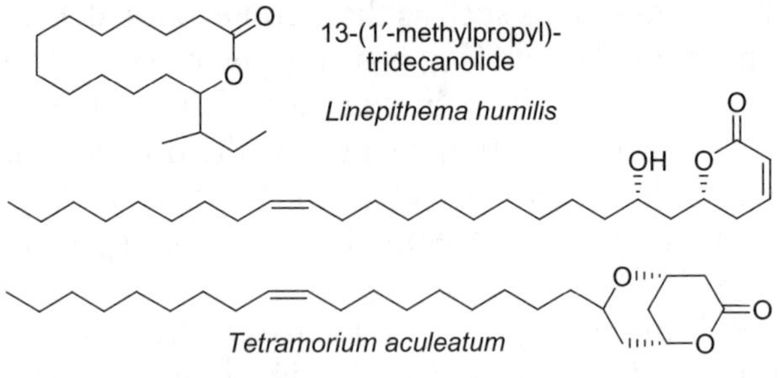

Figure 5.34 Macrocyclic lactone of the ant *Linepithemum humilis* and Dufour gland compounds of *Tetramorium aculeatum*.

5.6.3.1 Ant Venoms. Some ants of the genus *Solenopsis* have a very painful sting, and are consequently known as fire ants. Originating in South America, they have been accidentally spread to other areas, including the southern USA, where they have become a serious problem.

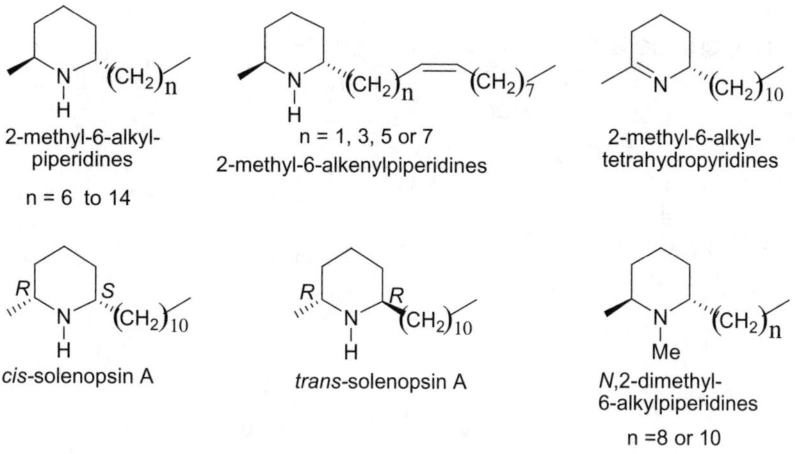

Aliphatic Compounds from Fatty Acids

Figure 5.35 2-Methyl-6-alkylpiperidine and tetrahydropyridine compounds and some N-methyl derivatives found in the venom of *Solenosis* fire ants. The *cis* and *trans* forms of solenopsins A form almost the total venom of *S. geminata*.

Their venom glands contain mixtures of 2-methyl-6-alkyl-piperidines and -tetrahydopyridines (Figure 5.35). The methyl and alkyl groups can be linked *cis* or *trans*; the *cis*-linked are always (*R,S*) and the *trans* are (*R,R*).

Because *Solenopsis geminata* makes essentially only *cis*- and *trans*-solenopsin A (2-methyl-6-undecylpiperidine), it was chosen for biosynthesis experiments (Leclercq *et al.*, 1996). Other species make more complex mixtures. Feeding experiments with 3000 to 4000 *S. geminata*, using sodium [1-^{14}C]acetate; [2-^{14}C]acetate; and [2-^{14}C]malonate plus unlabelled acetate, followed by extraction of basic compounds and dilution of the extract with unlabelled synthetic 2-methyl-6-undecylpiperidine, showed labelling from all three experiments. Degradation of the labelled compounds indicated that the compounds are biosynthesized as shown in Figure 5.36. The synthesis begins with the linking together of 12 acetate units, rather in the way already described for the coccinellines (Section 5.4.1), with keto groups either left in the alkyl chain or introduce later. A β-keto-acid is formed at some stage, and decarboxylated; addition of an amino-group to the chain, cyclization and reduction gives the solenopsins.

The solenopsins in *S. richteri* and *S. invicta* (Plate 9) are always *trans*. The alkyl groups are either saturated or have one double bond. A recent study of all the minor compounds and isomers in the venom of *S. richteri* (Chen and Fadamiro, 2009a) and *S. invicta* (Chen and Fadamiro, 2009b) led to a hypothesis that the biosynthesis proceeds mostly by route A in

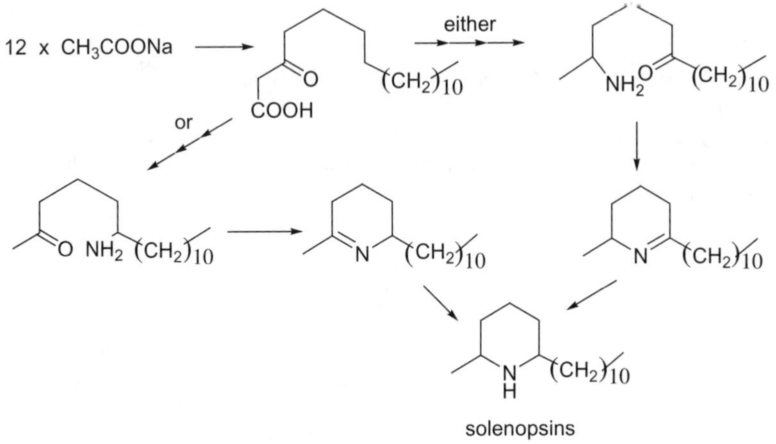

Figure 5.36 Labelling experiments with ^{14}C-labelled compounds indicate that the synthesis of solenopsins in *S. geminata* is from acetate units by one of the routes shown.

Figure 5.37. The presence of the double bond always at eight carbon atoms from the methyl end of the proposed fatty-acid chain and a total chain length of at least 17 carbons also suggests the unsaturated compounds are begun from pre-existing oleic acid molecules which are oxidized and decarboxylated, a fruitful area for further investigation.

Another group of *Solenopsis* species (Diplorhoptrum or thief ants) and some species of *Monomorium*, *Messor* and *Megalomyrmex* ants have alkyl-pyrrolidines and -pyrrolines (Figure 5.38). The sting of these ants does not seem to be so painful, but whether this is due to the different alkaloids or the protein part of the sting is not known. The biosynthesis of these compounds may be similar to the piperidines, but has not yet been studied. The terminal position of the double bonds indicates some difference. The venom of an Australian ant of the *Monomorium rothsteini* complex contains 14 2,5-dialkylpyrrolidines, but notably *trans*-2-ethyl-5-((Z)-tridecenyl)pyrrolidine (Figure 5.38), which links them to the piperidines, and also suggests a common origin (Jones et al., 2009). Alkylpyrrolidines and piperidines are also found in the defensive secretion of the beetle *Epilachna varivestis* (Section 5.4.4).

These unproven biosynthetic schemes can be compared with that of the untypical plant alkaloid coniine, which does not come from an amino acid (lysine is not incorporated) but [1-^{14}C]octanoic acid is efficiently converted into [6-^{14}C]coniine by young plants of *Conium maculatum*, [8-^{14}C]octanoic acid similarly gives [3'-^{14}C]coniine and [1-^{14}C]-5-oxo-octanal is well-incorporated (Leete, 1970). There is

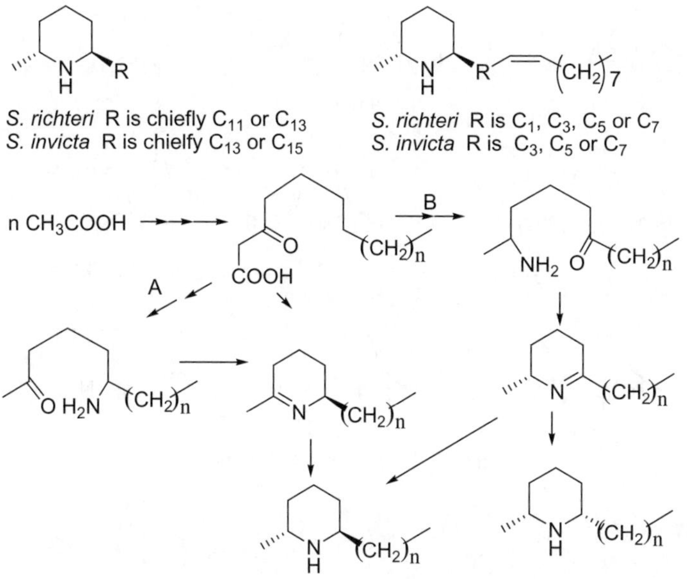

Aliphatic Compounds from Fatty Acids

Figure 5.37 Analysis of all the minor products formed by *Solenopsis richteri* and *S. invicta* suggests the alternative route shown, with route A more probable. The alkyl groups of these solenopsins are all arranged *trans*.

evidence that γ-coniceine is a precursor of coniine. The biosynthesis of coniine can then be summarized in Figure 5.39, a possible model for the insect route to piperidines.

5.7 ISOPTERA – TERMITES

Termites belong to the Hemimetabola group of insects, and provide the only other entirely social order. Termites lay trails to food sources with an abdominal sternal gland. Some 50 species have been examined for identification of their trail pheromones. Remarkably little diversity has been found in the compounds used. In the subfamily Macrotermitinae (Termitidae), some species use (Z)-3-dodecen-1-ol, and others use (3Z,6Z)-3,6-dodecadien-1-ol (Sillam-Dusses *et al.*, 2009) (Figure 5.40). It is suggested that these alcohols are formed from oleic acid and linoleic acids, respectively, by loss of three acetate units and reduction to the alcohol. Still other species of Macrotermitinae, all the Termitinae that have been examined, and some Rhinotermitidae and Coptotermitidae, and the mandibulate Nasutitermitinae use (3Z,6Z,8E)-3,6,8-dodecatrienol, which has a less obvious biosynthetic origin. The Kalotermitidae use the same dodecenol, dodecadienol and dodecatrienol. Other termites

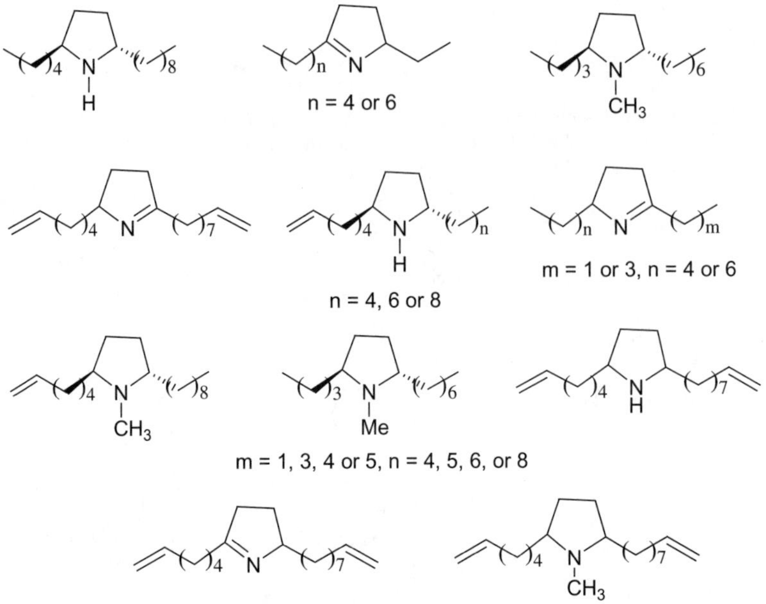

Figure 5.38 Alkyl- and alkenyl-pyrrolines, -pyrrolidines and N-methylpyrrolidines of unknown biosynthetic origin from *Solenopsis* thief ants and some *Monomorium* ants.

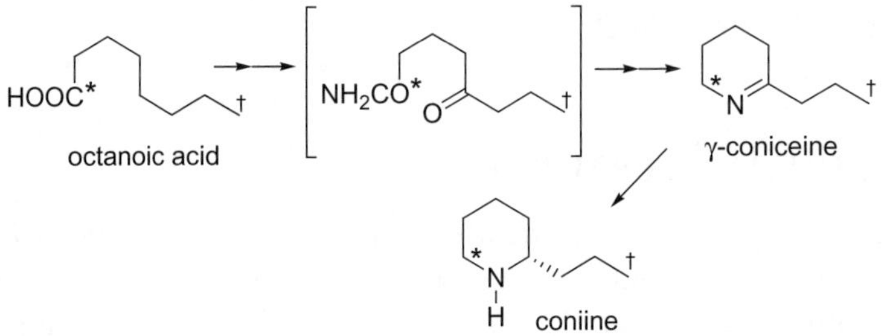

Figure 5.39 Coniine (in plants) is readily formed from octanoic acid. The asterisk and dagger indicate separate experiments where ^{14}C-labelled octanoic acid was incorporated.

use linear or cyclic terpenes. One group of C_{18} trail pheromones remains unidentified.

What is called *chemical parsimony* (frugality) is applied among termite pheromones. One compound serves as pheromone for a number of species. Moreover, dodecadienol and dodecatrienol, produced by females in some species at higher concentrations, act as sex pheromone

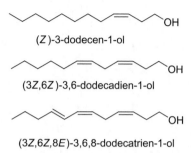

Figure 5.40 The C_{12} alcohols used by many species of termites as trail pheromones. Dodecadienol also acts as sex attractant (at higher concentration) in some species.

attracting males (about $0.1\,\mathrm{pg\,cm^{-1}}$ for trails and about 1 ng as sex pheromone) (Laduguie *et al.*, 1994; Peppuy *et al.*, 2001; Bland *et al.*, 2007).

The *Amitermes* genus of termites also use macrocyclic lactones in their defensive secretions. 15-Pentadecanolide, 21-heneicosanolide, 22-docosanolide and 24-tetracosanolide have all been identified along with the α-hydroxy- and β-hydroxytetracosanolides shown in Figure 5.41.

5.8 BLATTODEA – COCKROACHES

Cockroaches belong to the Family Blattodea, and have received special attention because of their association with human habitation. The female German cockroach *Blatella germanica* produces 3,11-dimethylnonacosan-2-one as a contact pheromone from a gland on the underside of the last abdominal tergite or pygidium. Because of the possibility of using the pheromone to control cockroaches in buildings, its biosynthesis has been studied in detail. It is produced with the aid of a fatty-acid synthase in microsomes. Its synthesis follows from the consideration of branched-chain hydrocarbons (Figure 5.42). Condensation of an acetate unit and a propionate unit gives a C_5 unit with the first branching point. The propionate is derived from valine, isoleucine or possibly methionine. Use of [1-^{13}C]propionate showed the methyl branch was incorporated early during chain growth. Three more acetate units are added before another propionate, and the chain is completed with nine more acetates. The sequence is completed by decarboxylation and oxidation of a CH_2 group in two stages to a ketone (Schal *et al.*, 2003). Minor components of the pheromone are formed by further oxidation at the far end of the alkyl chain. The fatty-acid synthase used to extend the alkyl chain is more efficient in

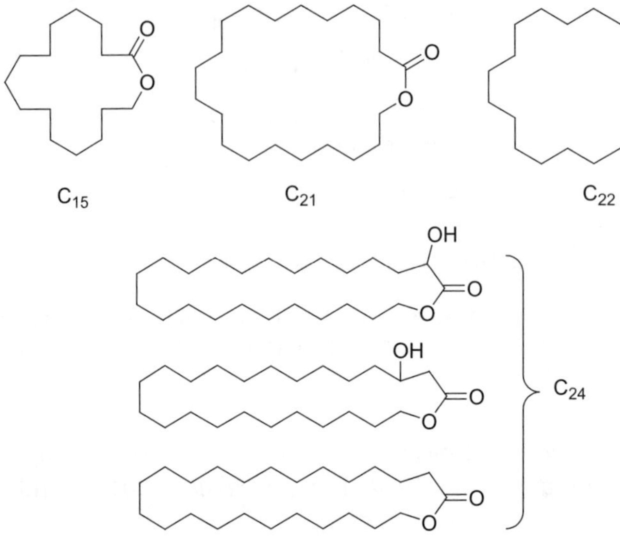

Figure 5.41 Macrocyclic lactones of the defensive secretion *Amitermes* termites. Note the odd number of carbon atoms in two compounds.

incorporating propionate units (as methylmalonate) than the cytosol fatty-acid synthase (Section 4.1.1). The production of pheromone in *B. germanica* is controlled by juvenile hormone (Section 7.6.1) but not in other cockroaches.

5.9 GREEN LEAF VOLATILES

Interesting by-products of plant metabolism are "green leaf volatiles", which are produced by hydroperoxide lyase enzymes from unsaturated fatty acids (Figure 5.43). These volatile compounds, which include (*Z*)- and (*E*)-3-hexenols and -3-hexenals, (*E*)-2-hexenal, (*E*)-2-hexen-1-ol, 1-hexanol and acetates are released by some plants naturally, and especially when damaged by insects. Their release can attract insect parasitoids to the damaging insects. The volatiles are also attractive to many insects and are used by others as pheromones. The leaf-footed bug *Leptoglossus zonatus* (Hemiptera: Coreidae) (Figure 5.44) uses an alarm pheromone consisting of (*E*)-2-hexenal, hexyl acetate, 1-hexanol, hexanal and hexanoic acid. (*E*)-2-Hexenal is a typical defensive compound of stink bugs, and is used as a male aggregation pheromone by *Podisus* bugs (Hemiptera: Pentatomidae).

There is no indication that these insects use the same biosynthetic route as plants. The Florida woods cockroach *Eurycotis floridana*

Aliphatic Compounds from Fatty Acids

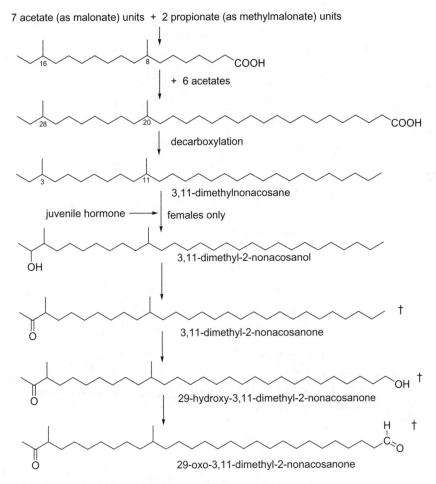

Figure 5.42 An outline of the biosynthesis of the contact pheromone of the German cockroach. All three compounds marked † are active pheromone components.

(Blattodea: Blattidae) emits a disagreeable odour when disturbed. Farine *et al.* (2000) showed that it consists of (*E*)-2-hexenal, (*E*)-2-hexenol and (*E*)-2-hexenoic acid. The biosynthesis of the secretion was studied by injection of ^{14}C-labelled acetic acid and fatty acids. They found that fatty acids were incorporated, but much more slowly than acetic acid, and although they were unable to firmly prove it, their findings suggested that the C_6 compounds were preferentially biosynthesized *de novo* from acetic acid.

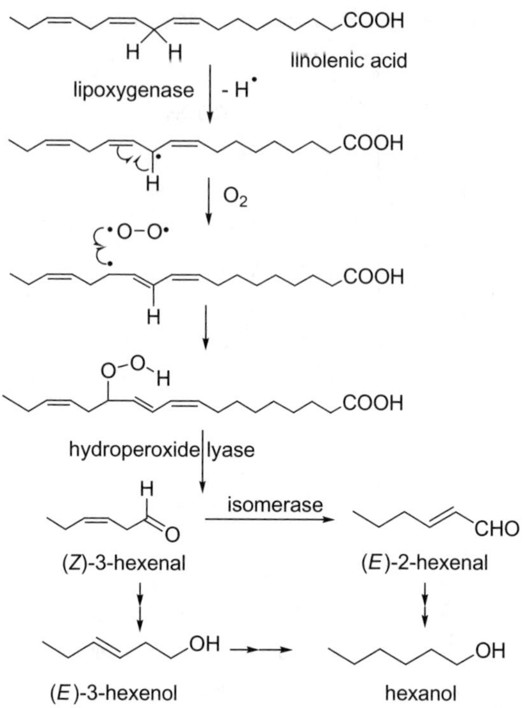

Figure 5.43 Formation of some of the "green leaf volatiles" by plants, which are also used by some insects. The biosynthesis in insects may not be by the same route.

Figure 5.44 The leaf-footed bug, *Leptoglossus zonatus*.

When pine trees are attacked by *Ips* and *Dendroctonus* bark beetles (Section 7.3.1) the green leaf volatiles can disturb the aggregation pheromones of the beetles.

BACKGROUND READING

Blomquist, G. J. and Vogt, R. G. (ed.) 2003. *Insect Pheromone Biochemistry and Molecular Biology*, Elsevier, London and San Diego, 745 pp.

Blum, M. S. 1978. Biochemical defenses of insects,. In: *Biochemistry of Insects* (Rockstein, M., ed.), Academic Press, New York and London, pp. 465–513.

Blum, M. S. 1987. Biosynthesis of arthropod exocrine compounds. *Annual Review of Entomology*, **32**, 381–413.

Francke W. and Schulz, S. 1999. Pheromones. In: *Comprehensive Natural Products Chemistry*, vol. 8 (Mori, K., ed.), Pergamon, Oxford, pp. 197–261.

Howse, P., Stevens I. and Jones, O. 1998. *Insect Pheromones and Their Use in Pest Management*, Chapman and Hall, London (Chapter 5, Chemical structures and diversity of pheromones, pp. 135–179; Chapter 6, Isolation and structure determination, pp. 180–225).

www-pherolist.slu.se for a comprehensive list of pheromones, species and pictures.

REFERENCES

Abdel-Latief, M., Garbe, L. A., Koch, M. and Ruther, J. 2008. An epoxide hydrolase involved in the biosynthesis of an insect sex attractant and its use to localize the production site. *Proceedings of the National Academy of Sciences, USA*, **105**, 8914–8919.

Ando, T., Kawai, T. and Matsuoka, K. 2008. Epoxyalkenyl sex pheromones produced by female moths in highly evolved groups: biosynthesis and its endocrine regulation. *Journal of Pesticide Science*, **33**, 17–20.

Attygalle, A. B., Blankespoor, C. L., Eisner, T. and Meinwald, J. 1994. Biosynthesis of a defensive insect alkaloid – epilachnene from oleic-acid and serine. *Proceedings of the National Academy of Sciences, USA*, **91**, 12790–12793.

Bjostad, L. B. and Roelofs, W. L. 1981. Sex-pheromone biosynthesis from radiolabeled fatty-acids in the redbanded leafroller moth. *Journal of Biological Chemistry*, **256**, 7936–7940.

Bland, J. M., Raina, A. K., Carpita, A. and Dickens, J. C. 2007. Comparative analysis of the sex/trail pheromone, 3,6,8-dodecatrien-1-ol, from three species of subterranean termites (Isoptera: Rhinotermitidae). *Sociobiology*, **50**, 535–551.

Blomquist, G. J. 2003. Biosynthesis and ecdysteroid regulation of housefly sex pheromone production. In: *Insect Pheromone Biochemistry and Molecular Biology* (Blomquist, G. J. and Vogt, R. G., ed.), Elsevier, London and San Diego, pp. 231–252.

Blomquist, G. J., Chu, A. J. and Remaley, S. 1980. Biosynthesis of wax in the honeybee *Apis mellifera* L. *Insect Biochemistry*, **10**, 313–321.

Butenandt, A., Beckmann, R., Stamm, D. and Hecker, E. 1959. Über den Sexual-Lockstoff des Seidenspinners *Bombyx mori*. Reindarstellung und Konstitution. *Zeitschrift für Naturforschung*, **14b**, 283–284.

Chen, L. and Fadamiro, H. Y. 2009a. Re-investigation of venom chemistry of *Solenopsis* fire ants. I. Identification of novel alkaloids in *S. richteri*. *Toxicon*, **53**, 469–478.

Chen, L. and Fadamiro, H. Y. 2009b. Re-investigation of venom chemistry of *Solenopsis* fire ants. II. Identification of novel alkaloids in *S. invicta*. *Toxicon*, **53**, 479–486.

Eigenheer, A. L., Young, S., Blomquist, G. J., Borgeson, C. E., Tillman, J. A. and Tittiger, C. 2002. Isolation and molecular characterisation of *Musca domestica* Δ-9-desaturase sequences. *Insect Molecular Biology*, **11**, 533–542.

Farine, J. P., Everaerts, C., Abed, D. and Brossut, R. 2000. Production, regeneration and biochemical precursors of the major components of the defensive secretion of *Eurycotis floridana* (Dictyoptera, Polyzosteriinae). *Insect Biochemistry and Molecular Biology*, **30**, 601–608.

Ferveur, J.-F., Savarit, F., O'Kane, C. J., Sureau, G., Greenspan, R. J. and Jallon, J.-M. 1997. Genetic feminization of pheromones and its behavioral consequences in *Drosophila* males. *Science*, **276**, 1555–1558.

Franke, S., Ibarra, F., Schulz, C. M., Twele, R., Poldy, J., Barrow, R. A., Peakall, R., Schiestl, F. P. and Francke, W. 2009. The discovery of 2,5-dialkylcyclohexan-1,3-diones as a new class of natural products. *Proceedings of the National Academy of Sciences, USA*, **106**, 8877–8882.

Görgen, G., Fross, C., Boland, W. and Dettner, K. 1990. Biosynthesis of 1-alkenes in the defensive secretion of *Tribolium confusum* (Tenebionidae) – stereochemical implications. *Experientia*, **46**, 700–704.

Goller, S., Szöcs, G., Francke, W. and Schulz, S. 2007. Biosynthesis of (3Z,6Z,9Z)-3,6,9-octadecatriene: the main component of the pheromone blend of *Erannis bajaria*. *Journal of Chemical Ecology*, **33**, 1505–1509.

Hoskovec, M., Luxová, A., Svatos, A. and Boland, W. 2002. Biosynthesis of sex pheromones in moths: stereochemistry of fatty alcohol oxidation in *Manduca sexta*. *Tetrahedron*, **58**, 9193–9201.

Jones, T. H., Andersen, A. N. and Kenny, J. C. 2009. Venom alkaloid chemistry of Australian species of the *Monomorium rothsteini* complex, with particular reference to the taxonomic implications. *Chemistry and Biodiversity*, **6**, 1034–1041.

Jurenka, R. A. 2003. Biochemistry of female moth sex pheromones. In: *Insect Pheromone Biochemistry and Molecular Biology* (Blomquist, G. J. and Vogt, R.G., ed.), Elsevier, London and San Diego, pp. 53–80.

Keeling, C. I., Slessor, K. N., Higo, H. A. and Winston, M. L. 2003. New components of the honeybee (*Apis mellifera* L.) queen retinue pheromone. *Proceedings of the National Academy of Sciences, USA*, **100**, 4486–4491.

Kodai, T., Umebayashi, K., Nakatani, T., Ishiyama, K. and Noda, N. 2007. Composition of royal jelly II. Organic acid glycosides and sterols of royal jelly of honeybees (*Apis mellifera*). *Chemical and Pharmaceutical Bulletin*, **55**, 1528–1531.

Laduguie, N., Robert, A., Bonnard, O., Vieau, F., Lequere, J. L., Semon, E. and Bordereau, C. 1994. Isolation and identification of (3Z,6Z,8E)-3,6,8-dodecatrien-1-ol in *Reticulitermes santonensis* Feytaud (Isoptera, Rhinotermitidae) – roles in worker trail-following and in alate sex-attraction behavior. *Journal of Insect Physiology*, **40**, 781–787.

Laurent, P., Braekman, J.-C., Daloze, D. and Pasteels, J. M. 2002. *In vitro* production of adaline and coccinelline, two defensive alkaloids from ladybird beetles (Coleoptera: Coccinellidae). *Insect Biochemistry and Molecular Biology*, **32**, 1017–1023.

Laurent, P., Braekman, J.-C., Daloze, D. and Pasteels, J. M. 2003. Biosynthesis of defensive compounds from beetles and ants. *European Journal of Chemistry*, **2003**, 2733–2743.

Laurent, P., Braekman, J.-C. and Daloze, D. 2005. Insect chemical defence. *Topics in Current Chemistry*, **250**, 167–229.

Leal, W. S., Zarbin, P. H. G., Wojtasek, H. and Ferreira, J. T. 1999. Biosynthesis of scarab beetle pheromones: enantioselective 8-hydroxylation of fatty acids. *European Journal of Biochemistry*, **259**, 175–180.

Leclercq, S., Braekman, J.-C., Daloze, D., Pasteels, J. M. and Vander Meer, R. K. 1996. Biosynthesis of the solenopsins, venom alkaloids of the fire ants. *Naturwissenschaften*, **83**, 222–225.

Leclercq, S., Braekman, J.-C., Daloze, D. and Pasteels, J. M. 2000. The defensive chemistry of ants. *Progress in the Chemistry of Organic Natural Products*, **79**, 115–229.

Leete, E. 1970. Biosynthesis of coniine from octanoic acid in hemlock plants. *Journal of the American Chemical Society*, **92**, 3835.

Lui, W., Jiao, H., McMurray, N. C., O'Connor, M. and Roelofs, W. L. 2002. Gene characterized for membrane desaturase that produces 9E-11 isomers of mono- and di-unsaturated fatty acids. *Proceedings of the National Academy of Sciences, USA*, **99**, 620–624.

Matoušková, P., Pichová, I. and Svatoš, A. 2007. Functional characterization of a desaturase from the tobacco hornworm moth (*Manduca sexta*) with bifunctional Z11- and 10,12-desaturase activity. *Insect Biochemistry and Molecular Biology*, **37**, 601–610.

Millar, J. G. 2000. Polyene hydrocarbons and epoxides: a second major class of lepidopteran sex attractant pheromones. *Annual Review of Entomology*, **45**, 575–604.

Moto, K., Yoshiga, T., Yamamoto, M., Takahashi, S., Okano, K., Ando, T., Nakata, T. and Matsumoto, S. 2003. Pheromone gland-specific fatty-acyl reductase of the silkmoth, *Bombyx mori*. *Proceedings of the National Academy of Sciences, USA*, **100**, 9156–9161.

Moto, K., Suzuki, M. G., Hull, J. J., Kurata, R., Takahashi, S., Yamamoto, M., Okano, K., Imai, K., Ando, T. and Matsumoto, S. 2004. Involvement of a bifunctional fatty-acyl desaturase in the biosynthesis of the silkworm, *Bombyx mori*, sex pheromone. *Proceedings of the National Academy of Sciences, USA*, **101**, 8631–8636.

Muñoz, L., Rosell, G., Quero, C. and Guerrero, A. 2008. Biosynthetic pathways of the pheromone of the Egyptian armyworm *Spodoptera litoralis*. *Physiological Entomology*, **33**, 275–290.

Oehlschlager, A. C., Pierce, A. M., Pierce, H. D. and Borden, J. H. 1988. Chemical communication in cucujid grain beetles. (Coleoptera, Cucujidoe) *Journal of Chemical Ecology*, **14**, 2071–2098.

Peppuy, A., Robert, A., Semon, E., Bonnard, O., Son, N. T. and Bordereau, C. 2001. Species specificity of trail pheromones of fungus-growing termites from northern Vietnam. *Insectes Sociaux*, **48**, 245–250.

Plettner, E., Slessor, K. N. and Winston, M. L. 1998. Biosynthesis of mandibular acids in honey bees (*Apis mellifera*): de novo synthesis, route of fatty acid hydroxylation and caste selective β-oxidation. *Insect Biochemistry and Molecular Biology*, **28**, 31–42.

Rafaeli, A. and Jurenka, R. A. 2003. PBAN regulation of pheromone biosynthesis in female moths. In: *Insect Pheromone Biochemistry and Molecular Biology* (Blomquist, G. J. and Vogt, R.G., ed.), Elsevier, London and San Diego, pp. 107–136.

Rule, G. S. and Roelofs, W. L. 1989. Biosynthesis of sex pheromone components from linolenic acid in Arctiid moths. *Archives of Insect Biochemistry and Physiology*, **12**, 89–97.

Ruther, J., Stahl, L. M., Steiner, S., Garbe, L. A., and Tolasch, T. 2007. A male sex pheromone in a parasitic wasp and control of the behavioral response by the female's mating status. *Journal of Experimental Biology*, **210**, 2163–2169.

Schal, C., Sevala, V. and Cardé, R. T. 1998. Novel and highly specific transport of a volatile sex pheromone by haemolymph lipophorin in moths. *Naturwissenschaften*, **85**, 339–342.

Schal, C., Fan, Y. and Blomquist, G. J. 2003. Regulation of pheromone biosynthesis, transport, and emission in cockroaches. In: *Insect Pheromone Biochemistry and Molecular Biology* (Blomquist, G. J. and Vogt, R. G., ed.), Elsevier, London and San Diego, pp. 284–322.

Schiestl, F. P., Peakall, R., Mant, J. C., Ibarra, F., Schulz, C., Franke, S. and Francke, W. 2003. The chemistry of sexual deception in an orchid-wasp pollination system. *Science*, **302**, 437–438.

Schroeder, F. R., Farmer, J. J., Smedley, S. R., Attygalle, A. B., Eisner, T. and Meinwald, J. 2000. A combinatorial library of macrocyclic polyamines produced by a ladybird beetle. *Journal of the American Chemical Society*, **122**, 3628–3634.

Serra, M., Piña, B., Bujons, J., Camps, F. and Fabriás, G. 2006. Biosynthesis of 10,12-dienoic fatty acids by a bifunctional Δ^{11} desaturase in *Spodoptera littoralis*. *Insect Biochemistry and Molecular Biology*, **36**, 634–641.

Serra, M., Piña, B., Abad, J. L., Camps, F. and Fabriás, G. 2007. A multifunctional desaturase involved in the biosynthesis of the processionary moth sex pheromone. *Proceedings of the National Academy of Sciences, USA*, **104**, 16444–16449.

Sillam-Dusses, D., Semon, E., Robert, A. and Bordereau, C. 2009. (Z)-Dodec-3-en-1-ol, a common major component of the trail-following pheromone in the termites Kalotermitidae. *Chemoecology*, **19**, 103–108.

Smedley, S. R., Schroeder, F. C., Weibel, D. B., Meinwald, J., Lafleur, K. A., Renwick, J. A., Rutowski, R. and Eisner, T. 2002. Mayolenes; labile defensive lipids from the glandular hairs of a caterpillar (*Pieris rapae*). *Proceedings of the National Academy of Sciences, USA*, **99**, 6822–6827.

Thom, C., Gilley, D. C., Hooper, J. and Esch, H. E. 2007. The scent of the waggle dance. *PLoS Biology*, **5**, 1862–1867.

Whittle, E. J., Tremblay, A. E., Buist, P. H. and Shanklin, J. 2008. Revealing the catalytic potential of an acyl-ACP desaturase: tandem selective oxidation of saturated fatty acids. *Proceedings of the National Academy of Sciences, USA*, **105**, 14738–14743.

CHAPTER 6
Aceto-propiogenins

6.1 ACETOGENINS

J. N. Collie, who proposed correctly that fatty acids were made from head-to-tail linking of acetate units (Section 4.1), also proposed that many simple cyclic compounds were also formed from head-to-tail chains of acetates. He found that boiling dehydroacetic acid with barium hydroxide, and acidifying the product, had converted it to orcinol, and suggested it had passed through a triketone stage (Figure 6.1). We now know that many compounds, other than the fatty acids and compounds derived from them, are made from chains of acetates, and we call these collectively acetogenins. Micro-organisms and insects extend this principle and often incorporate propionate units along with the acetates. Therefore, in considering insect biosynthesis, we need to speak of acetogenins and aceto-propiogenins, to include compounds made from a mixture of acetic acid and propionic acid. The syntheses are begun with acetyl-SCoA or propionyl-SCoA and extended with molecules of malonyl-SCoA (for C_2 units) or methylmalonyl-SCoA (for C_3 units) (Figures 4.16 and 4.17).

Collie's hypothesis on acetogenins was also confirmed by Birch and Donovan (1953) (Section 4.1), who used *Penicillium patulum* to show that ^{14}C-labelled acetic acid could be converted to 6-methylsalicylic acid (Figure 6.2). Degradation of the 6-methylsalicylic acid produced gave radioactivity distributed as predicted by his hypothesis (Figure 6.2). 6-Methylsalicylic acid belongs to the class of polyketides. 6-Methylsalicylic acid and its derivatives are also found in insects.

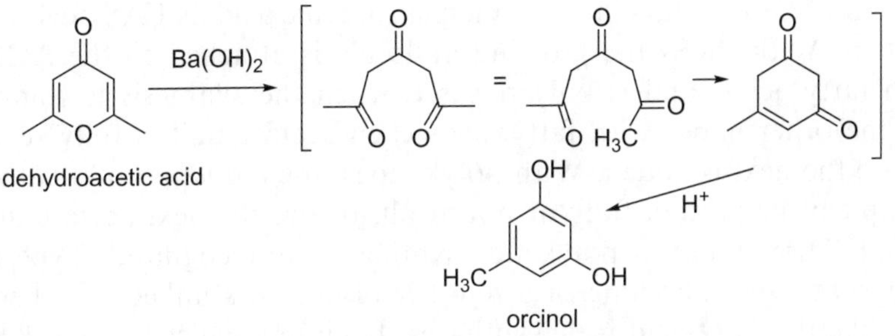

Aceto-propiogenins

Figure 6.1 Collie's conversion of dehydroacetic acid into orcinol as he conceived it.

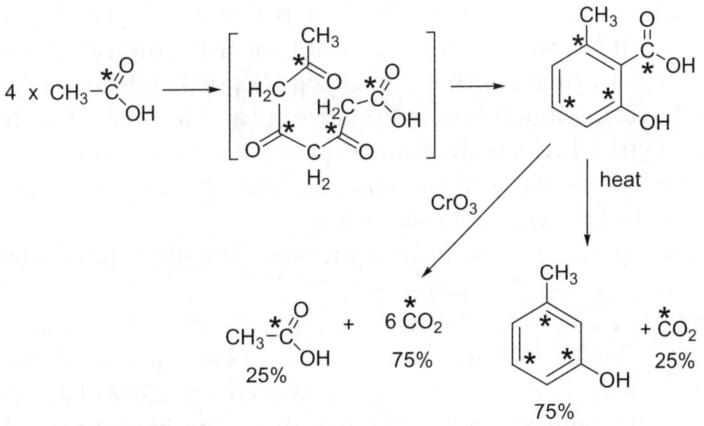

Figure 6.2 The biosynthesis of 6-methylsalicylic acid from radio-labelled acetic acid as hypothesized by Birch and the labelling of its degradation products that supported the hypothesis.

6.2 POLYKETIDES

The discovery of antibiotics led to intense investigation of natural products from micro-organisms, and the isolation of many new compounds which were shown, by labelling experiments similar to that of Birch (above), to consist of acetate units linked together. Characteristic enzyme complexes called polyketide synthases (PKS) have been identified through cloning of the corresponding genes, which enzymes are known to catalyse the synthesis of these compounds. Definitions have been turned around, so that now, polyketides are defined as compounds made through the catalysis of polyketide synthases. The PKS and fatty-acid synthases (FAS) have some characteristic in common. PKSs also use β-keto-synthases (KS) and keto-reductases (KR), and an acyl carrier protein (ACP), ending in the same pantothenic acid as FAS acyl carrier protein. With PKSs the growing molecule is attached to the ACP, as with fatty acids, and is only released when the synthesis is complete. On the other hand, with fatty acids each acetate unit is fully reduced before the next is added. With polyketides, the reduction of a carbonyl group can be partial, fully or not at all, before the next acetate unit is added. Three types of polyketide synthases are recognized. Type I are megasynthases consisting of a number of enzymes linked together, rather like the fatty-acid megasynthases. Each module contains a KS, an acyl transferase (AT) and ACP, and adds one building unit to the growing molecule. Other enzyme modules can be included, such as ketoreductase or dehydrase, and all enzymes are aligned in the order in which each of the steps of synthesis is performed. Type II polyketide synthases, which catalyse the formation of aromatic compounds, are clusters of individual enzymes, and type III polyketide synthases apply only to the formation of aromatic chalcones and flavones. A difficulty in studying Type I PKSs is the near total absence of isolatable intermediates, since the growing chain remains attached to the mega-enzyme complex until the synthesis is complete.

The formation of 6-methylsalicylic acid can then be written as in Figure 6.3.

It has generally been thought that polyketides can be synthesized by insects. That view was used in the first edition of this book (Morgan, 2004). However, Pankewitz and Hilker (2008) have carefully looked at the evidence now available about the biosynthesis of polyketides and come to the conclusion that there is no firm evidence that polyketide synthases exist in insects. Genes for polyketide synthases may exist only in micro-organisms, fungi, algae, dinoflagellates, some lower marine organisms and plants. It is possible that all the polyketides we find in insects may be made by micro-organisms living in or on insects. It is also possible that some insects possess the necessary genes through horizontal transfer from micro-organisms or fungi to insects.

6.3 BUTYRIC ACID COMPOUNDS

Some simple C_4 compounds, widely distributed, appear to be the product of condensation of just two acetate units. (R)-3-Hydroxybutyric acid, together with its dehydration-reduction product crotyl alcohol, provides the sex attractant of the female orb web spider *Agelenopsis aperta* (Araneae: Agelenidae), attracting males to the web. Once attracted, the dimer ($3R,3R'$)-3-($3'$-hydroxybutyroyl)butyric acid (Figure 6.4)

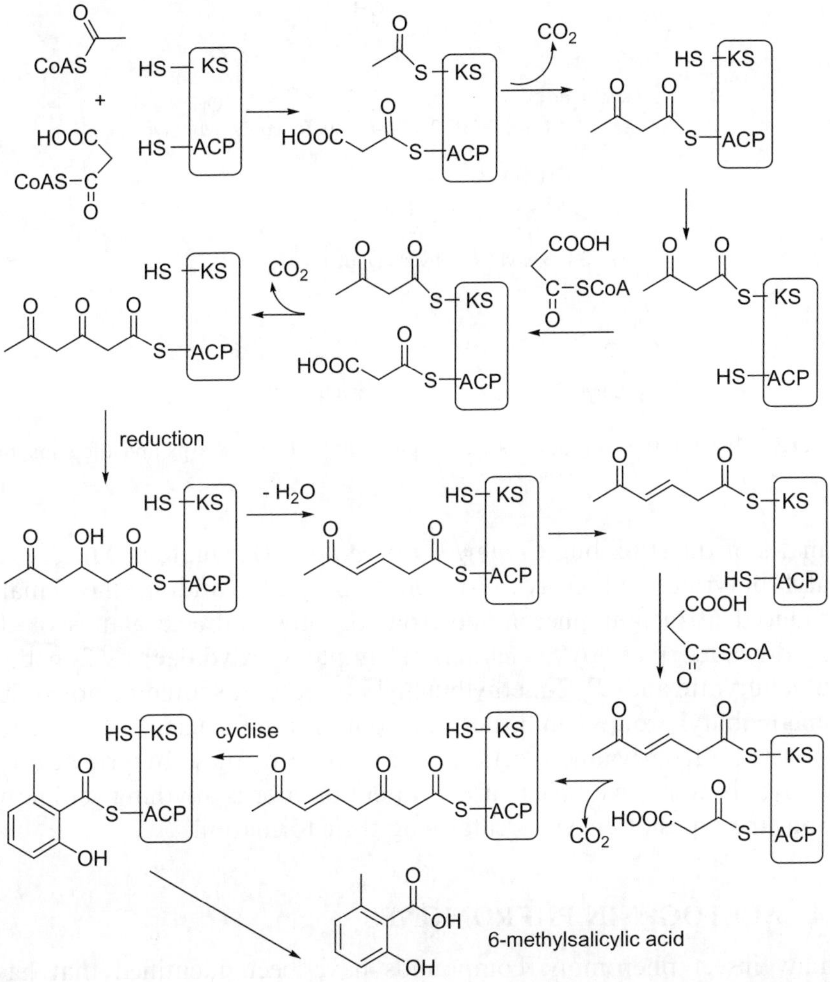

Figure 6.3 The biosynthesis of 6-methylsalicylic acid catalysed by its polyketide synthase as studied in the fungus *Penicillium patulum*. The two elements of the synthase complex shown are keto-synthase (**KS**) and acyl carrier protein (**ACP**). Other components are not shown.

stimulates the male to roll up the web to hide the odour and prevent other males being attracted (Schulz, 2004). These two compounds are found in the silk of virgin females of other species of *Linyphia*, *Microlinyphia* and *Neriene* spiders (all Araneae: Linyphiidae). The same (*R*)-3-hydroxybutyric acid occurs as the ester of (*Z*)-3-hexenol in the male marking secretion of three decorator wasps, *Eucerceris conata*, *E. rubripes* and *E. tricolor* (Hymenoptera: Crabronidae) (Figure 6.4). The fully reduced C_4 chain appears in the sex pheromone produced by

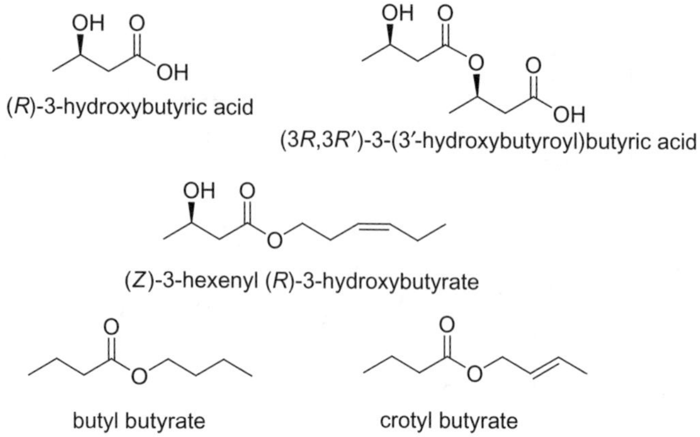

Figure 6.4 Butyrate compounds from opilionids, *Eucercis* wasps and the stink bug *Campyloma verbusei*.

females of the stink bug *Campylomma verbasci* (Hemiptera: Miridae) as butyl butyrate and crotyl (2-butenyl) butyrate, and in the female-produced attractant pheromone from the metathoracic glands of the broad-headed bug *Alydus eurinus* (Hemiptera: Alydidae) as 2-methylbutyl butyrate and (*E*)-2-methylbutenyl butyrate. It should be noted that 2-methylbutyl groups can be derived from the amino acid isoleucine, and isobutyrate from valine (Section 4.2.4). None of these butyric acid derivatives have received biosynthetic attention, nor is anything yet known about the types of enzymes catalysing their formation.

6.4 ACETOGENIN PHEROMONES

Many insect pheromone compounds have been identified that have short alkyl chains with an oxygen function (see *e.g.* Francke and Schulz, 1999). The chains may be straight or branched. Many of them can be visualized as composed of acetate and propionate units linked together and decarboxylated while still in the β-keto-acid stage. As an example, consider 2-heptanone and 2-heptanol. (*R*)-2-Heptanol and 2-heptanone comprise the pheromone of females of the caddis fly *Rhyacophila fasciata* (Trichoptera: Rhyacophilidae). 2-Heptanol is part of the aggregation pheromone of females of the bark beetle *Dendroctonus vitei* (Coleoptera: Curculionidae), and 2-heptanone is produced by the honeybee *Apis mellifera* (Hymenoptera: Apidae) in its mandibular glands and may be used by foraging bees to mark food sources. It is also present in the mandibular glands of the stingless bee *Oxytrigona mediorufa* (Hymenoptera: Apidae) but of unknown function there. A

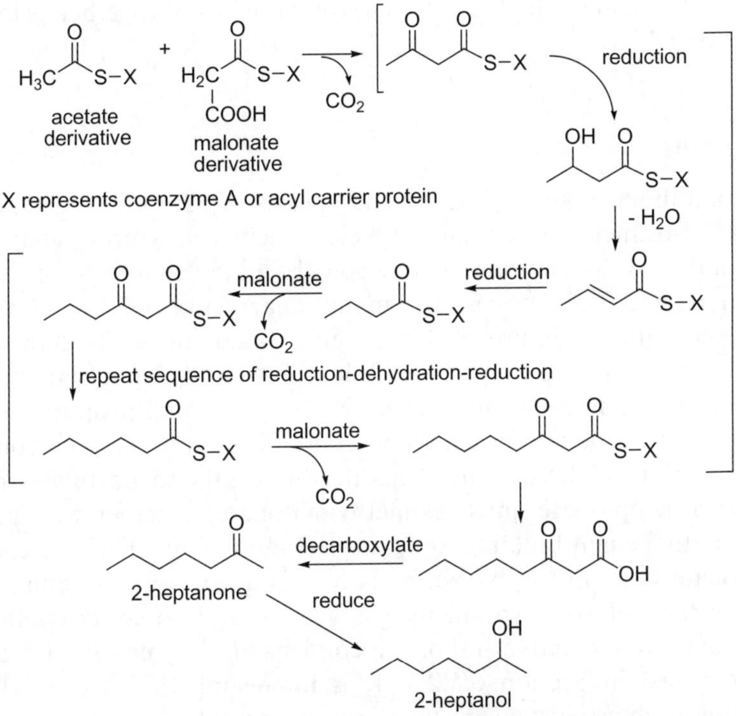

Aceto-propiogenins 151

Figure 6.5 The route proposed for the formation of 2-heptanone and 2-heptanol. It is not known whether the building blocks are handled as thioesters of co-enzyme A or whether they remain attached to an acyl carrier protein. Nor is the sequence of addition of a C_2 unit and reduction known, but with polyketide synthases the reduction usually precedes the addition of the next C_2 unit. Subsequent figures are simplified to show just un-derivatized acetic acid and propionic acid units without suggesting they react in that form.

probable formation of these compounds is outlined in Figure 6.5. The synthesis proceeds as in fatty-acid synthesis until four acetate units are linked together, but before the penultimate carbonyl is reduced, the resulting C_8 β-keto-acid is decarboxylated to the C_7 ketone, and this is reduced to the C_7 alcohol.

6.5 ACETO-PROPIOGENINS

There are many substances that, from appearance, must be made from acetate and propionate, some of which have been demonstrated to be so from labelling experiments, and yet these compounds are not closely related to fatty acids through synthesis. We must therefore call them aceto-propiogenins until we know just what type of enzymes catalyse their synthesis, and whether those enzymes are linked together as in FAS and Type I PKS.

6.5.1 Ants

The mandibular glands of at least 12 species of *Myrmica* ant (Hymenoptera: Formicidae) contain species-specific mixtures containing 3-octanol and 3-octanone together with minor amounts of related 3-alkanols and 3-alkanones. The mandibular glands of *Myrmica scabrinodis* contain a mixture of 3-hexanone, 3-heptanone, 3-octanone, 6-methyl-3-octanone, 3-nonanone, 3-decanone and 3-undecanone and the corresponding 3-alcohols. There is always a small proportion of 6-methyl-3-octanone and 6-methyl-3-octanol in the mixture. To conceive the formation of these compounds it is necessary to postulate the inclusion of propionate units (as methylmalonate) together with acetate units in the chain-building operation (Figure 6.6). The presence of propionate units in the biosynthesis of this group of compounds may indicate lack of total specificity for acetate and some acceptance of propionate by the active site of the condensing enzyme. Similar effects are described in Section 6.5.2.1. It is interesting that the polyketide synthases of micro-organisms tend to be of low substrate specificity, and can often incorporate "unnatural" building blocks into polyketides.

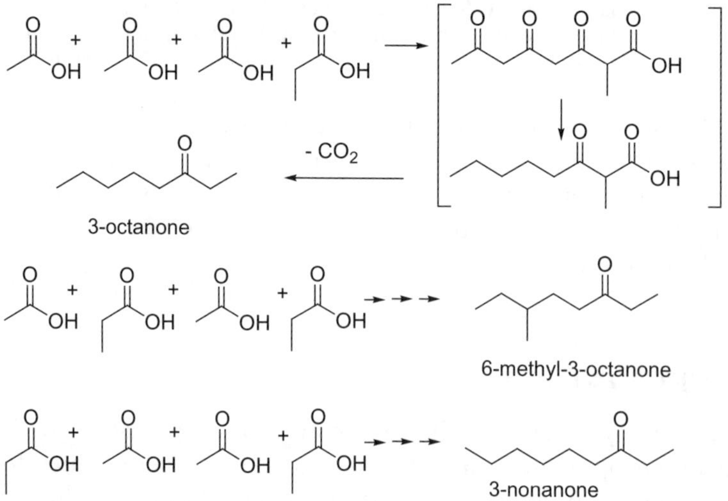

Figure 6.6 The formation of 3-octanone, 6-methyl-3-octanone and 3-nonanone suggested from acetate and propionate building blocks.

The related compound 4-methyl-3-heptanone is found in the mandibular glands of many other ants including *Atta texana*, *A. cephalotes* and *A sexdens*, where it serves as an alarm pheromone. (S)-4-Methyl-3-heptanone is found in the poison gland of *Aphaenogaster albisetosus*, where it is used as a trail pheromone. It is also found in wasps, caddisflies (Trichoptera) and the defensive secretion of three species of daddy longlegs or harvestmen (Opiliones) (Section 6.5.3). The compound is composed of three propionate units. The biosynthesis has been followed in *A. albisetosus* using ^{13}C- and ^{2}H-labelled methylmalonate, which was injected into crickets, which were then fed to the ants (Jarvis *et al.*, 2004). The labelling of the resulting 4-methyl-3-heptanone showed it indeed consists of three propionate groups, linked together and decarboxylated (Figure 6.7). The corresponding alcohol (3S,4S)-4-Methyl-3-heptanol is the trail pheromone of the ant *Leptogenys diminuta*, and is part of the aggregation pheromone of *Scolytus* bark beetles (Coleoptera: Scolitidae), while a 2 : 1 mixture of the (3S,4S)- and (3S,4S)-enantiomers is found in the caddisfly *Limnephilus lunatus* (Trichoptera: Limnephilidae) (Plate 15), but of unknown function.

The mandibular glands of the ant *Manica rubida* contain a mixture of unsaturated ketones that also illustrate the aceto-propiogenin theme. Manicone, the chief component, is accompanied by normanicone, homomanicone and bishomomanicone (Bestmann *et al.*, 1988). Their probable building blocks are emphasized by the thick black lines in Figure 6.8. Similarly, the mandibular glands of the ant *Pogonomyrmex salinus* contain chiefly 4-methyl-3-heptanone, but also 4-methyl-3-hexanone, 4-methyl-3-octanone, normanicone, manicone and homomanicone.

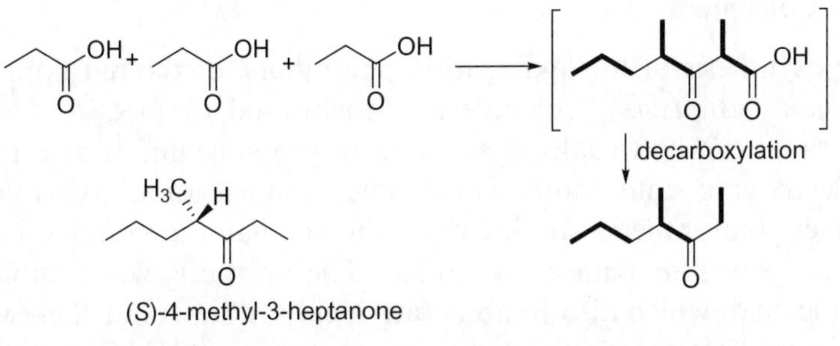

Figure 6.7 The formation of 4-methyl-3-heptanone, demonstrated through the use of deuterated and ^{14}C-labelled methylmalonic acid. The carbon skeletons of the propionate units are shown in heavy black lines.

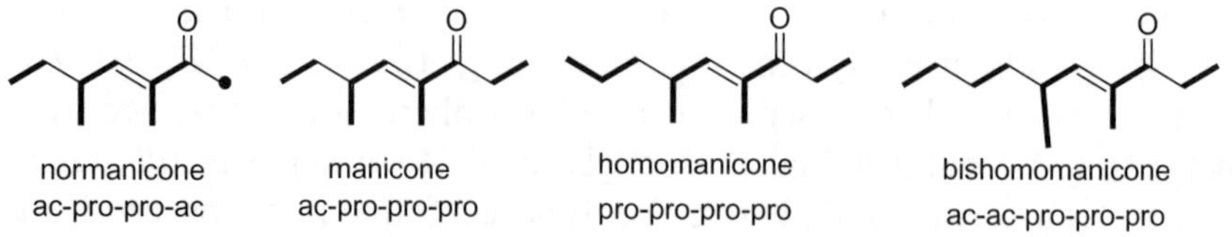

Figure 6.8 Compounds from the mandibular glands of the ant *Manica rubida*. The proposed building blocks of acetate and propionate are indicated by heavy black lines.

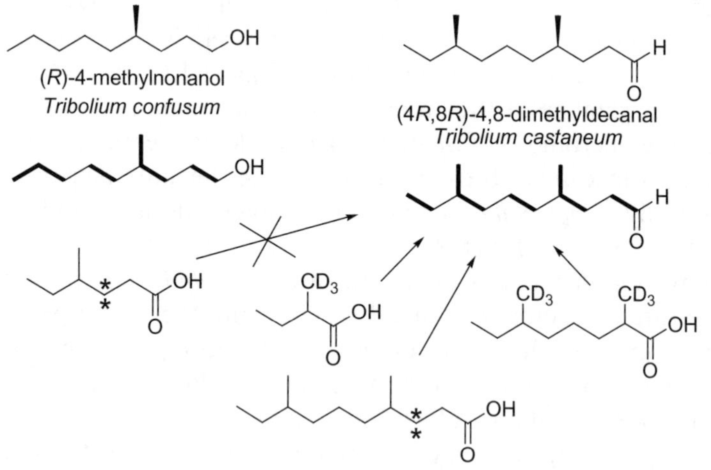

Figure 6.9 The aggregation pheromones of two *Tribolium* species. The biosynthetic studies on *T. castaneum* showed that three labelled intermediates were incorporated but another, 4-methylhexanoic acid, surprisingly, was not incorporated. Deuterium labelling is shown by D and *.

6.5.2 Coleoptera

The biosynthesis of the aggregation pheromone of the red flour beetle *Tribolium castaneum* (Coleoptera: Tenebrionidae) (4R,8R)-4,8-dimethyldecanal can be visualized as an aceto-propiogenin, or as a product of mevalonate and homomevalonate condensation (Chapter 7). However, biosynthetic studies have shown that the compound is an aceto-propiogenin (Kim *et al.*, 2005). The synthesis was inhibited by octynoic acid, which also inhibits fatty-acid synthesis but there was no inhibitory effect from mevastatin, an inhibitor of HMG-reductase in terpene synthesis (Section 7.2.1). Juvenile hormone JH III (Section 7.6.1), which activates the biosynthesis of the mevalonate pathway to

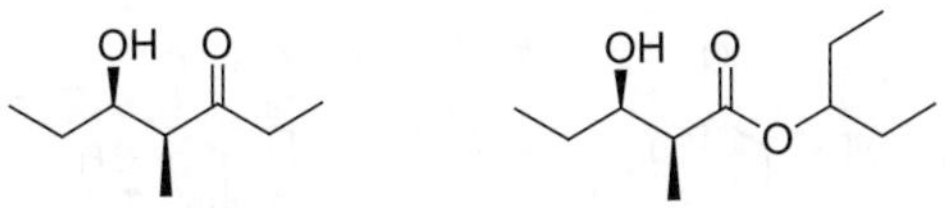

Aceto-propiogenins

(4*S*,5*R*)-5-hydroxy-4-methyl-3-heptanone (sitophinone)

3-pentyl (2*S*,3*R*)-3-hydroxy-2-methylpentanoate (sitophilate)

Figure 6.10 Male-produced attractants of grain weevils, sitophinone from *S. oryzae* and *O. zeamais* and sitophilate from *O. granarius*, starting from a common biosynthetic origin.

terpenes, had no effect. There was high incorporation of [1-^{13}C]acetate and [1-^{13}C]propionate and two deuterated intermediates. [3,3-^{2}H$_2$]-4,8-Dimethyldecanoic acid was reduced to the aldehyde, but [3,3-^{2}H$_2$]-4-methylhexanoic acid was not incorporated (Figure 6.9) suggesting a more complex route which was not investigated.

The aggregation pheromone shared by the confused flour beetle *Tribolium confusum* (Coleoptera: Tenebrionidae) and the yellow mealworm beetle *Tenebrio molitor* (Coleoptera: Tenebrionidae), (*R*)-4-methylnonanol, has the same (*R*)-4-methyl group as 4,8-dimethyldecanal, and appears to be made from two propionates and two acetates (Figure 6.9). The last stage of its synthesis has been studied in *T. molitor*, and shown to be the reduction of (*R*)-4-methylnonanoic acid CoA thioester. The reduction is a regulated step to pheromone production and performed only in mature females (Mangat *et al.*, 2006).

The weevil genus *Sitophilus* (Coleoptera: Curculionidae) presents an interesting example of divergence in aceto-propiogenins. The males of the rice weevil *S. oryzae* and of the maize weevil *S. zeamais* produce sitophinone (4*S*,5*R*)-5-hydroxy-4-methyl-3-heptanone, but in the grain weevil *S. granarius* (Plate 16) the growing chain is terminated in an ester (Figure 6.10). The male-produced attractant is 3-pentyl (2*S*,3*R*)-3-hydroxy-2-methylpentanoate, known as sitophilate. The systematic name is also given as 1-ethylpropyl (2*S*,3*R*)-3-hydroxy-2-methylpentanoate.

The bark beetle attractants multistriatin (which is an aceto-propiogenin) and *exo*- and *endo*-brevicomin (which may be fatty-acid derivatives or acetogenins) are considered in Chapter 7 with other bark beetle attractants because of their similar structure and biosynthesis (Section 7.3.1).

6.5.2.1 Nitidulid Beetles. Sap beetles (Coleoptera: Nitidulidae) exploit the aceto-propiogenin theme still further by allowing the incorporation of butyrate into their aggregation pheromones, which are

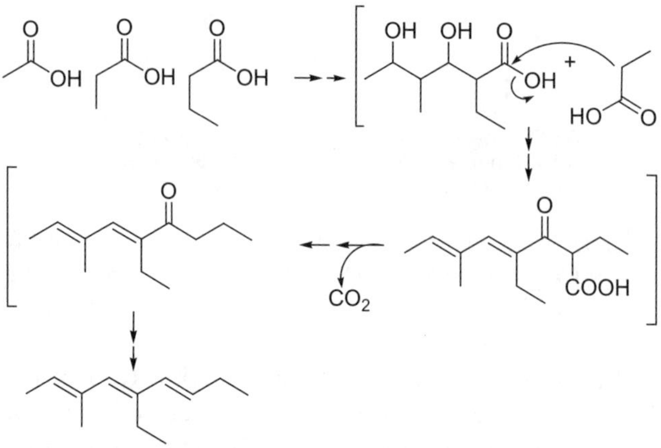

(2E,4E,6E)-5-ethyl-3-methyl-2,4,6-nonatriene

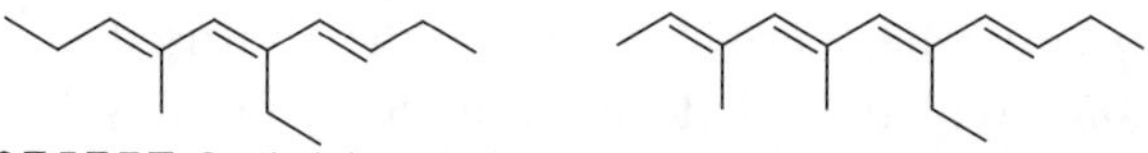

(3E,5E,7E)-6-ethyl-4-methyl-3,5,7-decatriene

(2E,4E,6E,8E)-7-ethyl-3,5-dimethyl-2,4,6,8-undecatetraene

Figure 6.11 Formation of the principal component of the aggregation pheromone of the beetle *Carpophilus freemani*, and two minor components where an acetate unit is replaced by a propionate and where an extra propionate has been added. The last unit added is the one that loses the carboxylate group.

poly-unsaturated hydrocarbons with methyl and ethyl branches. The secretion is produced only by males, from disc-shaped glands associated with the trachea in the posterior abdomen. Francke and Dettner (2005) list 23 compounds, all with three or four all-*trans* conjugated double bonds, from 10 species of *Carpophilus*. Petroski *et al.* (1994) studied the biosynthesis of (2E,4E,6E,8E)-5-ethyl-3-methyl-2,4,6-nonatriene (Figure 6.11) in *C. freemani* with deuterated acetic, propionic and butyric acids and ^{13}C-acetic acid. They expected to make the study with GC-MS alone and for that prepared deuterated versions of the final compound in order to understand the mass spectral fragmentation pattern. In practice, they found that they were able to add so much of the deuterated acids to the food and obtain sufficient secretion from the beetles that they had enough labelled material for NMR spectra, particularly for ^{13}C NMR, and they were able to gain much detailed information about the synthesis, and locate the origins of all the carbon atoms in the nonatriene. The labelling of [^{2}H]acetic acid and [^{13}C]acetic acid were considerably scrambled through metabolism, and therefore of little value, but from the other labelled compounds they could show that each molecule of the hydrocarbon consisted of one acetate, one propionate and two butyrate units, condensed as shown in Figure 6.11. The 5-ethyl-3-methylnonatriene was accompanied by smaller amounts of two other hydrocarbons (3E,5E,7E)-6-ethyl-4-methyl-3,5,7-decatriene and (2E,4E,6E,8E)-7-ethyl-3,5-dimethyl-2,4,6,8-undecatetraene, which represent, respectively, structures where the acetate unit has been replaced by another propionate and one where an extra propionate has been inserted.

An examination of the synthesis of 15 other such compounds in the Australian sap beetle *C. davidsoni* and the flower beetle *C. mutilatus* showed that, in each case, they were formed from the same three building blocks, and the last unit added is the one which loses its carboxylate group (Bartelt and Weisleder, 1996). Feeding the beetles with [1-^{13}C]propionic acid and [3,3,4,4,4-^{2}H$_5$]butyric acid made it possible to locate the positions of the one acetate, three propionates and one butyrate incorporated into the hydrocarbon molecules. In (2E,4E,6E)-3,5,7-trimethyl-2,4,6,8-undecatetraene, for example, the isotopic labelling made clear where each unit was placed. Each of the three propionates retained their labelled carboxylate carbons, but that of butyric acid was missing (Figure 6.12). Some further examples of nitidulid attractant hydrocarbons are shown in Figure 6.12.

Not all the compounds identified have shown pheromone activity. From a detailed study of the compounds present in the dried fruit beetle *C. hemipterus* (Figure 6.13), Bartelt (1999) found that the major

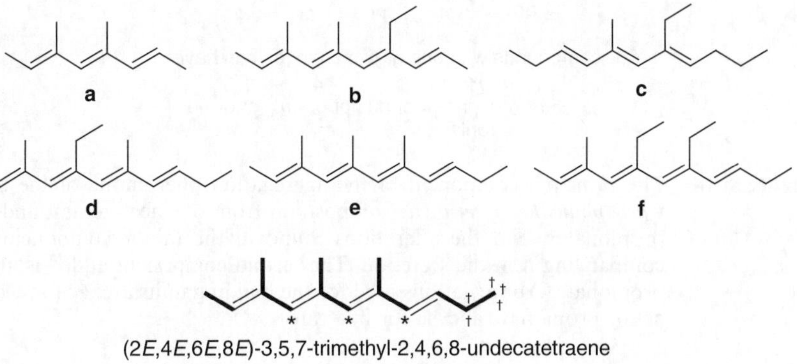

(2E,4E,6E,8E)-3,5,7-trimethyl-2,4,6,8-undecatetraene

Figure 6.12 The structures of some other aggregation compounds from *Carpophilus* beetles. All compounds except **c** start with acetate; **c** begins with propionate. The chains are then extended with propionate (to give methyl branches) or butyrate (for ethyl branches). Compounds **a**, **b** and **d** terminate with propionate and all the others with butyrate.

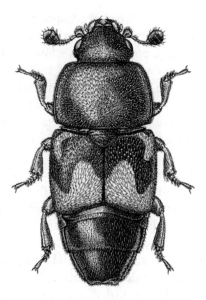

Figure 6.13 The dried fruit beetle *Carpophilus hemipterus*. © Her Majesty the Queen in Right of Canada, 2010.

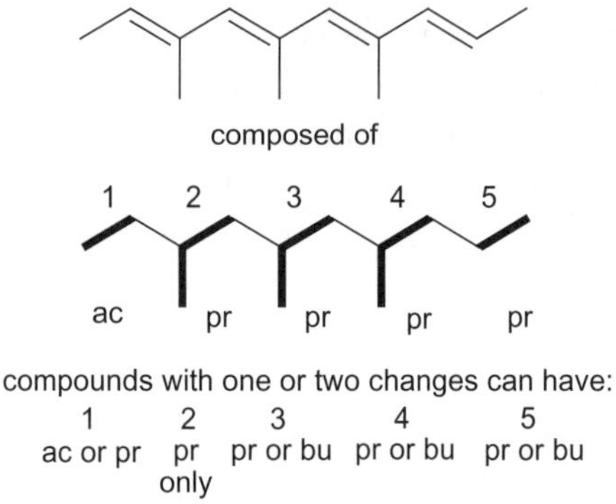

Figure 6.14 The principal component of the aggregation pheromone of the beetle *Carpophilus hemipterus*, its composition from one acetate unit and four propionates, and the alterations found in the minor components accompanying it in the secretion. The second component added is always propionate. Abbreviations used for the building units are: **ac** for acetate; **pr** for propionate and **bu** for butyrate.

compound was (2E,4E,6E,8E)-3,5,7-trimethyl-3,4,6,8-decatetraene. This compound was accompanied in the secretion by four compounds in which there was one substitution of propionate for acetate or butyrate for propionate and five compounds with two substitutions (Figure 6.14).

The synthesizing enzyme can apparently accommodate an extra carbon atom in the first, third and fourth units added, but the second unit is always propionate. When the abundance of the major component is assigned 100, the relative abundance of the compounds with one alteration varied from 13 down to 3, and only two of these had pheromone activity. The abundance of the compounds with two alterations was from 2 down to 0.2, and only the most abundant of these showed pheromone activity. The figures on abundance helped Bartelt to conclude that only one enzyme, with low specificity, able to accommodate a butyrate unit occasionally in place of a propionate, was responsible for making all the compounds (Bartelt, 1999).

6.5.3 Other Arthropods

Males of the brown stink bug *Euschistus heros* (Heteroptera: Pentatomidae), a major pest of soyabeans and other crops in Brazil, produce a pheromone attractant to females dominated by methyl 2,6,10-trimethyltridecanoate (presumably pro-pro-ac-pro-ac-pro, though not yet studied). Testing of the individual enantiomers on females in a choice olfactometer found that the (2R,6R,10S)-isomer (Figure 6.15) was the most attractive, although the isomeric composition of the natural secretion has not yet been determined. Another pentatomid stink bug *Stiretrus anchorago* (Hemiptera: Pentatomidae) produces 6,10,13-trimethyltetradecanol (Figure 6.15). The structure suggests an isovalerate starter unit, followed by pro-ac-pro-ac-ac. The same substance or the corresponding aldehyde or isovalerate ester has been found in large sternal glands in the abdomens of males of several other species (*Perillus bioculatus, Oplomus servus, O. dichrous, Mineus strigipes* and *Eocanthecona furcellata*), but neither their pheromonal purpose, their stereochemistry nor their biosynthesis have yet been studied (Millar, 2005).

Harvestmen or daddy longlegs (Opiliones, an order of Arachnida) possess well-developed exocrine glands that provide defensive secretions. In the suborder Palpatores the secretion consists of volatile

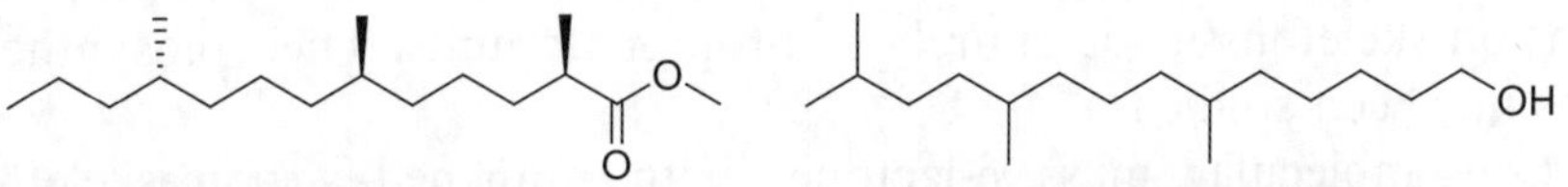

methyl (2R,6R,10S)-2,6,10-trimethyltridecanoate

6,10,13-trimethyltetradecanol

Figure 6.15 Male attractants of brown stink bugs.

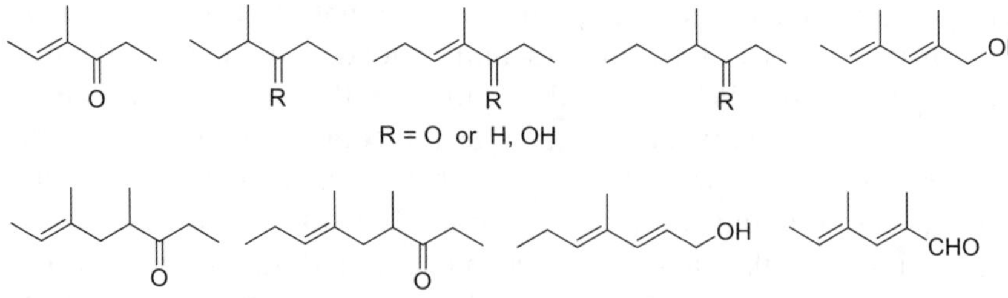

Figure 6.16 Defensive secretion compounds from opilionids. All can be conceived as made from acetate and propionate units.

branched-chain alcohols and ketones, many of them unsaturated (Figure 6.16). Their biosynthesis has not been studied, but their structures provide rich speculation on their origins.

The acarid mite *Lardoglyphus konoi* (Astigmata: Lardoglyphidae) produces an aggregation pheromone, (1R,3R,5R,7R)-1,3,5,7-tetramethyldecyl formate, which apparently is formed from five propionates condensed together, and another mite *Chortoglyphus arcutus* (Astigmata: Chortoglyphidae) produces (4R,6R,8R)-4,6,8-trimethyldecan-2-one (chortolure), an aggregation pheromone for both sexes. Its biosynthesis has been studied with deuterated propionic acid and shown to consist of a chain starting with one acetate unit followed by three propionates (Figure 6.17) (Schulz et al., 2004).

6.6 PYRONES AND LACTONES

It is noteworthy that the carbon skeleton of stegobinone, the dihydropyrone attractant of the drugstore beetle *Stegobium panicetum* (Coleoptera: Anobiidae), is the same as that of 3,5,7-trimethyl-2,4,6,8-decatetraene, from *Carpophilus hemipterus* (Coleoptra: Nitidulidae) (Figure 6.18). Stegobiol, also present in the secretion, appears less active. The only other anobiid beetle thus far investigated, the cigarette beetle *Lasioderma serricorne*, produces serricornin, (4S,6S,7S)-7-hydroxy-4,6-dimethylnonan-3-one, but also the less active α-serricorone, β-serricorone and serricorole. These are shown in Figure 6.18 with their proposed carbon skeletons made entirely of propionate units. Their biosynthesis has not been studied.

Low molecular mass δ-lactones with branched-carbon skeletons provide the trail pheromones of six ants of the *Camponotus* genus. Their biosynthetic studies were among the first on ant trail pheromones. Bestmann's group found that *Camponotus vagus* and *C. socius* use

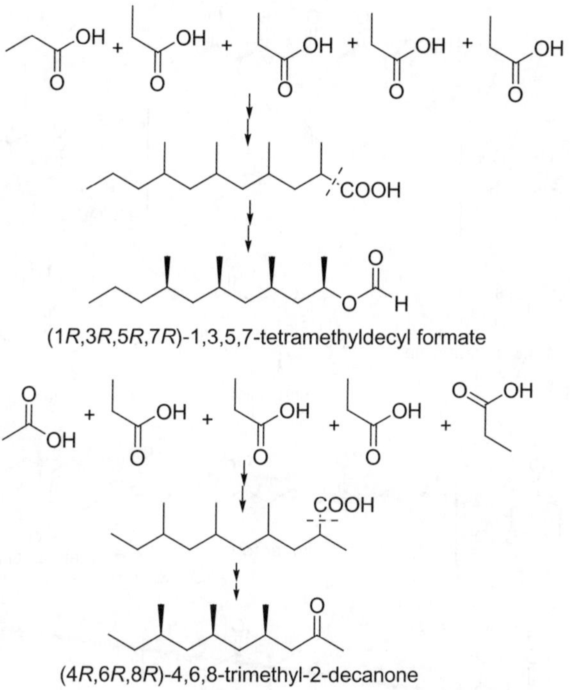

Figure 6.17 Two aceto-propiogenin aggregation pheromones of mites (Acaridae).

3,5-dimethylhexanolide (systematically 3,5,6-trimethyltetrahydro-2*H*-pyran-2-one, Figure 6.19) as pheromone. Feeding these ants with [3,3,3-^{2}H$_3$]propionic acid gave the lactone deuterated in the 3- and 5-methyl groups (Figure 6.17) (Bestmann *et al.*, 1997). *Camponotus atriceps* uses 3,5,7-trimethyloctanolide, which had three methyl groups labelled, indicating the incorporation of three propionic acid units. *Camponotus ligniperda* used a mixture of both these compounds plus 3,5,7-trimethylnonanolide with four labelled propionate units (Figure 6.19), while *C. herculeanus* uses a mixture of the hexanolide and nonanolide. Experiments with feeding [^{2}H$_3$-methyl]methionine to see if the methyl group could be incorporated found no evidence for its use in these compounds. 3,5-Dimethylhexanolide was already known as invictolide from the queen control secretion of the fire ant *Solenopsis invicta*.

6.7 CARBOCYCLIC COMPOUNDS

The formation of 6-methylsalicylic acid from a chain of acetate units has been described above (Sections 6.1 and 6.2). How it is made in insects is

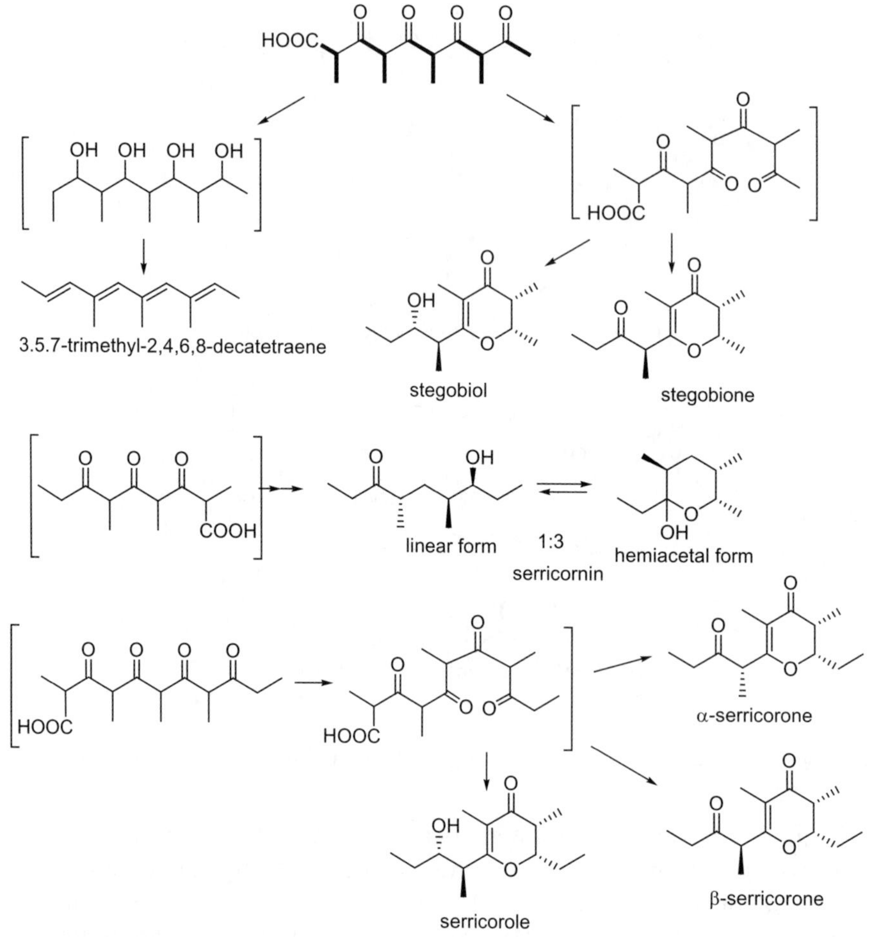

Figure 6.18 Stegobione and stegobiol, the attractants from *Stegobium panicetum*. They have the same carbon skeleton as the attractant trimethyldecatetraene of *Carpophilus hemipterus*. Serricornin, α-serricorone, β-serricorone and serricorole, composed entirely of propionate units, are attractants from *Lasioderma serricorne*.

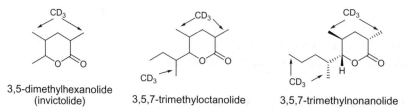

Figure 6.19 δ-Lactone trail pheromones from *Camponotus* ants. The positions of labelling when the ants were fed with $[3,3,3-{}^2H_3]$propionic acid are indicated.

Aceto-propiogenins

not known, but it and derivatives of it occur widely, especially in ants. For example, methyl 6-methylsalicylate forms the trail pheromone of the ant *Tetramorium impurum* and is also found in the secretion of several ponerine ants, and in the mandibular glands of *Gnamptogenys pleurodon* and some *Camponotus* species. 6-Methylsalicylaldehyde provides the pheromone of some species of mites. The decarboxylation product of 6-methylsalicylic acid, *m*-cresol, is a defensive compound in some beetles. 2-Hydroxy-6-methylacetophenone is found in the anal glands of various species of *Hypoclinia* ants, together with *m*-cresol and 2-acetoxy-6-methylacetophenone. In this section some aromatic compounds known to be made *via* acetogenins are described. Many other benzene derivatives are known from insects but they can be from aromatic precursors in the diet (typically phenylalanine or tyrosine), or through acetogenins or made by bacterial symbionts. These are discussed in Chapter 9.

The study of the biosynthesis of 2-hydroxy-6-methylacetophenone in the gaster of the ant *Rhytidoponera aciculate* has already been described in Section 3.2.3. Labelling studies showed that the compound was formed from acetate and the position of the ^{14}C label was consistent with an acetogenin origin. Pankewitz and Hilker (2008) suggest that insect chromones are very likely synthesized by polyketide synthases.

Mellein (Figure 6.20) is a compound found in many places in nature. In fungi, it is recognized as a polyketide, synthesized from a linear chain

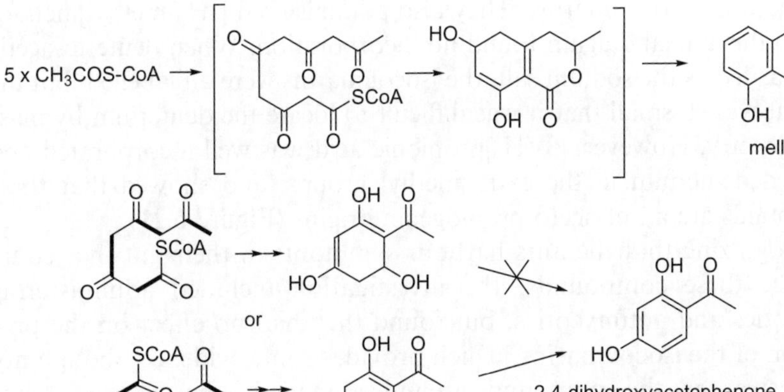

Figure 6.20 The formation of mellein and 2,4-dihydroxyacetophenone from acetogenins. The folding of the acetogenin to give dihydroxyacetophenone can occur in two ways. Labelling with $[1,2-^{13}C_2]$acetic acid made possible a choice between them. From the observed ^{13}C coupling constants, it was shown that the second example is the correct folding.

of acetate units (Holker and Simpson, 1981), but it also occurs in insects. It is present in the mandibular secretion of *Camponotus* ants, in the defensive secretion of termites and in the ant *Rhytidoponera chalybaea*. (*R*)-Mellein is present in the hairpencils (Section 11.3.1) of males of the oriental fruit moth *Grapholitha molesta* (Lepodoptera: Baculoviridae), part of a female-attracting pheromone. Sun and Toia (1993) studied its biosynthesis together with that of 2,4-dihydroxyacetophenone in *R. chalybaea* by feeding the ants with a solution containing sodium [1,2-$^{13}C_2$]acetate. The conclusions about the biosynthetic route generally agreed with those found earlier for the metabolite from fungus (Figure 6.20). More interesting was the discovery of 2,4-dihydroxyacetophenone also in the ant. There are two possible ways that this compound may be formed from acetate units (Figure 6.20). The ^{13}C-^{13}C coupling seen in the ^{13}C NMR spectrum (Section 3.3.5) made possible a choice between these, and gave the result shown (Figure 6.20).

Mellein also provides the trail pheromone of the formicine ants *Lasius fuliginosus* (Plate 17) and *Formica rufa*. Six other isocoumarin compounds provide the trail pheromone or accompany the trail pheromone in the rectal bladder of nine other formicine ants of the *Camponotus*, *Formica* (Plate 9) or *Lasius* genera. They all differ from mellein in having extra methyl groups. Bestmann *et al.* (1997) in their biosynthetic study synthesized deuterated mellein and fed it to ants to see if it would be incorporated into the other isocoumarins. They found the deuterated mellein in the rectum, but there was no incorporation of the deuterated mellein into the other isocoumarins. They also administered [$^2H_3$-methyl]methionine to the ants and again found no incorporation. When deuterio-acetic acid was fed as the sodium salt, the isocoumarins were all labelled, but the amount was so small that it was difficult to locate the deuterium by mass spectrometry. However, [3-$^2H_3$]propionic acid was well incorporated and revealed deuterium in the extra methyl groups, and showed that these compounds are all of aceto-propiogenin origins (Figure 6.21).

Recognizing that the ants harbour symbionts in their gut that could produce these compounds, the investigation included administering antibiotics and antimycotics, but found they had no effect on the production of the isocoumarins, which provides some evidence, though not conclusive, that the compounds are indeed made by the ant itself. More studies like those of Bestmann need to be carried out in the presence of antibiotics with other species and compounds to settle whether insects or their symbionts make polyketides.

A related compound is 8-hydroxyisocoumarin or centipedin, an antibiotic substance of unknown purpose from the centipede *Scolopendra subspinipes mutilans* (Chilopoda: Scolopendridae). It lacks the

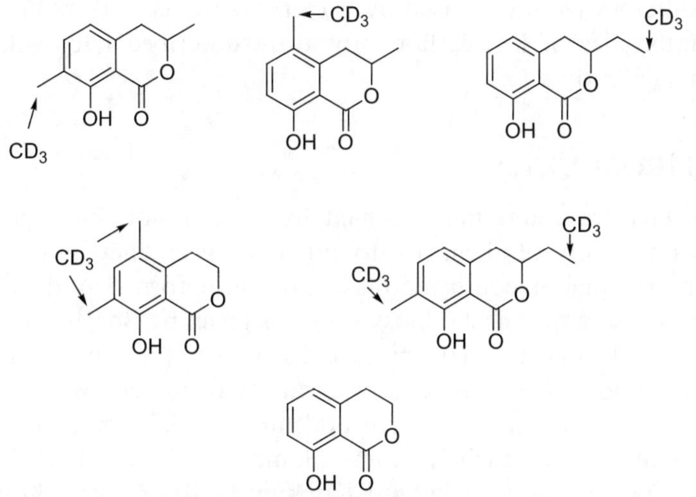

Aceto-propiogenins

Figure 6.21 Isocoumarins related to mellein that are trail pheromone substances, or associated with them in the rectal sac, of formicine ants, and centipedin. The extra methyl groups, not present in mellein and that were labelled with deuterium when the ants were fed with [3,3,3-^{2}H$_3$]propionic acid, are indicated.

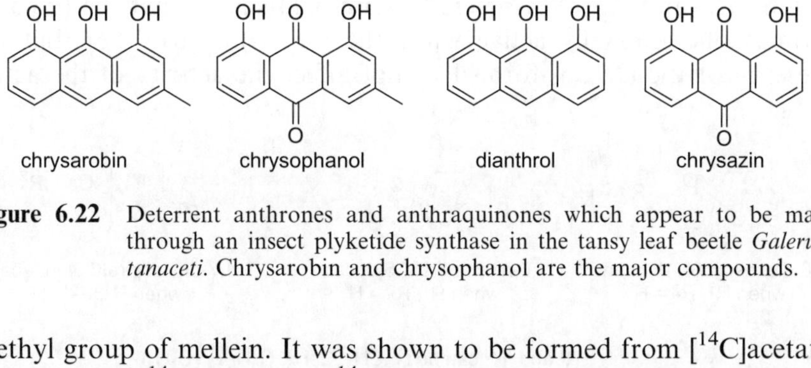

chrysarobin   chrysophanol   dianthrol   chrysazin

Figure 6.22 Deterrent anthrones and anthraquinones which appear to be made through an insect plyketide synthase in the tansy leaf beetle *Galeruca tanaceti*. Chrysarobin and chrysophanol are the major compounds.

methyl group of mellein. It was shown to be formed from [^{14}C]acetate, but not from [^{14}C]alanine or [^{14}C]tyrosine (Kim *et al.*, 1998).

A stronger case for an insect-produced polyketide is made by the deterrent anthrones and anthraquinones of Galerucini leaf beetles. The compounds chrysarobin and chrysophanol and two minor related compounds (Figure 6.22) are found in the larvae of the tansy leaf beetle *Galeruca tanaceti* (Coleoptera: Chrysomelidae), feeding on tansy *Tanacetum vulgare*, which does not contain these compounds. No endosymbionts have been identified, and it has been shown using isotopically labelled precursors that the polyketide is folded by the beetle in a way different from that used by bacteria (summarized by Piel, 2009). Anthraquinones of less certain origin are described later with insect pigments (Section 9.6.2).

6.8 SPIROACETALS

Spiroacetals are compounds formed from keto-diols by spontaneous cyclization. Many of them are found in nature. They can exist in a number of different conformations and chiral forms. A detailed discussion of insect spiroacetal biosynthesis is given by Booth et al. (2009). Volatile alkyl- and hydroxy-spiroacetals of the types shown in Figure 6.23 are found in the secretions of a variety of insects. With a few exceptions, insect spiroacetals have odd-numbered, unbranched carbon chains with 9 to 13 carbon atoms. Some contain an extra hydroxyl group. A review by Francke and Kitching (2001) records 30 different structures, without counting stereoisomers, in beetles, ants, bees, wasps, bugs and fruit flies. Some are pheromones; many are of unknown purpose.

The first spiroacetal identified was chalcogran, 2-ethyl-1,6-dioxaspiro-[4,4]nonane, the aggregation pheromone (together with methyl (2E,4Z)-2,4-decadienoate) of the European spruce bark beetle *Pityogenes chalcographus* (Coleoptera: Scolitidae). In the beetle it exists as a mixture of (2S,5R)- and (2S,5S)-isomers (Figure 6.24). Note that the (2S,5R)-form is pheromonally active while the (2S,5S)-form is not, but the presence of the (2S,5S)-form does not affect the activity of the active

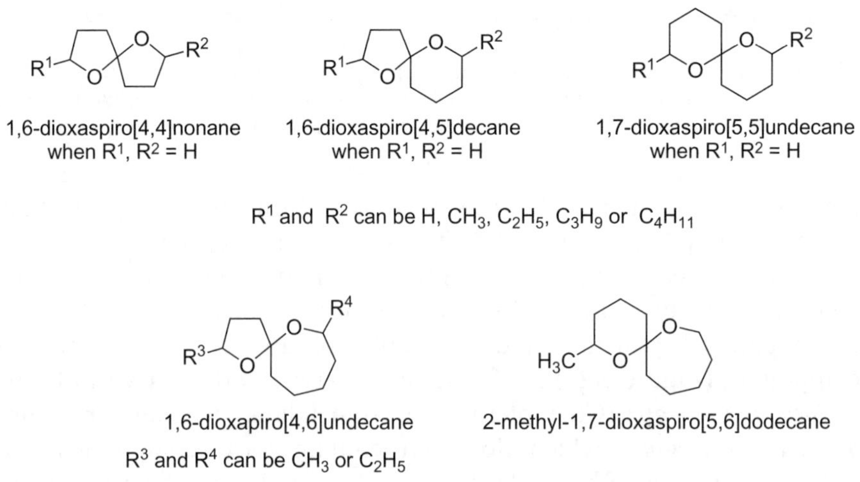

1,6-dioxaspiro[4,4]nonane when R^1, R^2 = H

1,6-dioxaspiro[4,5]decane when R^1, R^2 = H

1,7-dioxaspiro[5,5]undecane when R^1, R^2 = H

R^1 and R^2 can be H, CH_3, C_2H_5, C_3H_9 or C_4H_{11}

1,6-dioxapiro[4,6]undecane
R^3 and R^4 can be CH_3 or C_2H_5

2-methyl-1,7-dioxaspiro[5,6]dodecane

Figure 6.23 The types and system of naming of spiroacetals found in insects.

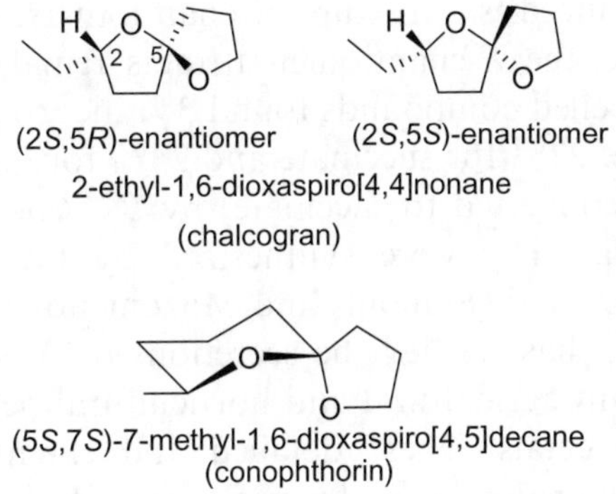

(2S,5R)-enantiomer   (2S,5S)-enantiomer
2-ethyl-1,6-dioxaspiro[4,4]nonane
(chalcogran)

(5S,7S)-7-methyl-1,6-dioxaspiro[4,5]decane
(conophthorin)

Figure 6.24 The structures of two spiroacetals from bark beetles, of unknown biochemical origin: chalcogran and conophthorin.

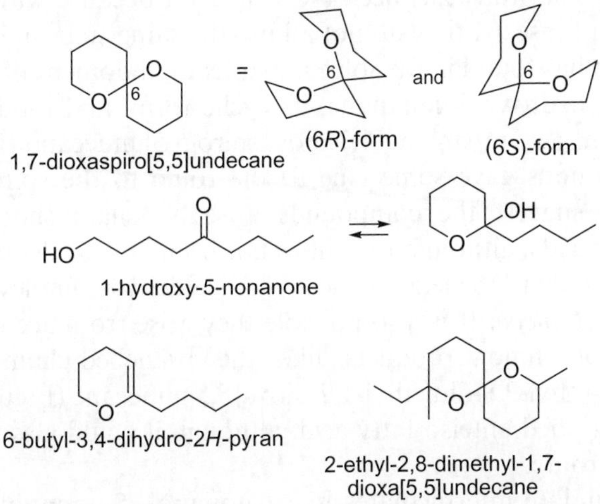

1,7-dioxaspiro[5,5]undecane   (6R)-form   and   (6S)-form

1-hydroxy-5-nonanone

6-butyl-3,4-dihydro-2H-pyran

2-ethyl-2,8-dimethyl-1,7-dioxa[5,5]undecane

Figure 6.25 Compounds found in the secretion of *Bactrocera oleae* and *B. cacuminata*, and the two forms of 1,7-dioxaspiro[5,5]undecane.

enantiomer. The racemic mixture is used commercially in pheromone traps for this forest pest. It has also been identified in other insects (Francke and Kitching, 2001). Other species of bark beetles and some pine cone borers produce conophthorin (5S,7S)-7-methyl-1,6-dioxaspiro[4,5]decane (Figure 6.24). The biosynthesis of these has not yet been investigated.

Biosynthetic studies of spiroacetals have concentrated on those of fruit flies (Tephrididae). The sex pheromone produced by females of the olive fly *Bactrocera* (*Dacus*) *oleae* (Diptera: Tephritidae), a Mediterranean pest, is 1,7-dioxaspiro[5,5]undecane (Figure 6.25). Although the compound as produced is a mixture of enantiomers, the R-enantiomer attracts males and the S-enantiomer attracts females. A preliminary study with ^{14}C-labelled compounds found 3% incorporation of acetate, 14% for malonate, 22% for succinate and 34% for glutamate (which is de-aminated and converted to succinate) by *B. oleae* into spiroacetals, which showed that they were synthesized by the insects, and not acquired with their food (Pomonis and Mazomenos, 1986).

Kitching's group has studied the secretions of Australian fruit flies, which are important agricultural and horticultural pests, and has identified many spiroacetals. These detailed and extensive studies have concentrated on the spiroacetals from *B. oleae*, *B. cacuminata*, the cucumber fly *B. cucumis* and the Queensland fruit fly *B. tryoni*. They have added to them a study of the spiroacetals of the giant ichneumon wasp *Megarhyssa nortoni nortoni* (Hymenoptera: Ichneumonidae) (Booth *et al.*, 2009). The study has been more difficult because within even this small group of insects, they do not all use the same steps in biosynthesis.

This group has found in the solvent extract of abdomens of *B. oleae* and *B. cucumis* 1-hydroxy-5-nonanone, its cyclic form, and 6-butyl-3,4-dihydro-2*H*-pyran, accompanying 1,7-dioxaspiro[5,5]undecane (Figure 6.25). These compounds gave some clue to the route to the spiroacetal. The investigators suggest the compounds arise by chain shortening from normal fatty acids, although they have not been able to demonstrate that by supplying either labelled linoleic acid, its aldehyde, linoleyl alcohol, or oleic acid to *B. oleae*. It is also possible they arise from acetogenins. The occurrence of minor products like the branched-chain spiroacetal (2*S*,6*R*,8*S*)-2-ethyl-2,8-dimethyl-1,7-dioxa[5,5]undecane (Figure 6.25) may suggest a degraded anteiso-fatty acid, although it could also be an aceto-propiogenin route.

Some candidate intermediates like 1-nonanol, 5-nonanol and 5-nonanone were not incorporated by these species into dioxaspiroundecane, neither were nonanoic acid, 5-keto-nonanoic acid nor their esters incorporated, which led to the discovery that the final stage before cyclization is the oxidation of 1-hydroxy-5-nonanone (Figure 6.26). Further evidence indicated this oxidation is probably catalysed by a cytochrome P450. Since these enzymes use molecular oxygen (Section 2.3.3), male insects of *B. oleae*, *B. cacuminata*, *B. cucumis* and *B. tryoni* were confined in an atmosphere containing $^{18}O_2$. In this condition, *both* oxygen atoms of dioxaspiroundecane in *B. oleae* and *B. cacuminata* are labelled with ^{18}O. It proved the intervention of a P450 mono-oxygenase for the final hydroxylation, but indicating that the OH group of 1-hydroxy-5-nonanone must also arise from an oxidation catalysed by cytochrome P450. A 1,2-diol in a trihydroxy-fatty-acid derivative was therefore

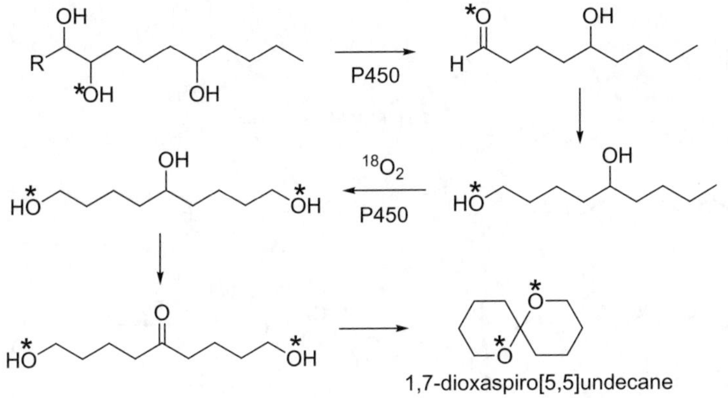

Figure 6.26 The final stages in the biosynthesis of 1,7-dioxaspiro[5,5]undecane by *Bactrocera oleae* and *B. cacuminata*. The * indicates atoms labelled by ^{18}O.

postulated (Figure 6.26), that on oxidation gives 5-hydroxypentanal. Reduction of the aldehyde and a second oxidation with $^{18}O_2$ gives 1,5,9-trihydroxynonane. Oxidation again gives the 1,9-dihydroxy-5-nonanone, which spontaneously cyclizes to the spiroacetal, which exists in two enantiomers, as shown. It also exists in several conformations, which are not shown in the figure (Booth *et al.*, 2009).

However, the ^{18}O labelling pattern of the spiroacetal of *B. cucumis* is different: there are three products formed, and a different route is proposed, summarized in Figure 6.27. Only one oxygen atom of the dimethylspiroacetals was labelled. Some candidate precursors, indicated in Figure 6.25, were not incorporated. Only 2-oxo-6-undecanone was reduced *in vivo* to 2-hydroxy-6-undecanone, which spontaneously cyclized to a hydroxytetrahydropyran. This was oxidized by $^{18}O_2$ at the ω-2, ω-1 or ω position to give the three possible spiroacetals shown (Figure 6.27). The cytochrome shows a preference for catalysing (*S*)-hydroxylation, but the hydroxylation reaction is selective but not stereospecific.

B. tryoni produces a more complex mixture of 11 C_9 to C_{13} spiroacetals; however, the major product (83%) is (2*S*,6*R*,8*S*)-2,8-dimethyl-1,7-dioxaspiro[5,5]undecane (Figure 6.28). The incorporation of a much wider range of intermediates, not used by other species, made a decision about the route used by the insect very difficult. Some intermediates that were incorporated into the major product are shown in Figure 6.28. Both oxygen atoms in the products were ^{18}O-labelled.

Taking an example away from Diptera for comparison, the Kitching team looked at the large parasitic wasp *Megarhyssa nortoni nortoni*,

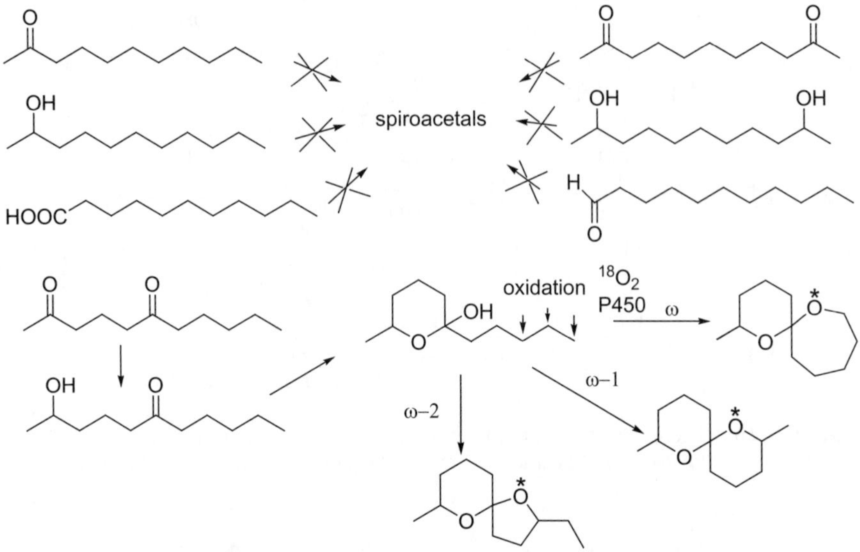

Figure 6.27 Some candidate precursors that were found not to be incorporated into spiroacetals by *Bactrocera cucumis* and the stages of biosynthesis that were established. The final oxidation can occur at any of three carbon atoms. The * indicates labelled by ^{18}O.

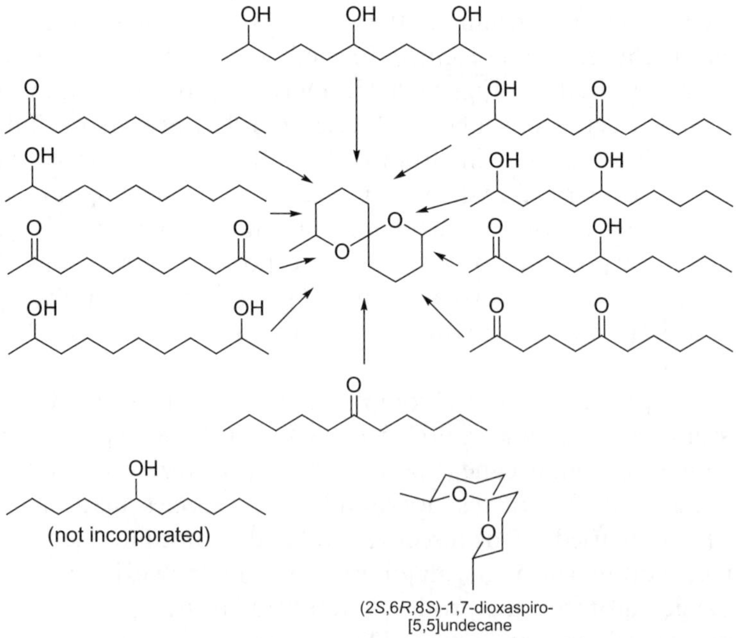

Figure 6.28 The principal spiroacetal produced by *Bactrocera tryoni*, and ketones and alcohols that were incorporated into it. Only 6-undecanol was not converted into the spiroacetal.

(2S,6R,8S)-1,7-dioxaspiro-[5,5]undecane

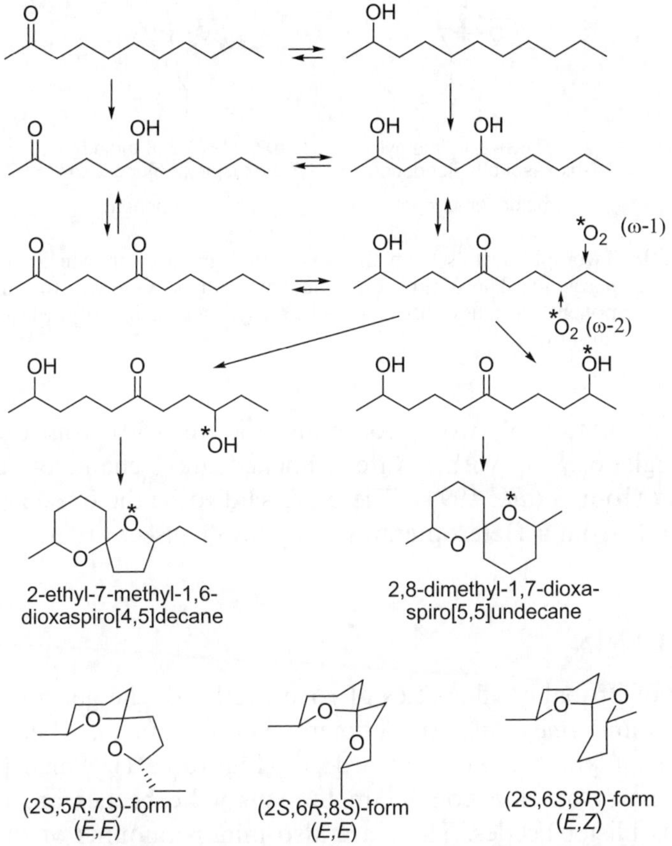

Figure 6.29 Final stages in the biosynthesis of the principal spiroacetals of *Megarhyssa nortoni nortoni*, and the stereochemical forms in which these compounds exist in the wasp. Several of the intermediates are interconvertible. The * indicates labelled by ^{18}O.

which produces a number of spiroacetals in its mandibular glands (Schwartz *et al.*, 2008). One atom of ^{18}O was incorporated into each molecule. Both 2-undecanone and 2-undecanol were incorporated into spiroacetals, and were inter-converted into each other, because some of the ketone and alcohol also occur in the secretion. An outline of the later stages is shown in Figure 6.29. Final oxidation can occur at two positions, ω-1 oxidation is unspecific and gives equal proportions of (*E,E*)- and (*E,Z*)-isomers of spiroacetals, while ω-2 oxidation is specific, and gives mainly the (*E,E*)-isomer.

As the unique example from the Heteroptera, both males and females of the shield bug *Cantao parentum* (Heteroptera: Scutelleridae) (Plate 18)

(2S,6R,8S)-2,8-dimethyl-1,7-dioxaspiro[5,5]undecane
minor component

(2S,4R,6R,8S)-2,,8-trimethyl-1,7-dioxaspiro[5,5]undecane
major component

Figure 6.30 Two spiroacetals from the abdominal gland of the shield bug *Cantao parentum*. The minor component was already known, the major component was a new discovery. The biosynthetic origin of the latter remains unknown.

produce a mixture of two spiroacetals (Figure 6.30) from dorsal abdominal glands, one with a rare branched alkyl chain, of unknown function (Moore *et al.*, 1994). These are said to be the only example of spiroacetals from a Heteropteran or any lower insect order.

6.9 PEDERIN

Females of probably all species of rove beetles of the genera *Paederus* (Plate 19) and *Paederidus* (Coleoptera: Staphylidinae), when crushed, exude haemolymph containing pederin (Figure 6.31), which produces severe blisters on skin contact in humans. The beetles are therefore known as blister beetles. There are also minor amounts of the closely related amides pederone and pseudopederin. It has now been shown with near certainty that pederin is made by symbiotic bacteria. Treating beetles with antibiotics removes their ability to produce pederin, and feeding of female larvae (but not adults) with beetle eggs, which harbour the bacteria, restores the ability of the larval beetles to produce it again (Kellner, 2003). The females that contain the bacterium pass it on *via* their eggs.

Some early studies suggested that pederin is largely synthesized from acetate and propionate units by a bacterial type I polyketide synthase. A biosynthetic scheme requires several methylations on oxygen and either participation of an isoprene unit (Section 7.2) or methylation on carbon. The total DNA was extracted from beetles known to contain pederin, and the DNA used in the polymerase chain reaction. A group of genes called the ped cluster, identified as possibly responsible for the biosynthesis of pederin, was isolated from the total DNA of *Paederus fuscipes* with typical bacterial appearance (Piel, 2002). The bacterium has not yet been isolated but the genetic data indicate it is close to

Plate 1 The life cycle of the red-humped caterpillar moth *Schizura concina*, the only lepiopteran known to produce formic acid. When disturbed, the larvae spray it from abdominal glands. Drawing attributed to R. E. Snodgrass.

Plate 2 Workers of the termite *Coptotermes formosanus*. A bacterioid inside bacteria in the termites fixes atmospheric nitrogen. USDA, ARS. Photo: Scott Bauer.

Plate 3 The bigheaded ground beetle *Scarites subterraneus* produces a defensive secretion consisting of C4 and C5 acids. Lynette Schimming, Earthlink.

Plate 4 *Drosophila melanogaster*, the experimental animal of many genetic studies, and the source of a complete genome. Prof. J. Holopainen, University of Eastern Finland.

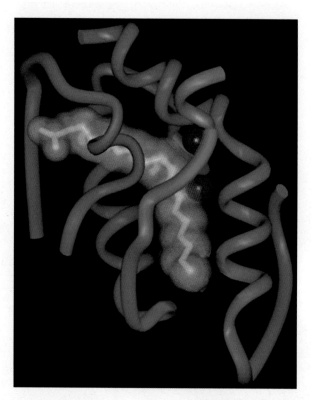

Plate 5 A model of the active site of the castor oil stearoyl-ACP desaturase containing a molecule of stearic acid (white stick model, grey space-filling outline) in the gauche conformation. In this conformation both C-9 and C-10 *pro*-R hydrogens are oriented towards the activated oxygen (not shown), which is bound to the di-iron complex. The protein chain is represented by the green wire, the two iron atoms by red spheres. J. Shanklin and W. McGrath, Brookhaven National Laboratory.

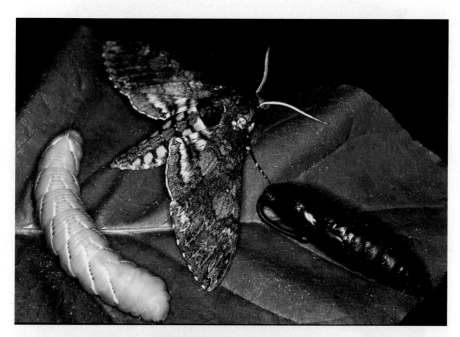

Plate 6 The tobacco hornworm *Manduca sexta*, the workhorse of many biosynthetic investigations. Larva, pupa and adult. The larva has a blue tinge from the haemocyanin pigment in its haemolymph, because it has been fed on a laboratory diet. On its normal food of tobacco leaves, it absorbs the carotenes from the leaves and appears green. Dr R. G. Vogt, University of Washington.

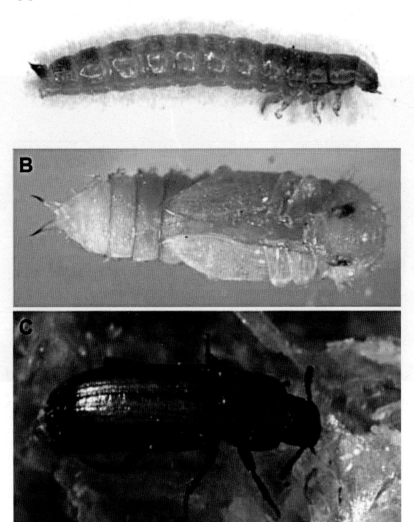

Plate 7 The red flour beetle *Tribolium castaneum* (larva, pupa and adult), a source of potent prostaglandin synthesis inhibitors. François Bonneton, ENS Lyon.

Plate 8 The American cockroach *Periplaneta americana*. Its cuticle contains 3-methylpentacosane. Prof. J. Holopainen, University of Eastern Finland.

Plate 9 The boll weevil *Anthonomis grandis* has long-chain compounds including 26,38-dimethylpentacontanol in its internal lipids. European and Mediterranean Plant Protection Organisation, EPPO.

Plate 10 (a) Queen of the primitive ant *Nothomyrmecia macrops*, which has methyl-branched alkenes in its cuticular lipids. (b) A worker of the fire ant *Solenopsis invicta* injecting alkylpyrrolidines with its venom. It grips the victim's skin with its mandibles while forcing in the sting lance. (c) A queen and workers of *Formic fusca*. The trail pheromone has been identified and its biosynthesis studied. Alex Wild, University of Illinois.

Plate 11 A male gypsy moth *Lymantria dispar*, with its broad antennae spread to detect female pheromone. © Entomart.

Plate 12 A female silkworm, *Bombyx mori*, with its pheromone glands in the abdominal tip displayed. Dr Kevin Wanner, Montana State University.

Plate 13 The alkaloids adaline and adalinine of the two-spotted ladybirds *Adalia punctata* have been shown to be synthesized through fatty acids. Andrei Lobanov, St. Petersburg. www.zin.ru/Animalia.

Plate 14 The mealybug destroyer *Cryptolaemus montrouzieri* larva and adult feeding on mealybugs (family Pseudococcidae). The larvae are covered in waxy excresences. Photo: Jack Kelly Clark, University of California Statewide IPM Project.

Plate 15 The caddisfly *Limnephilus lunatus*, a source of 4-methyl-3-heptanol, but of unknown function in this insect. Tim Ransom, Jersey.

Plate 16 The males of the grain weevil *Sitophilus granarius* produce the attractant sitophilate from a mixture of acetate and propionate units. Seabrooke Leckie, Ontario.

Plate 17 The trail pheromone of the formicine ant *Lasius fuliginosus* was identified by the group of Bestmann, who also studied its biosynthesis. Sebastian Lübcke, Berlin.

Plate 18 A female of the Australian scutellerid beetle *Cantao parentum* guarding her eggs. This beetle is the source of an unusual spiroacetal, but of unknown function. Robert Whyte, Save Our Waterways Now.

Plate 19 A cluster of larval stages of the pederin-producing blister beetle *Paederus riparius*. Henrik Hemplemann, Havelberg.

Plate 20 Two adults of the mustard beetle *Phaedon cochleariae* performing vital functions: eating and procreating. 8-Hydroxygeraniol has been found to be an inhibitor of HMG-CoA in this species. Prof. J. Holopainen, University of Eastern Finland.

Plate 21 The European spruce bark beetle *Ips typographus* has been used for several biosynthetic studies, including that of the attractant 2-methyl-3-buten-2-ol. Prof. J. K. Lindsey, Commanster.

Plate 22 A female of the parasitic wasp *Rhyssa persuasoria* inserting its ovipositor into wood. Its mandibular glands contain the attractant 3-hydroxy-3-methyl-2-butanone. Nigel Jones, Shropshire Invertebrate Group.

Plate 23 Osmeteria of *Papilio* butterfly larvae. The African citrus swallowtail *Papilio demodocus* (a) and the orchard swallowtail *Papilio aegeus* (b). The osmeterium is exposed with its deterrent chemicals when the caterpillar is disturbed. *P. demodocus*, Judy Burris and Wayne Richards, Butterfly Nature. *P. aegeus*, Robert Whyte.

Plate 24 The margined blister beetle *Epicauta pestifera* can contain 15% of its body weight as cantharidin, before mating and transferring it to the female for investing in its eggs. Charles Schurch Lewallen, Oklahoma.

Plate 25 A colony of *Ceroplastes destructor* on a branch of the bush *Bursaria spinosa*. Lotte von Richter, Botanic Garden Trust, Sydney.

Plate 26 The diving beetle *Ilybius fenestratus*, a producer of mammalian sterols for defence. N. Sloth, Biopix.dk.

Plate 27 The phenol and guiacol in the ventral abdominal gland of the Florida leaf-footed bug *Leptoglossus phyllopus* have been shown to be synthesized from the amino acid tyrosine. Charles Schurch Lewallen, Oklahoma.

Plate 28 The azalea lace bug *Stephanitis pyriodes* secretes from abdominal glands a defensive secretion containing long-chain phenolic compounds. Ashley Bradford, Bugline.

Plate 29 *Neanura muscorum*, the only collembola for which a pheromone has been identified. Shane Farrell, Cheshire moth recorder.

Plate 30 The bombardier beetle *Brachinus explodens* produces an explosive reaction between hydroquinones and hydrogen peroxide, and can aim the mixture in almost any direction when disturbed. Josef Dvorák, biolib.

Plate 31 The orange sulfur butterfly *Colias eurytheme* mixes pterin pigments with ultraviolet iridescence to make itself noticeable. Photo used with permission of www.laspilitas.com.

Plate 32 A colony of the cochineal bug *Dactylopius coccus*, which feeds on *Opuntia* (prickly pear) cactus and produces the food colouring cochineal and the anthraquinone pigment carminic acid, a deterrent for ants. Photo: Zyance, Image provided under the Creative Commons Attribution 2.5 licence. Original Image Source upload.wikimedia.org/wikipedia/commons/2/20/Cochenille_z02.jpg. Inset, a single female and a winged male. Dr P. J. Bryant, University of California Irvine.

Plate 33 The rose aphid *Macrosiphon rosae*, displaying its green colour from the pigment aphinin. Alessio Di Leo, Forlì, Italy.

Plate 34 The brimstone butterfly *Gonepterix rhamni* displaying the yellow pigment xanthopterin. Steve Ogden, Marsland Moths.

Plate 35 The wing tips of the males of the orange tip butterfly *Anthocharis* (*Euchloe*) *cardamines* contain red erythropterin. Simon Knott, simbird.com.

Plate 36 A cluster of nymphs of the milkweed bug *Oncopeltus fasciatus* displaying their red pterin pigments. Jean Hutchins and Roger Rittmaster.

Plate 37 The common bluebottle or triangle butterfly *Papilio* (*Graphium*) *sarpedon* contains the porphyrin pigment sarpedobilin. Todd Burrows, Australia.

Plate 38 The pink colour of the immature form of the desert locust *Schistocerca gregaria* is produced by ommachrome pigments. Jan Kašpar, Czech Republic.

Plate 39 The turnip sawfly, the hymenopteran *Athalia spinarum*, absorbs anthocynin pigments from its food, the turnip leaves. Kristin Viglander, Sweden.

Plate 40 The robber fly, the dipteran *Machimus* (*Asilus*) *chrysitis*, also displays anthocyanin pigments from its larval food. Fritz Geller-Grimm and Felix Grimm. Image provided under the Creative Commons Attribution 2.5 licence. Original Image Source upload.wikimedia.org/wikipedia/commons/thumb/9/93/Asilidae_fg07.jpg/180px-Asilidae_fg07.jpg.

Plate 41 The venom of the Australian funnel-web spider *Atrax robustus* contains spermine and spermidine. Colin Halliday, Australia.

Plate 42 Adults and larvae of the poplar leaf beetle *Chrysomela tremulae*. It produces a secretion containing a nitropropionic ester of an oxazolinone glucoside. Andrei Lobanov, St. Petersburg. www.zin.ru/Animalia.

Plate 43 The cherry fruit fly *Rhagoletis cerasi* leaves a complex involatile oviposition deterrent on the fruit after it has laid its egg there. B. Hamers, The Netherlands.

Plate 44 A firefly of the *Photinus* genus from Georgia photographed on a mirror, showing its bioluminescent abdominal organ. Dr D. B. Fenolio, Atlanta Botanic Garden.

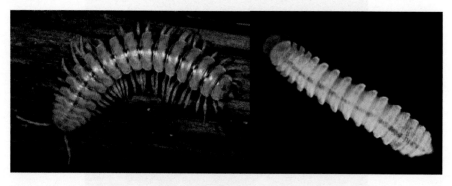

Plate 45   The bioluminescent millipede *Motyxia sequoia alia*, in normal light and its own illumination. The luminescence is probably warning of its content of cyanogenic glycosides. A sort of insect lighthouse. Dr D. B. Fenolio, Atlanta Botanic Garden.

Plate 46   The colours of the ornate moth *Utetheisa ornatrix* warn potential predators of its content of pyrrolizidine alkaloids, part of which it degrades to hydroxydanaidal to attract females. Matt Edmonds, Wonders of Nature.

Plate 47 The tiger beetle *Platyphora boucardi* advertises its content of pyrrolizidine alkaloids stored in abdominal glands. It can absorb both tertiary amine and *N*-oxide forms from plants. Dr Arthur Anker.

Plate 48 The aposematic colour of the grasshopper *Zonocerus variegata* directs the researcher to investigate its alkaloid content and why its colour is so variable. Photo: Luekk. Image provided under the Creative Commons Attribution 2.5 licence. Original Image Source wapedia.mobi/thumb/8b7914599/de/fixed/470/362/Harlekinschrecke.jpg.

Plate 49 The seven-spotted ladybird *Coccinella septempunctata* eating an aphid, from which it absorbs its alkaloids. There are yellow droplets of secretion on the cornicals of the aphid. Cornical secretions often contain alarm pheromones that induce the aphids to disperse. Bruce Marlin, Cirrusimage.com.

Plate 50 The burnet moth *Zygaena trifolii* both makes cyanogenic glycosides and absorbs them from the plant *Lotus cornicuatus*. J. L. Calleiras Vieitez, Pontevedra, Spain.

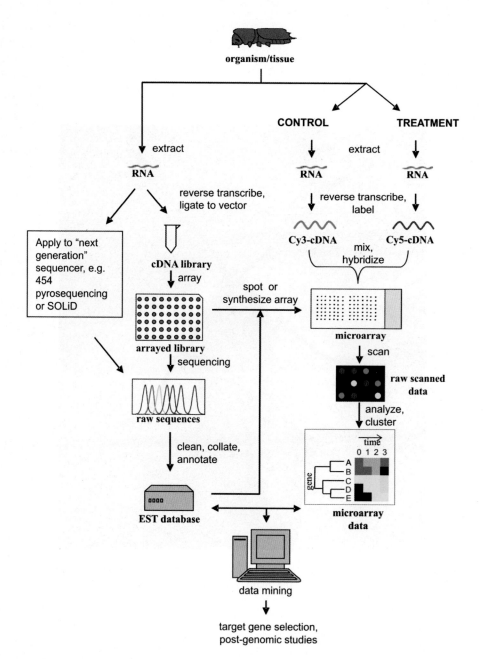

Plate 51 Scheme of a study using functional genomics. Figure modified from Tittiger, C. 2004. Functional genomics and insect chemical ecology. *Journal of Chemical Ecology*, **30**, 2342. Reproduced with permission of Springer.

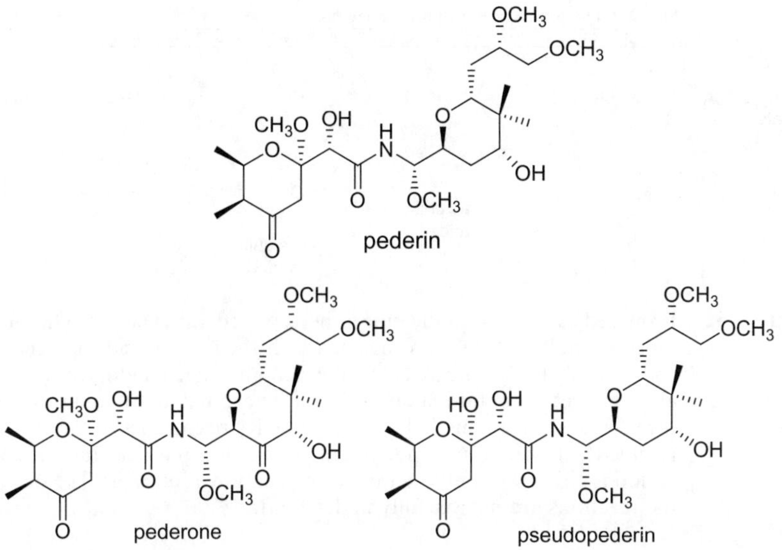

Figure 6.31 The defensive compounds pederin, pederone and pseudopederin from rove beetles.

Pseudomonas aeruginosa. The ped cluster is part of a genomic island acquired by the bacterium. A genomic island is a large region of DNA transferred horizontally from another organism. Such transfers can enhance metabolic and colonizing ability so that bacteria can evolve by sudden leaps (Piel et al., 2004a). The *ped* genes code for a hybrid type I polyketide synthase, not neatly arranged in a cluster as expected for a type I polyketide synthase (Section 6.2), but distributed among three distinct genomic regions, although the greater part of genes coding for the pederin synthesis is found in one locus. Other fragments have been found subsequently (Piel et al., 2004b). The *ped* genes are found only in female beetles capable of producing pederin and their eggs. The *pedF* region is bounded on both sides by pseudogenes of transposons, that is, inactive fragments genes that were once involved in gene mobility, which also indicates that the pederin-synthesizing ability has been acquired by transfer into the genome.

The synthesis, while recognizably that of a polyketide synthase has some rare and unusual steps, but the architecture exactly matches a large portion of the pederin structure. Each module typically contains at least a keto-acyl synthase, acyl transferase and acyl carrier protein. Each module attaches one building block, usually an acetyl group, to the growing polyketide chain. Various other enzymes are involved to make modifications, such as keto-reductase, dehydratase or cyclase. In the first

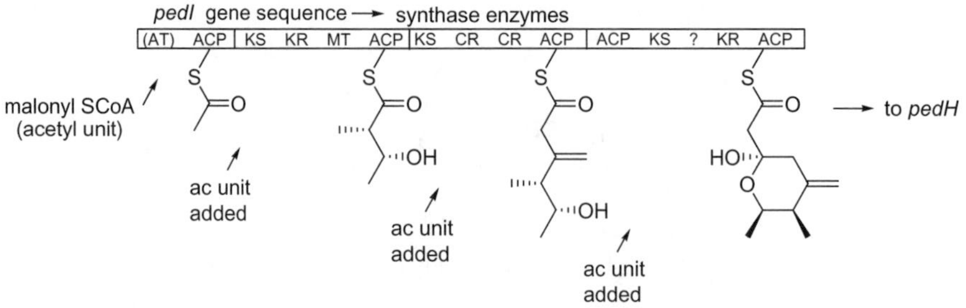

Figure 6.32 Proposed early steps in the biosynthesis of pederin. The catalytic enzyme units are indicated in the order in which they occur and the genes that code for them are arranged in the genome. Each keto-synthase adds another acetyl unit (as malonyl-SCoA). The initial unit is an unusual acyl transferase. MT is a methyl transferase, CR are enzymes of the crotonase family, their exact function here unknown. One unit near the end of this region is either without function, or of unknown function. The details of the reactions are not yet fully understood.

step an acetyl-S-ACP is condensed with a second acetate unit, followed by reduction of the keto-group and introduction of a branch methyl (Piel et al., 2004b) (Figure 6.32). A keto-synthase that catalyses the condensation of the third acetyl unit is followed by two unusual enzyme units of the crotonase family. Their functions are not clear, but the result is addition of an exo-methylene group. After the addition of the next acetate unit, there follows an enzyme structure of unknown or no function and a keto-reductase. The final product on this enzyme cluster is a C_{10} fragment attached to an acyl carrier protein, which are transferred to the larger enzyme cluster called *pedF*.

After addition of another acetate unit on *PedF*, a non-ribosomal peptide synthase introduces a glycine unit. (Non-ribosomal peptide synthases are enzymes catalysing the incorporation of amino acid units. The formation of such synthases is not coded by ribosomes as for other parts of polyketide synthases.) A methyl transferase catalyses the insertion of a dimethyl group. After the introduction of a double bond, a Michael addition closes the tetrahydropyran ring (Piel, 2002). The product formed on the *PedF* enzyme may be either extended further on another polyketide synthase enzyme called *PedH* to give a product like the onnamides found in sponges, and then cleaved to a pederin precursor, or directly cleaved and methylated to pederin (Figure 6.33). An interesting point is that nowhere is a propionate unit required.

Not a single symbiotic bacterium has been isolated for any of a large number of natural products found in invertebrates, and this is the first example known of an insect defensive compound, shown to be produced

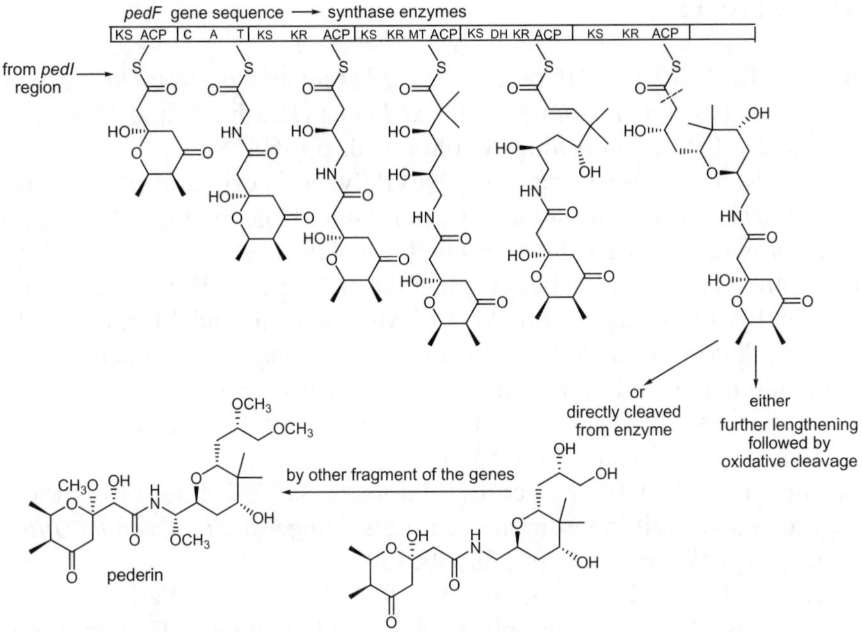

Figure 6.33 The polyketide fragment constructed by the enzymes coded by *pedI* is transferred to those of the large fragment called *pedF*. The nearly complete pederin produced by *pedF* can be either further elongated on another polyketide synthase fragment called *pedH*, or directly cleaved from the enzyme and methylated. The abbreviations for enzyme regions are: KS, keto-synthase; ACP, acyl carrier protein; C, non-ribosomal peptide synthase (NRPS) condensation domain; A, NRPS adenylation domain; T NRPS thiolation domain; KR, keto-reductase; MT, methyl transferase; DH, dehydratase. (Based on Fig. 3 of J. Piel, *Proceedings of the National Academy of Sciences, USA*, 2002, **99**, 14002–14007.)

by a bacterial symbiont. This is important work and will surely be followed by more examples where the biosynthesis is traced through the total genetic code of the insect to a symbiont.

BACKGROUND READING

Francke, W. and Dettner, K. 2005. Chemical signalling in beetles. *Topics in Current Chemistry*, **240**, 85–166.

Piel, J. 2009. Metabolites from symbiotic bacteria. *Natural Product Reports*, **26**, 338–362.

Staunton, J. and Weissman, K. J. 2001. Polyketide biosynthesis: a millennium review. *Natural Products Reports*, **18**, 380–416.

REFERENCES

Bartelt, R. J. 1999. Sap beetles. In: *Pheromones of Non-lepidopteran Insects Associated with Agricultural Plants* (Hardie, J. and Minks, A. K., ed.), CABI Publishing, Wallingford, pp. 69–89.

Bartelt, R. J. and Weisleder, D. 1996. Polyketide origin of pheromones of *Carpophilus davidsoni* and *C. mutilatus* (Coleoptera: Nitidulidae). *Bioorganic and Medicinal Chemistry*, **4**, 429–438.

Bestmann, H.-J., Attygalle, A. B., Glasbrenner, J., Riemer, R., Vostrowsky, O., Constantino, M. G., Melikian, G. and Morgan, E. D. 1988. Pheromones 65. Identification of the volatile components of the mandibular gland secretion of the ant *Manica rubida*, structure elucidation, synthesis, and absolute configuration of Manicone. *Liebigs Annalen der Chemie*, **1988**, 55–60.

Bestmann, H.-J., Übler, E. and Hölldobler, B. 1997. First biosynthetic studies on trail pheromones in ants. *Angewandte Chemie, International Edition in English*, **36**, 395–397.

Birch, A. J. and Donovan, F. W. 1953. Studies in relation to biosynthesis. 1. Some possible routes to derivatives of orcinol and phloroglucinol. *Australian Journal of Chemistry*, **6**, 360–368.

Booth, Y. K., Kitching, W. and De Voss, J. J. 2009. Biosynthesis of insect spiroacetals. *Natural Product Reports*, **26**, 490–525.

Francke, W. and Dettner, K. 2005. Chemical signalling in beetles. *Topics in Current Chemistry*, **240**, 85–166.

Francke, W. and Kitching, W. 2001. Spiroacetals in insects. *Current Organic Chemistry*, **5**, 233–251.

Francke, W. and Schulz, S. 1999. Pheromones. In: *Comprehensive Natural Products Chemistry*, vol. 8 (Mori, K., ed.), Elsevier, Oxford, pp. 197–261.

Holker, J. S. E. and Simpson, T. J. 1981. Studies on fungal metabolites 2. C-13 Nuclear magnetic resonance biosynthetic studies on pentaketide metabolites of *Aspergillus melleus*; 3-(1,2-epoxypropyl)-5, 6-dihydro-5-hydroxy-6-methylpyran-2-one and mellein. *Journal of the Chemical Society – Perkin Transactions I*, 1397–1400.

Jarvis, A. P., Liebig, J., Hölldobler, B. and Oldham, N. J. 2004. Biosynthesis of the insect pheromone (*S*)-4-methyl-3-heptanone. *Chemical Communications*, 1196–1197.

Kellner, R. L. L. 2003. Stadium-specific transmission of endosymbionts needed for pederin biosynthesis in three species of *Paederus* rove beetles. *Entomologia Experimentalis et Applicata*, **107**, 115–124.

Kim, K. T., Hong, S. W., Lee, J. H., Park, K. B. and Cho, K. S. 1998. Mechanism of antibiotic action and biosynthesis of centipedin purified from *Scolopendra subspinipes multilans* L. Koch (centipede). *Journal of Biochemistry and Molecular Biology*, **31**, 328–332.

Kim, J., Matsuyama, S. and Suzuki, T. 2005. 4,8-Dimethyldecanal, the aggregation pheromone of *Tribolium castaneum*, is biosynthesised through the fatty acid pathway. *Journal of Chemical Ecology*, **31**, 1381–1400.

Mangat, J., Langedock, C. and Vanderwel, D. 2006. In vitro assay for sex pheromone biosynthesis by the female yellow mealworm beetle and identification of a regulated step. *Insect Biochemistry and Molecular Biology*, **36**, 403–409.

Millar, J. G. 2005. Pheromones of true bugs. *Topics in Current Chemistry*, **240**, 37–84.

Moore, C. J., Hubener, A., Tu, Y. Q., Kitching, W., Aldrich, J. R., Waite, G. K., Schulz, S. and Francke, W. 1994. A new spiroketal type from the insect kingdom. *Journal of Organic Chemistry*, **59**, 6136–6138.

Morgan, E. D. 2004. *Biosynthesis in Insects*, Royal Society of Chemistry, Cambridge, Chapter 4, pp. 57–68.

Pankewitz, F. and Hilker, M. 2008. Polyketides in insects: ecological role of these widespread chemicals and evolutionary aspects of their biogenesis. *Biological Reviews*, **83**, 209–226.

Petroski, R.J., Bartelt, R. J. and Weisleder, D. Q. 1994. Biosynthesis of (2E,4E,6E)-5-ethyl-3-methyl-2,4,6-nonatriene: the aggregation pheromone of *Carpophilus freemani* (Coleoptera: Nitidulidae). *Insect Biochemistry and Molecular Biology*, **24**, 69–78.

Piel, J. 2002. A polyketide synthase-peptide gene cluster from an uncultured bacterial symbiont of *Paederus* beetles. *Proceedings of the National Academy of Sciences, USA*, **99**, 14002–14007.

Piel, J. 2009. Metabolites from symbiotic bacteria. *Natural Products Reports*, **26**, 338–362.

Piel, J., Hofer, I. and Hui, D. Q. 2004a. Evidence for a symbiosis island involved in horizontal acquisition of pederin biosynthetic capabilities by the bacterial symbiont of *Paederus fuscipes* beetles. *Journal of Bacteriology*, **186**, 1280–1286.

Piel, J., Wen, G., Platzer, M. and Hui, D. Q. 2004b. Unprecedented diversity of catalytic domains in the first four modules of the putative pederin polyketide synthase. *Chembiochem*, **5**, 93–98.

Pomonis, J. G. and Mazomenos, B. E. 1986. Biosynthesis of a pheromone, 1,7-dioxaspiro[5,5]undecane, from C-14 substrates in vivo and by explanted female rectal glands of the olive fruit-fly, *Dacus oleae* (Gmel.) – a preliminary study. *International Journal of Invertebrate Reproduction and Development*, **10**, 169–177.

Schulz, S. 2004. Semiochemistry of spiders. In: *Advances in Insect Chemical Ecology* (Carde. R. T. and Miller, J. G., ed.), Cambridge University Press, Cambridge, pp. 110–150.

Schulz, S., Fuhlendorff, J., Steidle, J. L. M., Collatz, J. and Franz, J. T. 2004. Identification and biosynthesis of an aggregation pheromone of the storage mite *Chortoglyphus arcuatus*. *Chembiochem*, **5**, 1500–1507.

Schwartz, B. D., Moore, C. J., Rahm, F., Yates, P. Y., Kitching, W. and De Voss, J. J. 2008. Spiroacetal biosynthesis in insects from Diptera to Hymenoptera: the giant ichneumon wasp *Megarhyssa nortoni nortoni* Cresson. *Journal of the American Chemical Society*, **130**, 14853–14860.

Sun, C. M. and Toia, R. F. 1993. Biosynthetic-studies on ant metabolites of mellein and 2,4-dihydroxyacetophenone from $(1,2\text{-}^{13}C_2)$acetate. *Journal of Natural Products*, **56**, 953–956.

CHAPTER 7
Lower Terpenes

7.1 INTRODUCTION

The largest group of all natural products in plants and animals is the terpenes, which are all based on a five-carbon unit. Ruzicka in 1922 proposed that the basic building block of the terpenes is an isoprene (2-methyl-1,3-butadiene) unit (Figure 7.1), and later Robinson proposed that these isoprene units are usually joined head-to-tail. The terpenes are sub-divided into groups by their number of isoprene units. Monoterpenes (C_{10} compounds) contain two isoprene units, sesquiterpenes (C_{15} compounds) contain three isoprenes, diterpenes (C_{20}) four units, triterpenes (C_{30}) six units and tetraterpenes (C_{40}) eight units. There are more than 1000 monoterpenes known, more than 7000 sesquiterpenes, 3000 diterpenes, 7000 triterpenes and 700 tetraterpenes. There is a smaller number of five-isoprene-unit compounds called sesterterpenes. Some simple examples of terpenes are shown in Figure 7.1, with their structures dissected into isoprene units. Although plants are their richest source, terpenes are frequently found in insects, as pheromones, defensive secretions and as hormones. Our understanding of terpenoid biosynthesis was greatly helped by the efforts of Cornforth, Popjak, Lynen and others on the biosynthesis of cholesterol in the 1950s and 1960s.

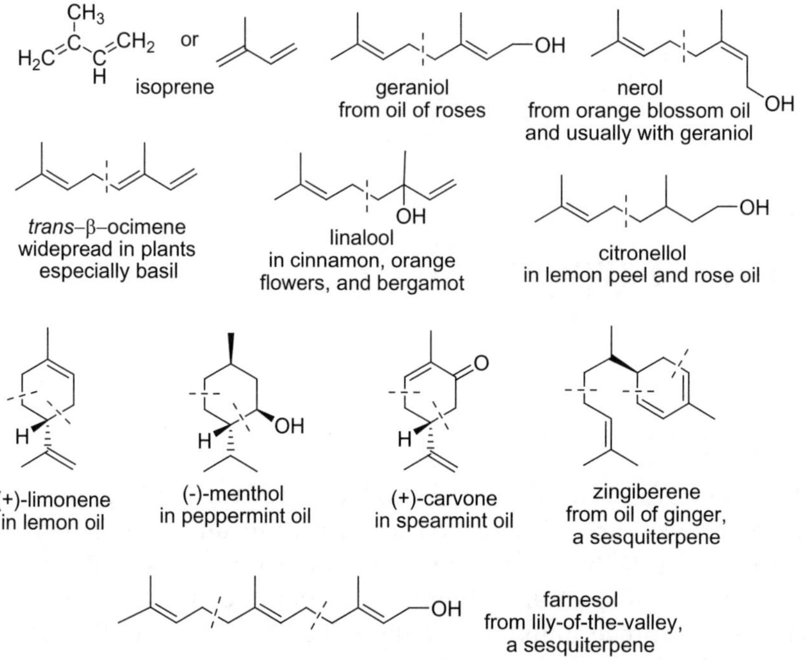

Figure 7.1 Isoprene and some examples of simple terpenes, dissected into their isoprene units. Note that some of these substances are chiral.

7.2 MONOTERPENE BIOSYNTHESIS

In 1937 it was shown that labelled acetate gave labelled terpenes, but evidently by a different route from the fatty acids and acetogenins. In 1956 it was accidentally found that a substance called mevalonic acid (Figure 6.2) was an intermediate between acetate and terpenes, and this gave the clue needed to study their biosynthesis.

7.2.1 The Mevalonate Pathway

In the first step, two molecules of acetyl-co-enzyme A thioester undergo a Claisen condensation, catalysed by acetoacetyl-CoA synthase, an enzyme in the cytosol, to give acetoacetyl-CoA. In this reaction there is no first conversion of acetate to malonate to aid the condensation, as in the case of fatty-acid synthesis. The equilibrium in this reaction is slightly in favour of the starting materials, but the condensation product is quickly removed by the subsequent reactions, which drives the reactions forward. One acetyl-CoA is first transferred to a cysteine-SH group on the enzyme, and brought into contact with the second acetyl-CoA for the Claisen

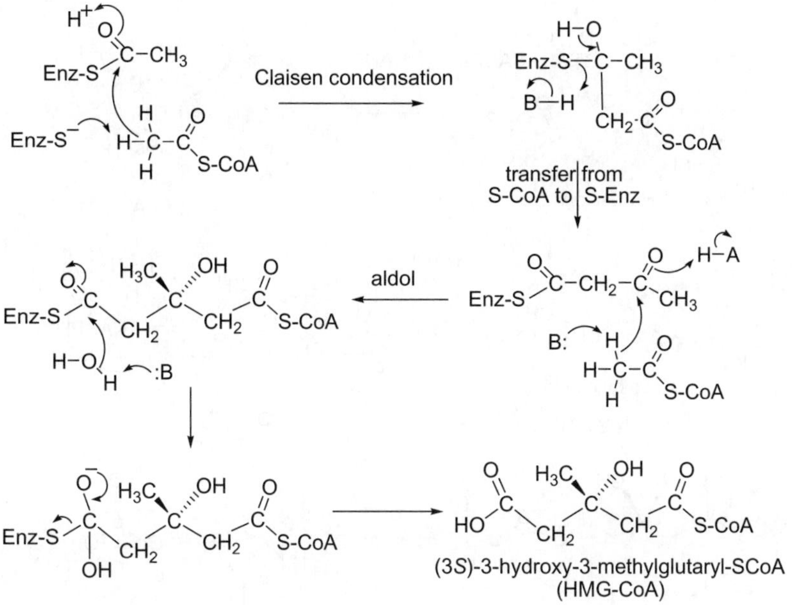

Lower Terpenes

Figure 7.2 The first stages in terpene biosynthesis, two condensations to give the C_6 intermediate HMG-CoA.

condensation. The next step is an aldol-type of reaction, in which the substrate is again transferred to a cysteine-SH. A third acetyl-CoA is added to give β-hydroxy-β-methylglutaryl co-enzyme A (HMG-CoA), catalysed by HMG-CoA synthase, also in the cytosol (Figure 7.2).

HMG-CoA is reduced by two successive molecules of NADPH, in the presence of HMG-CoA reductase, to give (3*R*)-3,5-dihydroxy-3-methylvaleric acid, known as mevalonic acid (MVA), in the rate-determining step for the whole sequence of reactions that builds up the terpenes (Figure 7.3). The first addition of NADPH gives an intermediate aldehyde, and release of co-enzyme A. The second reduces this aldehyde to an alcohol. Mevalonic acid is converted to the 5-diphosphate and further phosphorylated by diphosphomevalonate decarboxylase to the 3-phosphate-5-diphosphate followed by *trans* elimination of carbon dioxide and phosphate, to give isopentenyl diphosphate (IPP), the first building block of the terpenes (Figure 7.3). These reactions all occur in the cytosol, but HMG-CoA reductase is held in the endoplasmic reticulum with its active site facing the cytosol.

HMG-CoA reductase is a very highly regulated enzyme. Researchers have long been interested to know how that regulation is effected in insects. Burse *et al.* (2008) have found that 8-hydroxygeraniol is a

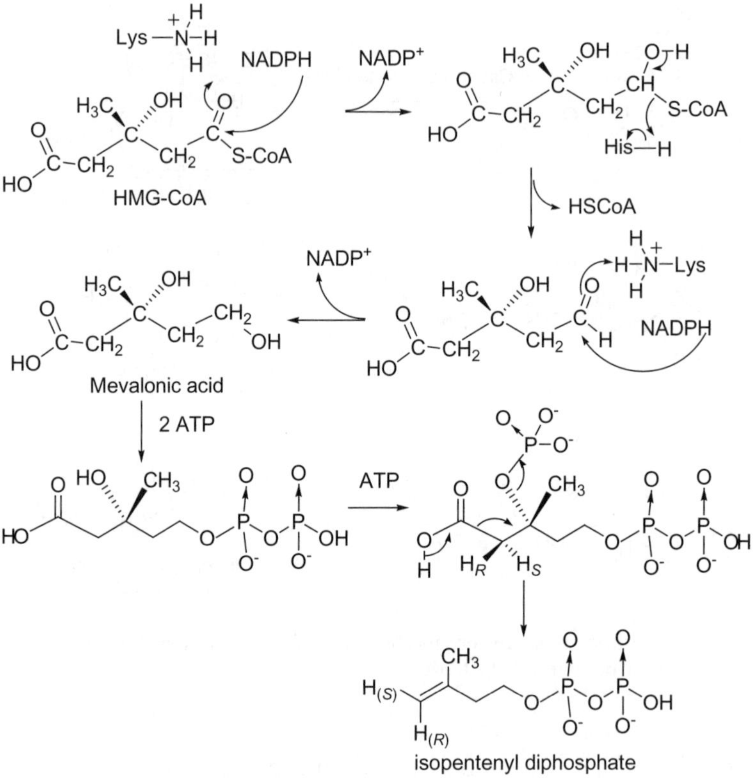

Figure 7.3 The formation of mevalonic acid by reduction of HMG-CoA, the rate-determining step in terpene synthesis, and the conversion of mevalonic acid to isopentenyl diphosphate (IPP). Here (H_S) and (H_R) are not pro-chiral, the labels are used to trace atoms between structures. The diphosphate groups are shown in full, but in subsequent figures they are represented by OPOP.

competitive inhibitor of HMG-CoA reductase in the mustard beetle *Phaedon cochleariae* (Coleoptera: Chrysomelidae) (Plate 20) (Section 11.7). Inhibition was also observed for other insect HMG-CoA reductases, including that of *Drosophila melanogaster*.

One further reaction, catalysed by isopentenyl diphosphate isomerase, is required to make dimethallyl diphosphate (DMAPP), the starter unit for terpenes. Labelling experiments with deuterium and tritium have shown that the *pro*-chiral H_R of isopentenyl diphosphate (from the back of the molecule as drawn) is removed in this step (Figure 7.4). The condensation of isopentenyl diphosphate (IPP) and dimethallyl diphosphate (DMAPP), catalysed by geranyl diphosphate synthase (prenyl transferase) gives geranyl diphosphate, the parent of all the monoterpenes. The IPP is added

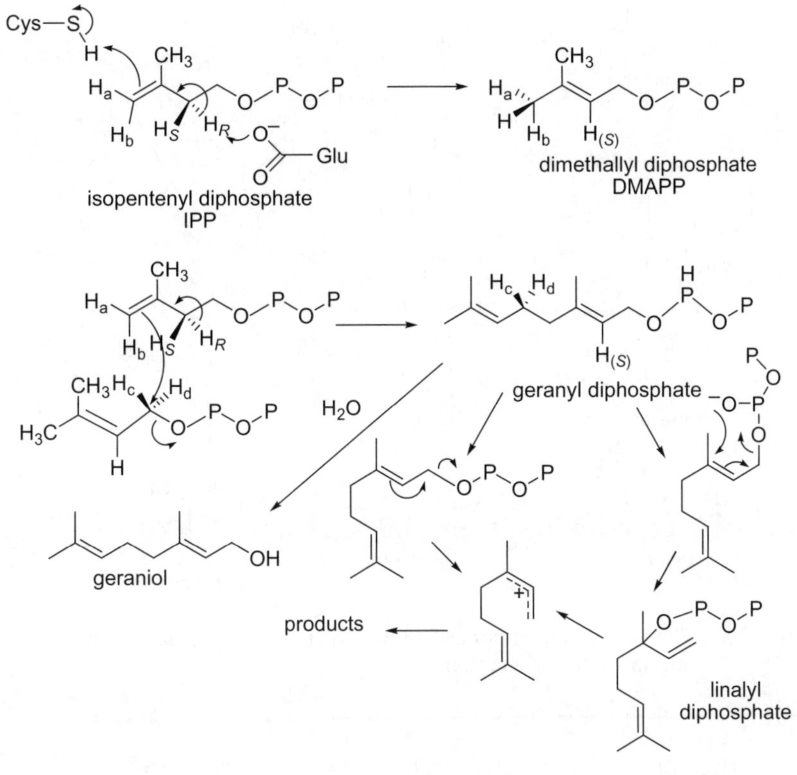

Lower Terpenes 183

Figure 7.4 The isomerization of IPP to dimethallyl diphosphate (DMAPP) and the condensation of these two units to give geranyl diphosphate, its rearrangement to linalyl diphosphate, and formation of the carbocation that is the source of further monoterpenes. Hydrogen atoms are labelled a, b, c and d to distinguish them; they are not all pro-chiral.

from above to the DMAPP as phosphate on DMAPP is eliminated from below, so its *pro*-chiral centre is inverted. The *pro*-chiral H_R on IPP is eliminated from the resulting carbocation (Figure 7.4). From acetyl-CoA to geranyl diphosphate requires eight enzymes. Hydrolysis of the diphosphate with diphosphatase gives geraniol (Figure 7.4) directly. Rearrangement gives linalyl diphosphate, and cleavage of the carbon–oxygen bond gives a carbocation and a diphosphate ion. Geraniol, linalool and this carbocation can all undergo various changes that provide the wealth of monoterpene compounds (Figure 7.5). The structure of the terpene formed from the carbocation is largely determined by the shape in which it is folded in the enzyme pocket, and the rules of rearrangement chemistry.

The idea of one enzyme – one product is losing ground. Some enzymes are less than specific for a single product. In the case of simple monoterpenes and sesquiterpenes plurality is rather common. One way to test that only

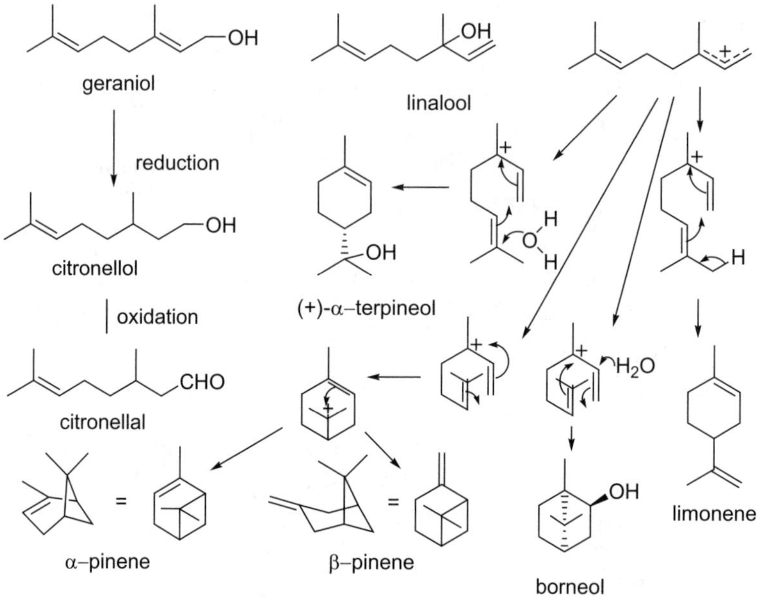

Figure 7.5 The derivation of typical monoterpenes from geraniol, linalool or an intermediate carbocation.

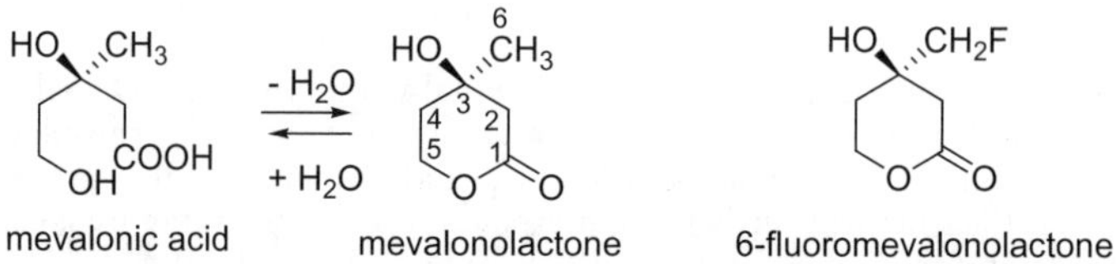

Figure 7.6 The key compounds mevalonic acid and its lactone, with fluoromevalonolactone, an inhibitor of the terpene pathway.

one enzyme is involved is to have a pure enzyme (Section 3.5.3), but there is a simpler chemical test. If deuterium atoms are introduced at specific points in the substrate, the ratio of amounts of the products formed will be altered by the kinetic isotope effect (Section 3.3.1). A specific example is found in the ratio of α- and β-pinene and myrcene formed from geranyl diphosphate. With a CD_3 group in geranyl diphosphate, the proportions of myrcene fell from 9 to 4%, that of α-pinene rose from 26 to 38% and β-pinene fell from 21% to 13% (Croteau et al., 1987).

Mevalonic acid, the key intermediate in solving the terpene biosynthesis pathway, is an oily liquid, but in solution it is in equilibrium with mevalonolactone (Figure 7.6), a crystalline solid, so the latter is more convenient to use in biosynthetic studies. Fluoromevalonolactone,

7.2.2 The Methylerythritol Phosphate Pathway

and the corresponding acid, are powerful inhibitors of terpene formation. If addition of fluoromevalonate to a biosynthesizing system blocks the formation of a compound, then it can be concluded that that compound has a terpene origin.

The scientific world was surprised in 1993 to learn that there was another completely different route to terpenes. M. Rohmer showed that micro-organisms, green algae and plastids (membrane-bound organelles of plants, *e.g.* chloroplasts) use a different pathway to mevalonic acid, as outlined in Figure 7.7 (Rohmer *et al.*, 1996). Bacteria use both pathways. It is not surprising that in plants, where sugars are abundant, the starting materials are sugar derivatives. The process begins with pyruvic acid being decarboxylated using thiamine diphosphate (Chapter 2) and the intermediate being condensed with glyceraldehyde 3-phosphate (Figure 2.20) to give 1-deoxy-D-xylulose 5-phosphate (DOXP), catalysed

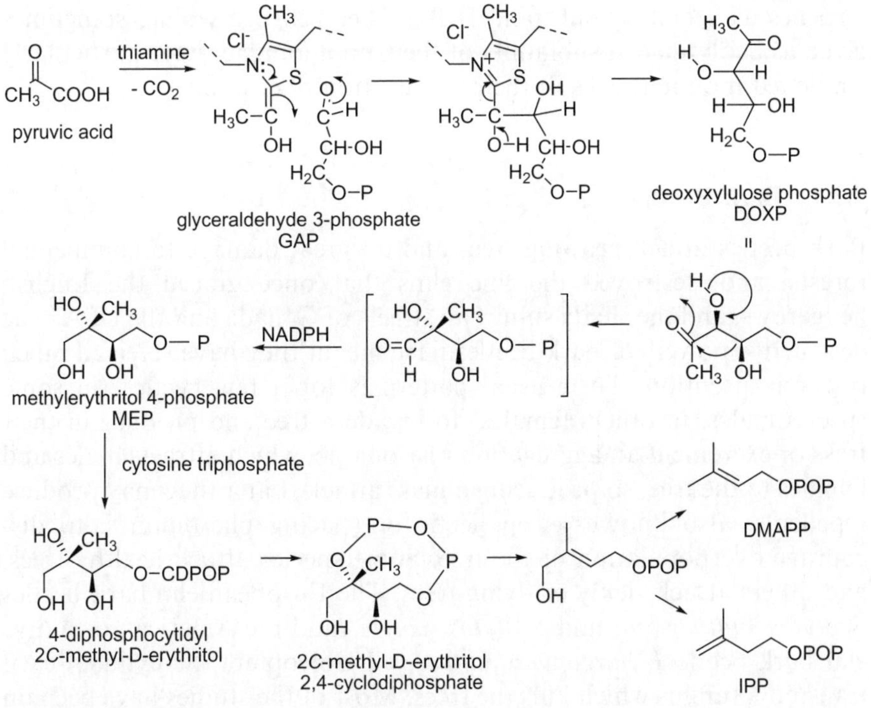

Figure 7.7 The methylerythritol or non-mevalonate pathway to terpenes used by micro-organisms and plastids of plants.

by deoxyxylulose phosphate synthase. DOXP undergoes rearrangement and reduction in the presence of DOXP reductase to methylerythritol 4-phosphate (MEP). The final stages of the synthesis are not fully explored, but involve a cyclic diphosphate, which eliminates to give isopentenyl diphosphate and dimethallyl diphosphate. In all, nine enzymes are required. This route to terpenes is known variously as the MEP/DOXP pathway, the non-mevalonate pathway or the mevalonic acid-independent pathway. Incidentally, DOXP is also an intermediate in the synthesis of both thiamine and pyridoxal (Chapter 2).

7.3 MONOTERPENE PHEROMONES

A large number of insect pheromone compounds are either simple terpenes or are derived from them through further reactions. Lists of insect monoterpene pheromones can be found in reviews such as that of Francke and Schulz (1999). The few examples of monoterpenes given in Figure 7.6 illustrate the variety of structures that can be made from geranyl diphosphate or linalyl diphosphate without much further reaction. Both the Coleoptera and the Hymenoptera make frequent use of terpenes as secretory substances. Bark beetles (Scolytidae; sometimes given as Scolytinae, a subfamily of the Curculionidae) have particularly employed monoterpenes in their aggregation pheromones.

7.3.1 Bark Beetles

Bark beetles attack growing trees and do great damage to commercial forests, and destroyed the fine elms that once graced the English hedgerows, and the shady suburban streets of Canada and the USA. The destructive powers of bark beetles have meant they have received much research attention. Their usual pattern is for a few beetles (in some species males, in others females) to invade a tree and produce in their frass or excrement an aggregation pheromone, which attracts males and females to the tree, so producing a mass attack. Later they may produce repellents, also known as epidiectic or spacing pheromones, to discourage overpopulation of the tree. Some species attack healthy trees, and others attack sickly or dying trees. The European elm bark beetles *Scolytus multistriatus* and *Scolytus scolytus* and the American or native elm bark beetle *Hylurgopinus rufipes* (all Coleoptera: Scolytidae) each transmit a fungus which kills the trees. Most of the studies have been on species that attack commercial coniferous trees. A detailed account of bark beetle compounds is given by Francke and Dettner (2005).

The first of their pheromones identified were ipsenol and ipsdienol, compounds used by a number of *Ips* species, and some others. At first it was thought, and there was supporting evidence, that such terpenoid pheromones were derived from the terpenes in the trees, but further studies have overturned this. Terpenes were found to accumulate in the midgut of beetles of various species that had no contact with plant terpenes (Seybold *et al.*, 1995). The synthesis of the aggregation pheromone could be located in the midgut (Seybold and Tittiger, 2003). The gene coding for the enzyme for catalysing the *de novo* synthesis has been isolated and expressed in bacteria (Gilg *et al.*, 2009). More surprising is the discovery that the geranyl diphosphate synthase in some *Ips* species is a bi-functional enzyme, and catalyses the formation of both geranyl diphosphate and myrcene. The *Ips* beetles require myrcene to make ipsdienol with the aid of a cytochrome P450 enzyme (Figure 7.8). The gene for this enzyme, for the oxidation, has also been cloned from the pinyon ips, *I. confusus*, and expressed in *E. coli*. It was found to

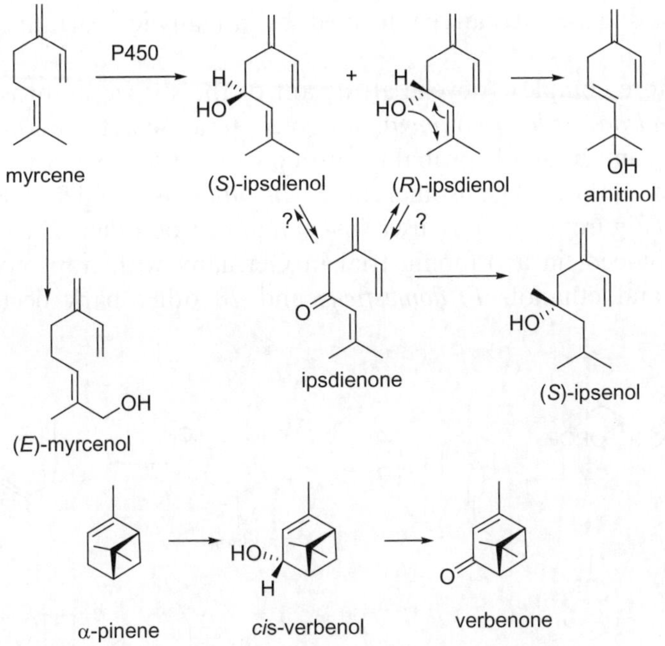

Figure 7.8 Formation of ipsdienol, ipsenol and amitinol from myrcene in bark beetles. The ratio of (*S*)- to (*R*)-ipsdienol differs between that catalysed by the P450 enzyme and that found in the emitted attractant. The ratio may be changed through the reversible formation of ipsdienone. α-Pinene from the tree is oxidized by beetles to *cis*-verbenol and verbenone, and forms part of the attractant.

produce more of (*S*)-(−)-ipsdienol than (*R*)-(+)-ipsdienol (Sandstrom *et al.*, 2008). The ratio of these two enantiomers emitted by the beetles varies with species and population, but *I. confusus* emits an attractant which is about 90% (*S*)-(+)-ipsdienol and *I. pini* about 95% (*R*)-(−)-ipsdienol, which shows that the cytochrome P450 does not control the stereochemistry of the final mixture. The explanation for the difference between the enzyme ratio and the attractant ratio is not yet known, but it is probably caused by a reversible oxidation through ipsdienone (Figure 7.8). These beetles also use ipsenol in their attractant, which is probably made through reduction of the ipsdienone.

Both males and females oxidize α-pinene (derived from the tree resin) to *cis*-verbenone, which adds to the attractancy of the blend, and to *cis*-verbenol (Figure 7.8). The host tree as the source of α-pinene has been confirmed by several experiments, including the use of deuterated pinene. It is sufficient for the beetles to absorb the vapours of pinene and other compounds. (*E*)-Myrcenol in the attractant of male *I. pini* and *I. duplicatus* is probably formed by beetle oxidation of its own myrcene. Amitinol, from the attractant of *Ips amitinus*, is probably insect-synthesized, since it can be formed by an allylic rearrangement of ipsdienol.

The more complex tricyclic attractant of the striped ambrosia beetle *Trypodendron* (*Xyloterus*) *lineatum* (Coleoptera: Scolytidae), lineatin, is derived from geraniol, probably through grandisol (Section 7.3.2) by further oxidation. The isomer in *T. lineatum* is (1*S*,4*R*,5*S*,8*S*)-(+)-lineatin (Figure 7.9). It is used by a number of other *Trypodendron* species; indeed, in a trapping trial in Germany with traps containing lineatin and ethanol, *T. domesticus* and 28 other bark beetles were caught.

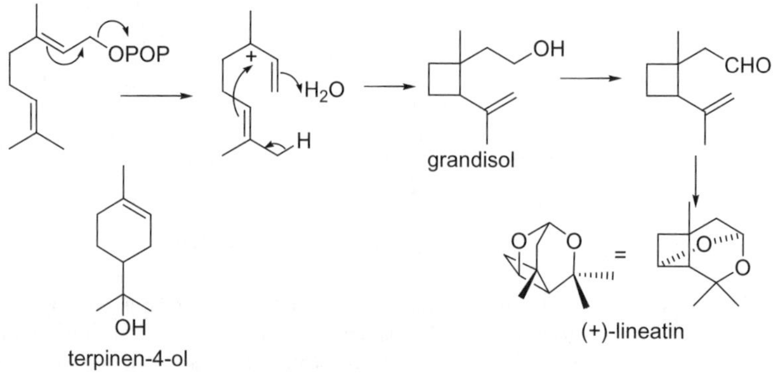

Figure 7.9 The likely route from geranyl diphosphate to lineatin through grandisol, and the structure of terpinen-4-ol.

Lower Terpenes

There still remain a number of bark beetle aggregation pheromones for which their biosynthesis remains conjecture, partly because of the small amounts the beetles produce or the tiny size of the beetles that produce them. Monoterpenes are found in the attractants of nearly all bark beetles attacking coniferous trees but it is not always clear how much of the synthesis has been done by the beetle and how much by the tree. (4*R*)-Terpinen-4-ol in *Polygraphus polygraphus* (Scolytidae) may be made by insect oxidation of tree terpinene, like verbenol, its dehydration product verbenene, and verbenone, all of which are found in bark beetles.

7.3.1.1 Hemiterpenes. Bark beetles attractants also contain some unusual C_5 hemiterpene compounds. 2-Methyl-3-buten-2-ol (Figure 7.10) is the principal substance in the attractant of *Pteleobius vittatus* (Scolytidae), along with *cis*-pityol and *cis*-vittatol (see below). It is also present in the attractant of several *Ips* species, and its biosynthesis has been demonstrated in *Ips typographus* (Plate 21) from labelled mevalonolactone (Lanne *et al.*, 1989). 3-Methyl-3-buten-1-ol, the direct hydrolysis product of isopentenyl diphosphate, is found in the attractant of the larch bark beetle *Ips cembrae*. The doubly oxygenated compound 3-hydroxy-3-methyl-2-butanone is found in several species of ambrosia beetle, but also in the attractant (for its own species) in the mandibular glands of the parasitic wood wasp *Rhyssa persuasoria* (Hymenoptera: Ichneumonidae) (Plate 22). The reduced derivative 3-methylbutanol (isopentanol) is found in the attractant of another ambrosia beetle *Platypus flavicornis* (Coleoptera: Platypodidae), along with hexanol and sulcatol (Section 7.3.1.2). Only 3-methyl-2-butenol, the hydrolysis product of dimethallyl diphosphate, is not found in beetle attractants.

7.3.1.2 Degraded Monoterpenes. There are other bark beetle attractants that contain isoprene units but have been degraded from monoterpenes

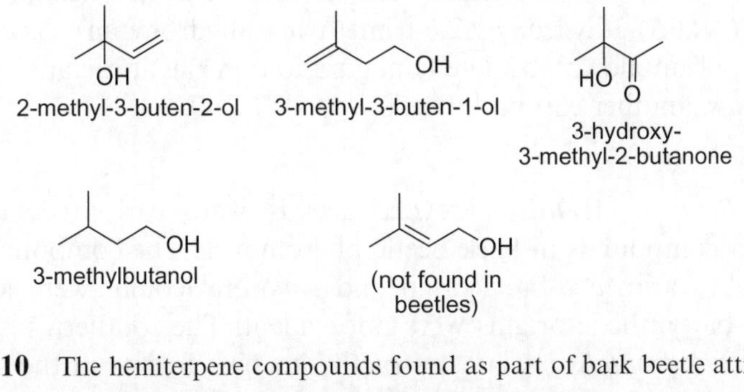

Figure 7.10 The hemiterpene compounds found as part of bark beetle attractants.

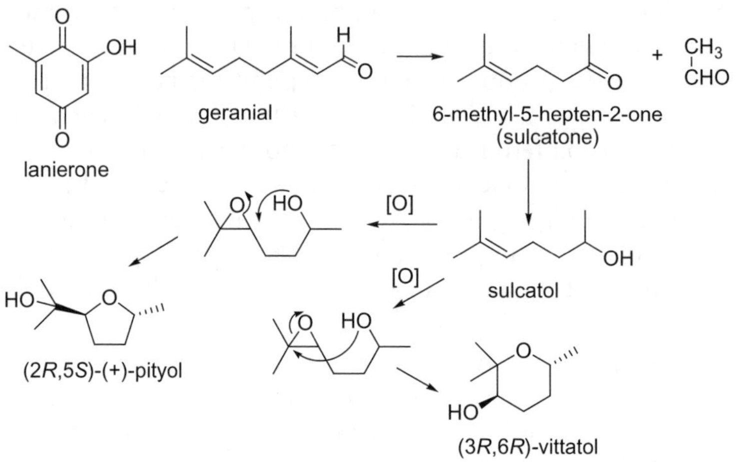

Figure 7.11 Some degraded monoterpenes from bark beetles, and the formation of cis-pityol and cis-vittatol from geranial, probably via sulcatol.

in some way. Examples are lanierone, pityol, sulcatol and vittatol (Figure 7.11). Lanierone is another component of the male attractant of *Ips pini*. Its formation is unexplored and not obvious. A widely distributed substance is 6-methyl-5-hepten-2-one, which can be formed by oxidation of geraniol (or geranyl diphosphate), or through a retro-aldol reaction from geranial. It is also found in ants, bees, other beetles and male butterflies. Sulcatol is evidently formed by reduction of 6-methyl-5-hepten-2-one (Figure 7.11). Sulcatol exists in two enantiomeric forms, and the blend of enantiomers is different for different species of *Gnathotrichus* (Scolitidae) ambrosia beetles. In *Platypus mutatus* (*sulcatus*) sulcatol is accompanied by sulcatone. Ring closure of sulcatol gives pityol (2-(2'-hydroxy-2-propyl)-5-methyltetrahydrofuran), probably through an intermediate epoxide. Pityol can be made in the laboratory from sulcatol. It is found in several species of *Conophthorus* and *Pityophthorus* as (2R,5S)-(+)-pityol, in some species made by males and in others by females; vittatol also is derived from sulcatol, and cis-vittatol ((3R,6R)-3-hydroxy-2,2,6-trimethyltetrahydropyran) accompanies cis-pityol (of unknown absolute configuration) in the attractant of *Pteleobius vittatus*, another elm bark beetle.

7.3.1.3 Bicyclic Acetals. Bicyclic acetals were early recognized as important compounds in bark beetle pheromones. The compounds frontalin, multistratin, *exo*-brevicomin and *endo*-brevicomin were identified but their biosynthetic origins were not evident. The southern pine beetle *Dendroctonus frontalis* is very destructive to pine forests in the southern

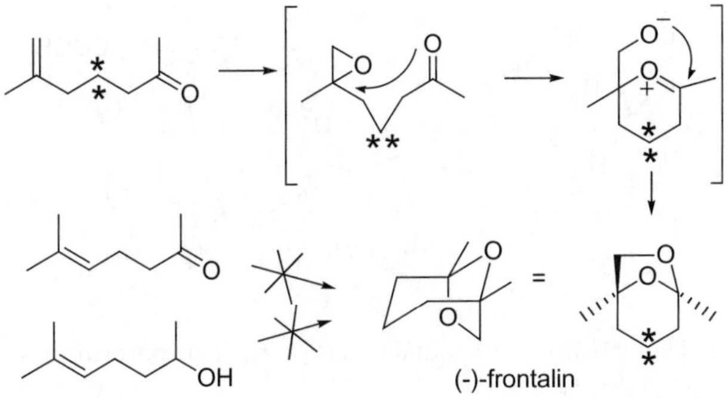

Lower Terpenes

Figure 7.12 The biosynthesis of frontalin by *Dendroctonus rufipennis* and *D. ponderosae*: the deuterium labelling used is shown by *.

USA, and therefore its behaviour and control have received much attention. Females of this species invade the tree and first emit the attractant which is chiefly (−)-frontalin (1S,5R)-1,5-dimethyl-6,8-dioxabicyclo [3,2,1]octane (Figure 7.12). The compound is produced by at least seven species of *Dendroctonus*. Brand *et al.* (1979) suggested that 6-methyl-6-hepten-2-one could be a precursor of frontalin. This was demonstrated by Perez *et al.* (1996) by confining beetles of five species with [4-^{2}H$_2$]6-methyl-6-hepten-2-one and showing that they produced [3-^{2}H$_2$]frontalin. They used first the spruce beetle *Dendroctonus rufipennis* and the mountain pine beetle *D. ponderosae*, which normally produce frontalin. These species produced deuterated frontalin, but they did not convert either sulcatol or sulcatone (Figure 7.11) to frontalin. Perez *et al.* (1996) also tried the experiment on the California five-spined ips *Ips paraconfusus*, the pine engraver *Ips pini*, *Ips tridens* and a curculionid beetle, the West Indian sugarcane weevil *Metamasius hemipterus sericeus* (Coleoptera: Curculionidae). None of these normally produce frontalin, but supplied with the required precursor, they appeared to use a non-specific mono-oxidase to produce frontalin (Figure 7.12).

The early idea that frontalin was a modification of a tree terpene was finally disproved by Barkawi *et al.* (2003). They injected several radio-labelled potential precursors into Jeffrey pine bark beetles *Dendroctonus jeffreyi* and found they all were incorporated into frontalin. Using [1-^{14}C]acetate, [2-^{14}C]mevalonolactone, [1-^{14}C]isopentenol, [1-^{14}C]:[1-^{3}H]isopentenol and [4,5-^{3}H$_7$]leucine all gave labelled frontalin (Figure 7.13). The fatty-acid synthesis inhibitor 2-octynoic acid gave a slight increase in incorporation of acetate into frontalin, probably a result of diverting it away from fatty-acid synthesis towards isoprenoids. They used [1-^{14}C]:[1-^{3}H]isopentenol to show there was no significant difference in the

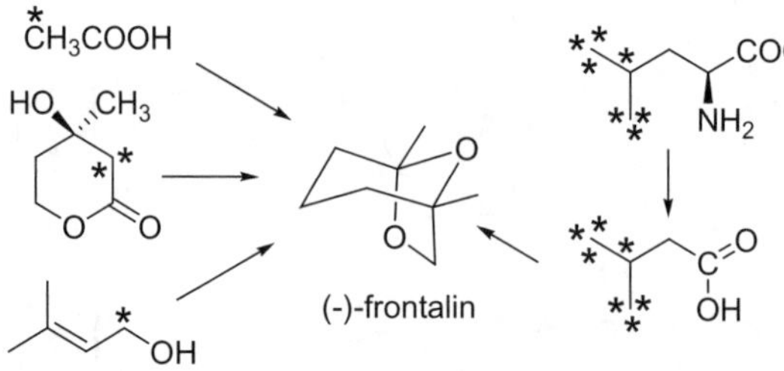

Figure 7.13 The labelled compounds that were incorporated into frontalin by *D. jeffreyi*.

ratio of isotopes incorporated, and those in the frontalin produced, to rule out any hybrid pathway. Acetate was also incorporated in the same way using two other species, *D. rufipennis* and *D. simplex*.

It is now understood that bark beetles can convert any suitable intermediate unsaturated ketone to a bicyclic acetal. Two other well-studied compounds; brevicomin and α-multistriatin, are not derived from terpenoid intermediates. True to the versatility of beetles, α-multistriatin is likely derived from an aceto-propiogenin, 4,6-dimethyl-7-octen-3-one and the brevicomins from two isomers of 6-nonen-2-one, which may be either a fatty-acid derivative or an acetogenin.

The mechanism for the biosynthesis of brevicomin from (Z)-6-nonen-2-one was studied in detail by Vanderwel and Oehlschlager (1992), using the mountain pine bark beetle *Dendroctonus ponderosae* and $^{18}O_2$ and $H_2^{18}O$. They showed that in an atmosphere containing $^{18}O_2$ the beetles produced (+)-*exo*-brevicomin labelled in both oxygen atoms. By considering the ways in which the double bond of the unsaturated ketone could be opened up to give brevicomin, they deduced from the labelling pattern that it occurred through a keto-epoxide, and the epoxide was not converted to an intermediate diol (Figure 7.14). They also deduced that both epoxide enantiomers give brevicomin; one epoxide enantiomer (6S,7R) is favoured by males, and the other (6R,7S) by females. Males of *D. ponderosae* produce (1R,5S,7R)-*exo*-brevicomin; females do not normally release brevicomin, but supplied with the unsaturated ketone they have the ability to epoxidize it. Mild heat or a trace of acid *in vitro* is sufficient to convert the epoxy-ketone to brevicomin. (+)-*exo*-Brevicomin is used as a pheromone by at least 12 species of bark beetle and also, incidentally, by the African elephant, while frontalin is found in both African and Asian elephants.

In addition to the female-produced pheromone frontalin of *D. frontalis*, it has recently been discovered that (+)-*endo*-brevicomin

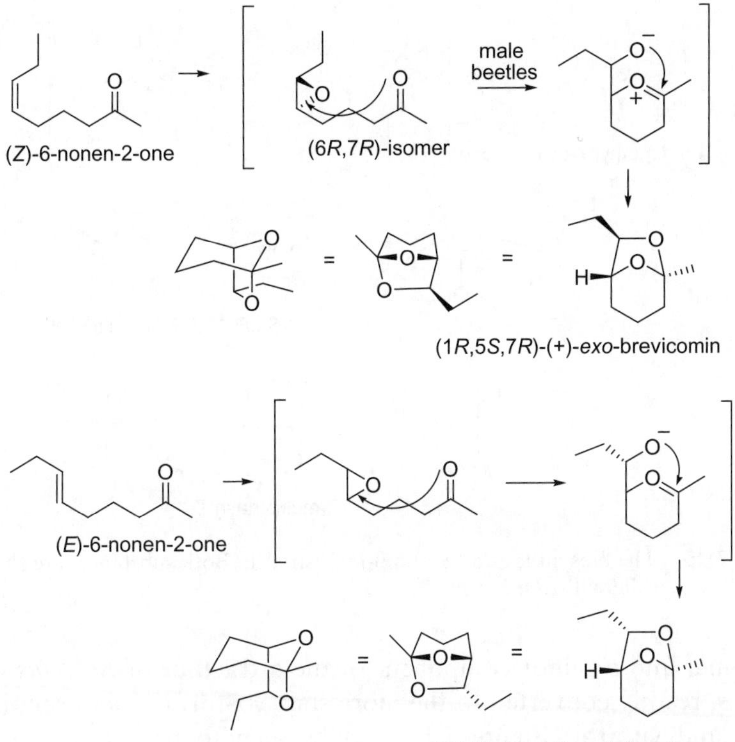

Lower Terpenes 193

Figure 7.14 The formation of *exo-* and *endo-*brevicomin. Both structures are shown from different perspectives, as used in different publications.

(Figure 7.14) is important for that species (Sullivan *et al.*, 2007). *Endo*-brevicomin is released by the later-arriving males of *D. frontalis* as part of the attack on the tree. It is produced in very small quantity, and was thought of little importance, but the study has shown that the beetles are very sensitive to the compound. By comparison with *exo*-brevocomin, the precursor of *endo*-brevicomin has been demonstrated to be (*E*)-6-nonen-2-one (Vanderwel *et al.*, 1992a) (Figure 7.14).

In the above discussions, the same reaction of attack of a ketone on an epoxide operates in each example. α-Multistriatin from the European elm bark beetle *Scolytus multistriatus* has not received the same biosynthetic attention as frontalin and brevicomin, but Francke and Schulz (1999) have suggested it is formed from 4,6-dimethyl-7-octen-3-one *via* the epoxy-ketone (Figure 7.15). Unlike the other bark beetle examples, in *S. scolytus* and *S. multistriatus* pheromone production, storage and release is from accessory glands on the vaginal palps. β-Multistriatin is

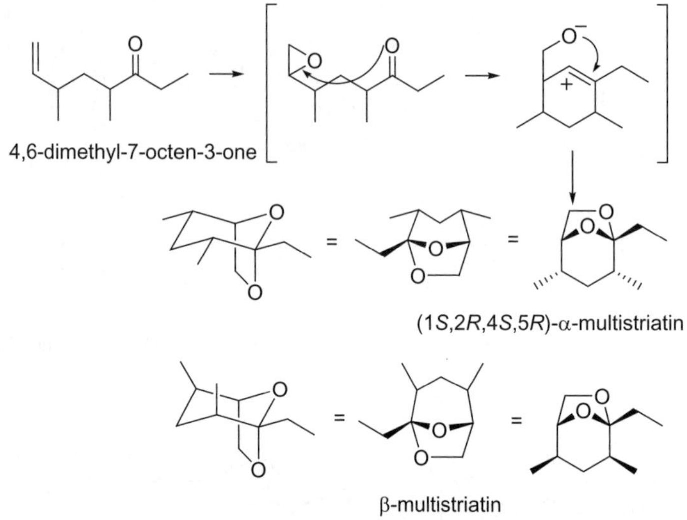

Figure 7.15 The biosynthesis of α- and β-multistriatin. Both substances are shown in different perspectives.

an isomer and a minor component of the attractant of *Scolytus multistriatus*, readily converted to the more stable α-form. The formation of pityol and vittatol (Figure 7.11) can be seen to belong to the same general type of reaction that gives frontalin, multistriatin and brevicomin.

Aggregation pheromone production in some bark beetles is regulated by another terpene, juvenile hormone JH-III (Seybold and Vanderwel, 2003). The Blomquist group is studying how JH-III induces the synthesis of bark beetle attractants. They find that different species, even those of the *Ips* genus (*I. pini* and *I. paraconfusus*), respond to JH differently (Tillman *et al.*, 2004). JH rapidly induces the transcription of many genes. They have found a likely primary responder gene to JH III in the midgut of *Ips pini* (Bearfield *et al.*, 2008).

7.3.2 Weevils

The males of the cotton boll weevil *Anthonomus grandis* (Coleoptera: Curculionidae) (Plate 8), a serious pest of cotton, produce a sex pheromone to attract females. The pheromone, known as grandlure, consists of a mixture of at least four monoterpenes, (1R,2S)-grandisol and (Z)-ochtodenol, the major components, and (E)- and (Z)-ochtodenal minor components (Figure 7.16). Whereas it was once thought the weevils modified cotton terpenes, it has now been shown that the males

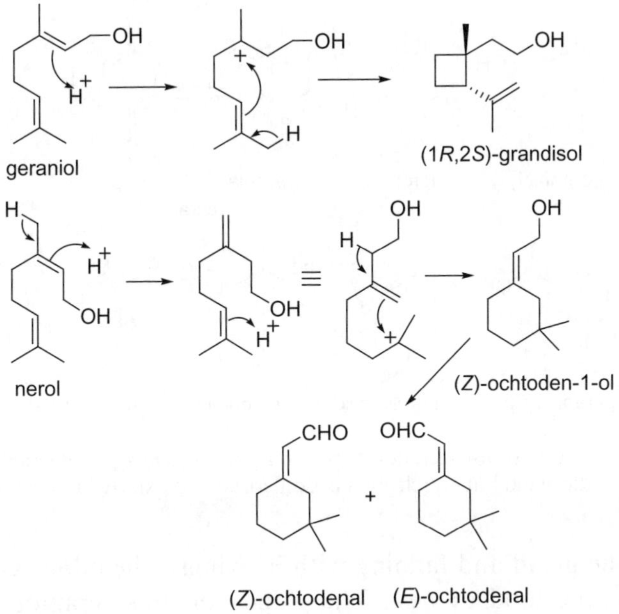

Lower Terpenes 195

Figure 7.16 The presumed routes to grandisol and the other compounds of the male attractant pheromone of the boll weevil *Anthonomus grandis*. The four compounds together make up the commercial attractant grandlure.

of *A. grandis* make these compounds from acetate in specialized cells in their gut (Taban *et al.*, 2006). The chief component, grandisol, is made from geraniol, as already outlined in Figure 7.9. The three cyclohexane components are derived from nerol, the *cis*-isomer of geraniol (Figure 7.16). Pheromone production is about 3.4 µg per day. At least ten weevil species use grandisol in their attractants. The pecan weevil *Curculio caryae* (Curculionidae) uses the same mixture but with (1*S*,2*R*)-grandisol added, while the pepper weevil *Anthonomis eugenii* has the three cyclohexane compounds, plus geraniol and geranic acid but no grandisol. The formation and release of attractant in boll weevils, as in bark beetles, is stimulated by juvenile hormone JH-III (Seybold and Vanderwel, 2003).

7.3.3 Other Pheromones

The Nasonov gland, part of the sting apparatus, in the abdomen of honeybees provides an attractant pheromone for marking food sources and the nest. The gland secretes into a groove on the seventh abdominal segment. The secretion is dispersed by the bee raising its abdomen,

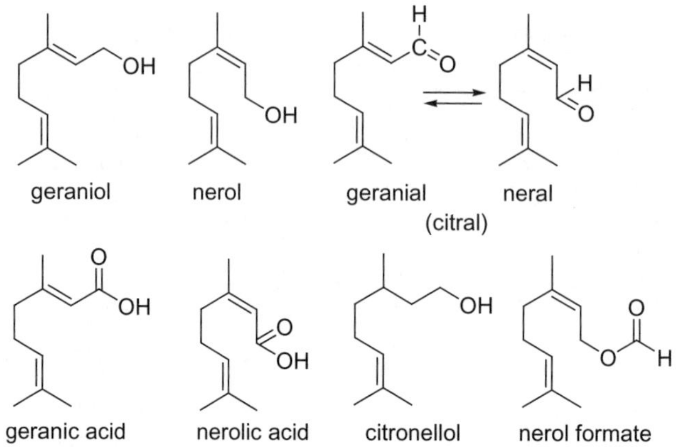

Figure 7.17 The Nasonov secretion of the honeybee *Apis mellifera*. Geraniol is stored in the gland but on release it is isomerized and oxidized to the mixture shown.

exposing the gland and fanning with its wings. The odour guides other workers to the food or nest. The gland contains geraniol and (*E*,*E*)-farnesol (Section 7.5). The compounds are enzymatically oxidized on release (Figure 7.17) so that the emitted pheromone also contains nerol, geranial and neral [(*E*)- and (*Z*)-citral], geranic and nerolic acids and corresponding products from farnesol (Pickett *et al.*, 1980).

Citronellol (Figure 7.17) is a sex attractant produced by male spider mites. The cheese or house mite *Tyrophagus putrescentiae* (Astigmata: Acaridae) normally forms clusters, but if one of them is crushed, the formate ester of nerol (Figure 7.17) is released and they disperse.

The large number of known monoterpene pheromones remaining has not received biosynthetic attention but their frequently simple structures make it not difficult to predict how they may be formed.

7.4 MONOTERPENE DEFENSIVE COMPOUNDS

Insects from beetles to termites use terpenes in defensive secretions. Beetles characteristically use a wide range of different compounds in defence. Such compounds can be directly sequestered from plants on which they feed, or partially modified from food compounds or totally synthesized by the beetle.

7.4.1 Iridoids

In 1949 Pavan isolated the substance he called iridomyrmecin from the Argentine ant *Iridomyrmex humilis* (now *Linepithema humile*). Later

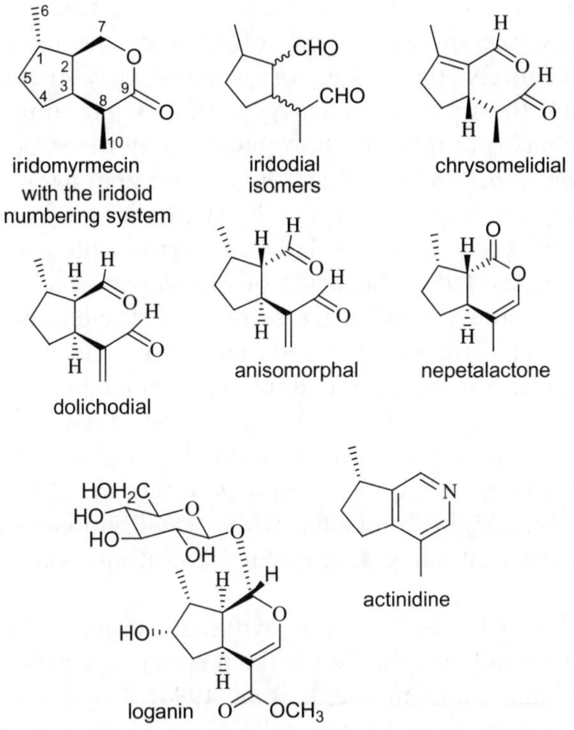

Figure 7.18 Some examples of iridoids. Iridomyrmecin and dolichodial are from ants, chrysomelidial from a leaf beetle larva, anisomorphal from a stick insect, nepetalactone from aphids and the catnip plant, loganin an example of a plant iridoid glycoside, from *Strychnos* fruits and *Hydrangea* bark, and the alkaloid actinidine, which accompanies iridoids in many species.

more compounds of related structure were isolated. Iridomyrmecin, dolichodial and other monoterpenes with a methylcyclopentanoid structure are called iridoids (Figure 7.18). They act as defensive compounds in ants, stick insects, rove beetles and leaf beetle larvae. Some of the group (*e.g.* the isomer of nepetalactone shown in Figure 7.18) are also sex pheromones in aphids. Iridoids (including iridomyrmecin and nepetolactone), often in the form of glycosides, were subsequently found in many plants, where they presumably provide defence against insects. They have a variety of physiological effects in humans and, to us, they have a bitter taste. Nepetalactone from the catnip plant *Nepeta cataria* also excites members of the cat family (domestic cats, lions and jaguars but not tigers). Loganin from *Strychnos nux-vomica* is an example of a plant iridoid glycoside.

Since many insects that have iridoids do not feed on iridoid-containing plants, the question arose of whether the insects make the compounds entirely themselves or whether they transform simple monoterpenes to iridoids. In early work [^{14}C]mevalonolactone was shown to be incorporated into anisomorphal by the stick insect *Anisomorpha buprestoides* (Phasmida: Pseudophasmatidae). Then using [4,4,6,6,6-^{2}H$_5$]mevalonolactone the biosynthesis was studied in leaf beetle larvae (Figure 7.19). The labelled intermediate was first painted onto a cabbage leaf fed to the larvae of *Phaedon amoraciae* (Coleoptera: Chrysomelidae) and the defensive secretion collected and analysed by GC-MS. The mass spectrum showed that at least one unit of deuterated mevalonolactone was incorporated into chrysomelidial, confirming it as a product of the beetle metabolism (Veith *et al.*, 1994). When a droplet of water containing the labelled compound was placed directly onto a pair of defensive glands of *P. amoraciae* (there are nine pairs on abdominal segments), chrysomelidial with a small amount of the deuterated compound containing seven deuterium atoms was seen (Oldham *et al.*, 1996).

To study some of these reactions, synthetic norgeraniol (Figure 7.20), rather than geraniol, was used to make it easier to separate the labelled products and analyse them (Veith *et al.*, 1994).

The early stages of oxidation take place in the fat body in four species of leaf beetles (*Phaedon amoraciae*, *P. cochleariae*, *Gastrophysa viridula* and *Plagiodera versicolora*) and follow the same route as in plants, that is, the terminal methyl of geraniol is oxidized to an alcohol.

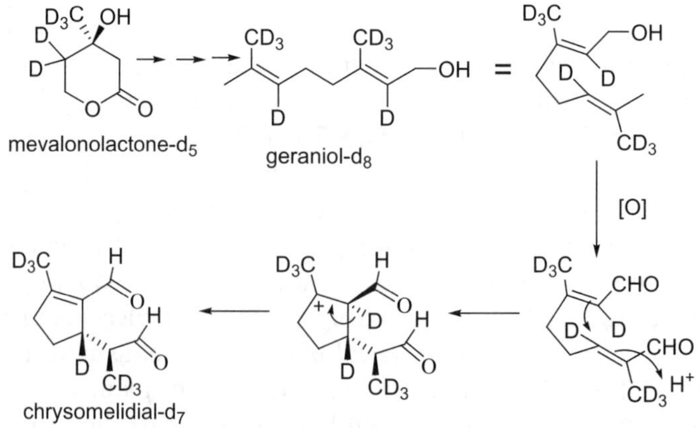

Figure 7.19 The formation of D$_7$-chrysomelidial from D$_5$-mevalonolactone in *Phaedon* larvae defensive glands.

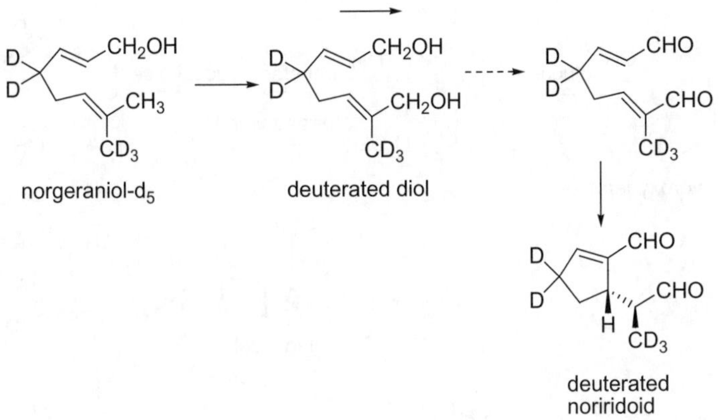

Lower Terpenes

Figure 7.20 The incorporation of deuterated norgeraniol and its corresponding diol into deuterated noriridoid in leaf beetles.

The molecule is then conjugated to glucose and transported through the haemolymph as 8-hydroxygeraniol-8-*O*-β-D-glucoside, into the gland reservoir. Because of the ease of hydrolysis of glucosides, model experiments were made with thioglucosides, which are not hydrolysed by glucosidases. In the gland reservoir the glucoside is hydrolysed back to 8-hydroxygeraniol and then both hydroxyls are oxidized by oxygen, catalysed by an oxidase to 8-oxocitral by removal of the *pro-R* hydrogens at C-1 and C-8 (Figure 7.21). The oxocitral in *G. viridula* is cyclized directly to chrysomelidial; in *P. versicolora* it gives plagiodial, but in *P. amoraciae* and *P. cochleariae* it is first converted to plagiodial and then the double bond is isomerized to chrysomelidial.

Rove beetles of the Staphylinina and Philonthina are carnivorous insects with paired abdominal glands that also produce iridoids similar to those of leaf beetles, and follow a comparable overall biosynthetic route, but with the opposite stereochemistry on cyclization (Figure 7.22), that was studied with deuterated norgeraniol (Weibel *et al.*, 2001). The rove beetles are also able to use saturated intermediates like citronellol.

Beetles are very versatile in making defensive compounds, but also in taking potential defensive materials from their food, and sequestering them in glands (see Section 11.3.2). Some leaf beetles can make their iridoids, but can also collect them from plants. In the latest twist to this unfolding story of biosynthesized *versus* sequestration, Kunert *et al.* (2008) have found that the leaf beetles *Plagiodera versicolora* (feeding on *Salix fragilis*) and *Phratora laticollis* (both Chrysomelidae) (feeding on *Populus canadensis*) are able to make iridoids *de novo* from acetate, or

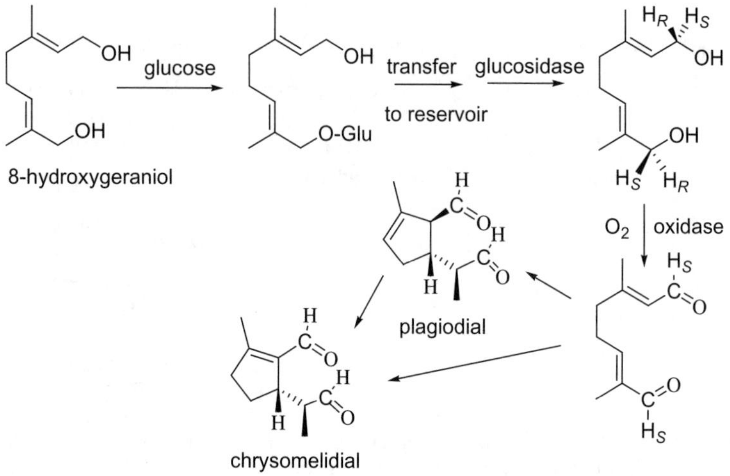

Figure 7.21 Summary of the steps in conversion of 8-hydroxygeraniol in the fat body to a glucoside, carried by haemolymph to the defensive glands, hydrolysis of the glucoside, removal of the *pro*-R hydrogens and cyclization to iridoids.

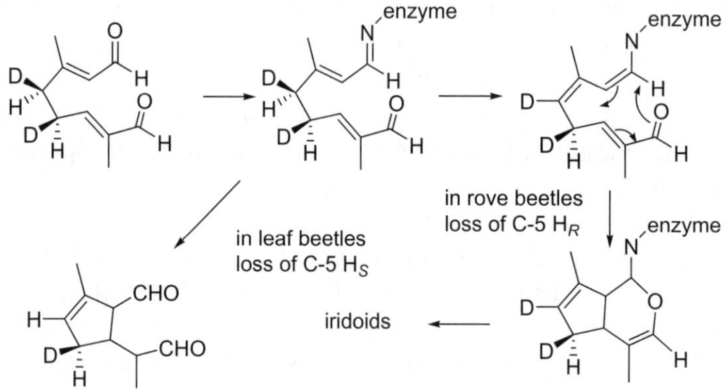

Figure 7.22 Iridoid formation in leaf beetles and rove beetles is very similar, but with a different stereochemistry in the loss of C-5 hydrogen.

they can sequester them from food. Larvae were fed on leaves impregnated with terpenoid precursors in the food plant labelled with ^{13}C, the precursors were injected into the haemolymph and their distribution followed in the haemolymph, the defensive secretion and the faeces. Of various compounds, only 8-hydroxygeraniol-8-*O*-β-D-glucoside is passed to the gland reservoir and converted to iridoids. Other potential defensive compounds, such as salicin, used by other beetles were not sequestered.

Iridoids are not confined to beetles and plants. Larvae and adults of the thrips *Callococcithrips fuscipennis* (Thysanoptera: Phlaeothripidae) produce droplets of secretion, which contain dolichodial and another iridoid close to anisomorphal (Figure 7.18), in a solvent mixture of hydrocarbons and esters (Tschuch *et al.*, 2008). Chrysomelidial is found in the opisthonotal glands of the ground-dwelling oribatid mite *Oribotritia berlesi* (Acarini: Oribatidae). In adults, chrysomelidial is accompanied by β-springene and a mixture of hydrocarbons as solvent. Juveniles contain only chrysomelidial and epi-chrysomelidial (Raspotnig *et al.*, 2008).

7.4.2 Termite Defensive Secretion

Most termite (Isoptera) species have a soldier caste. In the families Rhinotermitidae, Termitidae and Serritermitidae the soldiers have a modified head with a frontal gland. The secretion contains a range of compounds from alkanes, monoterpenes, sesquiterpenes and diterpenes. Sometimes other compounds such as lactones, vinyl ketones and keto-aldehydes are present. Terpenes are commonly found. In six of seven European species of *Reticulitermes* the diterpene geranyllinalool is the major compound, and in the seventh it is geranylfarnesol (Quintana *et al.*, 2003). Among the monoterpenes are myrcene, ocimene and limonene, α-pinene and β-pinene (Figure 7.23).

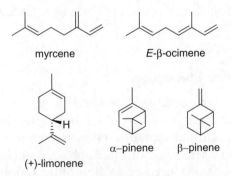

Figure 7.23 Some monoterpenes from termite defensive secretions.

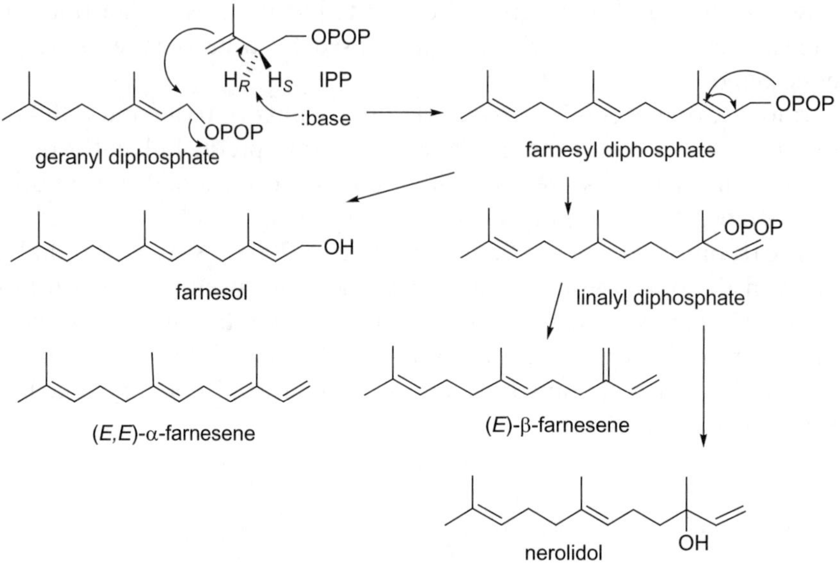

Figure 7.24 The formation of farnesyl diphosphate by addition of another molecule of isopentenyl diphosphate (IPP) to geranyl diphosphate, and some simple sesquiterpenes derived from it.

7.5 SESQUITERPENES

Addition of another isopentenyl diphosphate (IPP) to geranyl diphosphate, catalysed by prenyl transferase enzyme, in the same way that gave geranyl diphosphate, gives farnesyl diphosphate, the parent of the sesquiterpenes (Figure 7.24). Hydrolysis, or rearrangement followed by hydrolysis, or dehydration leads to a number of products as for the monoterpenes. By suitable folding of the intermediate diphosphate or carbocation, further cyclic products are obtained. A farnesyl diphosphate synthase has recently been isolated and characterized from the boll weevil *Anthonomis grandis* (Coleoptera: Curculionidae) (Taban et al., 2009).

7.5.1 Sesquiterpene Pheromones

Sesquiterpenes are frequently encountered in insects. For the simpler ones, their biosynthesis probably does not differ from that already studied in plants, and has therefore not attracted attention. (*E*)-β-Farnesene is an alarm pheromone of some aphids, and (*E*,*E*)-α-farnesene is an alarm pheromone produced by soldiers of the termite *Prorhinotermes canalifrons* (Isoptera: Heterotermitidae), from the frontal gland. Trail pheromones of about 50 species of termites have been identified, most of them of aliphatic origin, but recently discovered is the norsesquiterpene *trans*-2,6,10-trimethyl-5,9-undecadien-1-ol (Sillam-Dussès *et al*., 2007) (Figure 7.25). Simple sesquiterpenes like farnesols, dihydrofarnesols and nerolidols, and esters of these, are used as pheromones. More complex are the germacrene derivatives periplanones A and B, sex attractants from the females of the American cockroach *Periplaneta americana* (Blattodea: Blattidae) (Figure 7.25), produced in the hind gut. Other species of cockroach use these and related periplanones. Some stink bugs (Pentatomidae) produce tricyclic sesquiterpene pheromones (Millar, 2005) that, as yet, have received no biosynthetic attention. The structures of many insect sesquiterpene compounds together with their biological properties are given by Francke and Schulz (1999) and Hick *et al*. (1999).

The male-produced cucujolides, macrocyclic lactone aggregation pheromones of *Cryptolestes* (Cucujidae) and *Oryzaephilus* (Silvanidae) beetles, have been discussed (Section 5.3.1). One was omitted. Cucujolide I is unlike the others in having methyl branches, which suggested a terpenoid origin. This was confirmed by incorporation of radio-labelled acetate and mevalonate into cucujolide I. Using both *Cryptolestes ferrugineus* and *Oryzaephilus mercator* it was shown that farnesol labelled on C-1 with both deuterium and ^{18}O was converted to cucujolide I (Vanderwel *et al*., 1992b). Apparently the terminal isopropylidene group is oxidized to a carboxyl group. The ^{18}O and deuterium are retained in the lactone (Figure 7.26).

Females of the oleander scale insect *Aspidiotus nerii* (*A. hedera*) (Homoptera: Diaspididae) (Plate 21), which are immobile, secrete a sex attractant from pygidial glands and release it through the rectum to

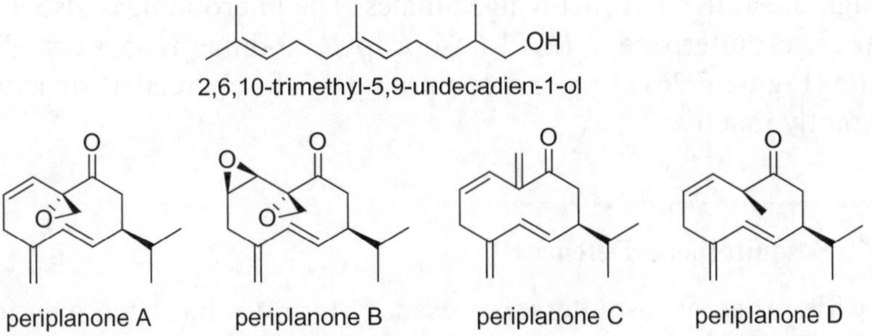

2,6,10-trimethyl-5,9-undecadien-1-ol

periplanone A   periplanone B   periplanone C   periplanone D

Figure 7.25 Sesquiterpene pheromones: the norsesquiterpene trail pheromone of the termite *Mastotermes darwiniensis*, and periplanones A and B, sex attractants of the cockroach *Periplaneta americana* with two related compounds from other cockroach species.

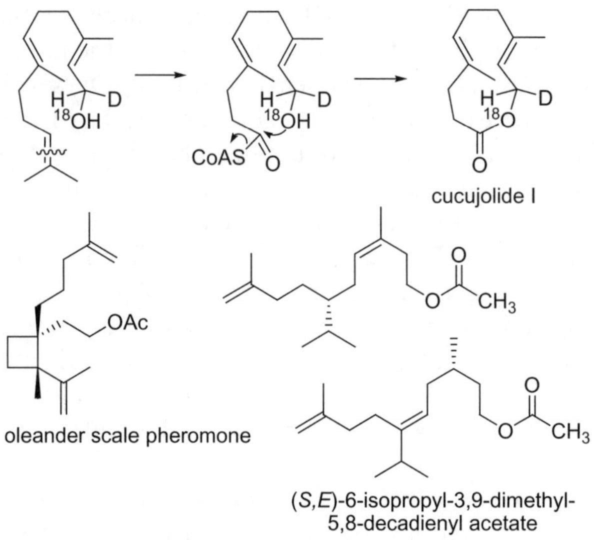

Figure 7.26 The formation of cucujolide I from farnesol as indicated by labelling experiments; the structure of oleander scale insect pheromone, and two non-linear sesquiterpene acetates from the California yellow scale *Aonidiella citrina*. The last compound is the sex pheromone.

attract winged males. The compound can be seen as a sesquiterpene analogue to grandisol (Figures 7.9 and 7.16), with the same stereochemistry around the cyclobutane ring (Einhorn *et al.*, 1998), but if the accepted biosynthesis of grandisol is correct, the oleander scale pheromone does not have a linear sesquiterpene origin (Figure 7.26).

The females of the California yellow scale *Aonidiella citrina* (Homoptera: Diaspididae), another sessile insect, secrete a sex pheromone through the anus to attract winged males. The pheromone is also a non-linear sesquiterpene (*S,E*)-6-isopropyl-3,9-dimethyl-5,8-decadienyl acetate (Figure 7.26), a second compound of closely related structure is apparently inactive.

7.5.2 Sesquiterpene Defences

The wide range of sesquiterpenes used by termites has been mentioned (Section 7.4.2). Interesting, but biosynthetically unsolved, structures are ancistrodial and ancistrofuran, the chief compounds in the repellents of the minor and major soldiers, respectively, of *Ancistrotermes cavithorax* (Isoptera, Macrotermitinae) (Figure 7.27) (Baker *et al.*, 1978). These and

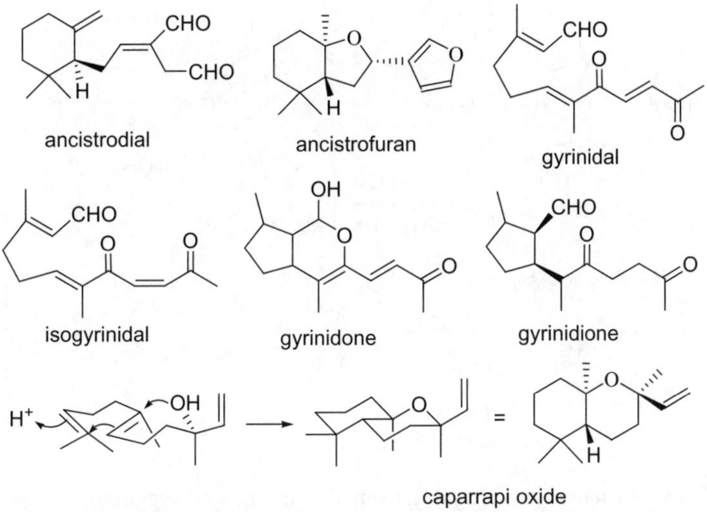

Figure 7.27 Some sesquiterpene defensive compounds of yet uncertain biosynthesis, and a suggestion for the possible formation of caparrapi oxide from farnesol.

caparrapi oxide, of unknown function from *Amitermes* (Isoptera: Termitidae), are evidently derived from farnesol by oxidation and cyclization. A probable route to caparrapi oxide is shown in Figure 7.27. Whirligig beetles (Gyrinidae) have a pair of defensive glands opening on the tip of the abdomen. The glands of *Gyrinus* and *Dinuetes* species secrete norsesquiterpene ketones, gyrinidal, isogyrinidal and gyrinidione while gyrinidone is found only in *Dinuetes*. These latter compounds look rather like sesquiterpene relations of iridoids. Fish have been shown to reject the beetles or gyrinidal.

Dendrolasin is a defensive compound from the mandibular glands of the ant *Lasius fuliginosus*, discovered by M. Pavan 50 years ago. It is accompanied by β-farnesene and a monoterpenefuran, perillene (Figure 7.28). In early experiments with labelled sodium acetate, sodium mevalonate and glucose, all were incorporated into dendrolasin by the ant, but the label was scattered over all the carbon atoms, so that firm conclusions about the biosynthetic route could not be made. A clue to the biosynthesis of dendrolasin and perillen may be the presence of farnesal and citral accompanying them in the mandibular glands. Low yields of dendrolasin and perillen have been obtained by autoxidation of the enol acetates of farnesal and citral (Tada *et al.*, 1982). Comparison with ancistrodial and ancistrofuran, and also gyrinidione and gyrinidone, suggest an oxidation-dehydration route. Dendrolasin has also been found in plants and marine molluscs.

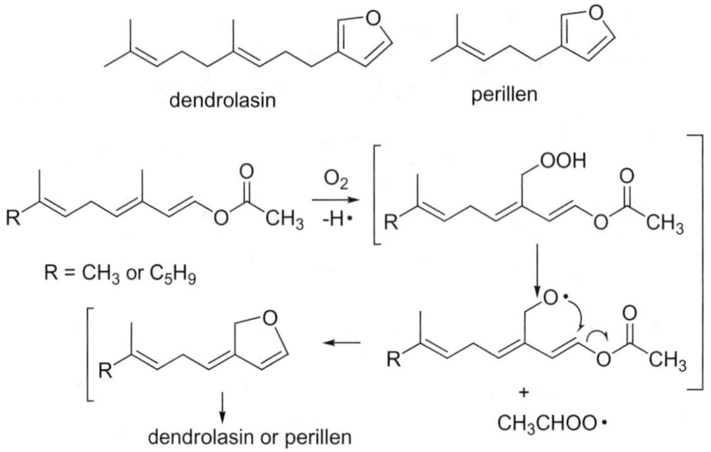

Figure 7.28 Defensive compounds from the ant *Lasius fuiginosus*, and a possible route to them by autoxidation of enol acetates.

7.5.2.1 Papilio Osmeterial Defence. The larvae of swallowtail (Papilionid) butterflies have a horn-like organ, called an osmeterium, on the back of their heads, normally hidden, but everted when disturbed (Plate 23). They attempt to smear its secretion onto the disturber. K. Honda has made a close study of the secretions. The younger larvae usually have a secretion consisting of mono- and sesquiterpenes. In six species of *Papilio* and two of *Chilasa* fourth instar larvae contained typically α-pinene, β-myrcene, sabinene, limonene, β-caryophylline, germacrene-A and various farnesene isomers. The osmeterial secretion of one species, *Papilio memnon*, contains caryophyllene oxide (Figure 7.29), which is also found in the oil of some tropical plants and there acts as a deterrent to leaf-cutting ants. *Papilio protenor* produces germacrenes A and B (Figure 7.29) in its osmeterium. When the organ of *P. protenor* was treated with [^{2}H$_4$]acetic acid, deuterium was incorporated into both compounds, and [1,2-^{13}C]acetic acid was incorporated into β-myrcene and (*E*)-β-farnesene by *Papilio protenor*, *P. helenus* and *Luehdorfia puzibi* (Honda, 1990). In the fifth and final instar the secretion changes from terpenoids to lower mass acids and esters (butyric, isobutyric and acetic acids and their methyl esters and methyl 3-hydroxybutyrate). The authors suggest the change reflects shifting predator pressure as the larvae age and grow (Omura *et al.*, 2006).

7.5.2.2 Cantharidin. Cantharidin is a defensive secretion of blister beetles (Meloidae) and false blister beetles (Oedemeridae). It forms

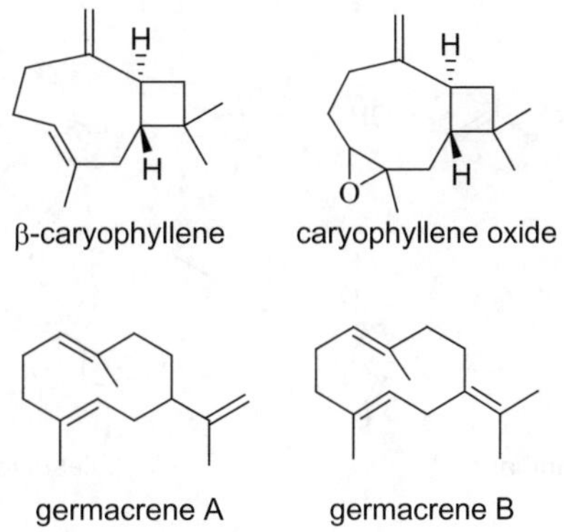

Figure 7.29 Sesquiterpenes from the osmeterial secretion of some *Papilio* butterfly larvae.

about 0.25–0.5% of the body weight, and is stored in the haemolymph and male genitalia. The haemolymph of the oil beetle *Meloe proscarabaeus*) contains about 25% cantharidin. In the margined blister beetle *Epicauta pestifera* (Meloidoe) (Plate 24) a male before copulation has an average of 23.8 mg of the substance, equivalent to 15% of its body weight, most of it stored in its accessory reproductive glands. In meloid beetles, it is present in all life stages. When disturbed, adult insects bleed as a reflex from the leg joints, while early larvae regurgitate a milky secretion from the mouth. Cantharidin is highly toxic in humans and an extreme irritant to all tissues; it is known to inhibit protein phosphatases. In the Meloidae, cantharidin is synthesized by both sexes as larvae, but only by the male adult beetles. Females acquire it from males through frequent copulation and it passes thence to eggs. In the Oedemeridae, both sexes of adults are reported to produce cantharidin.

The full biosynthesis route to cantharidin has still not been established. McCormick and Carrell (1987) give a detailed account of the extended biosynthesis studies, which began with radio-labelled materials, largely before heavy isotope labelling came into general use. The discovery that [2-^{14}C]farnesol could be incorporated into cantharidin was a significant advance. In further studies with labelled mevalonate it was found that carbon atoms 1 and 5 to 7 of farnesol were lost (Figure 7.30) (McCormick and Carrell, 1987). Specific labelling of the terminal methyl groups of farnesol should distinguish them, but by a curious effect they become equivalent and the label is divided between two atoms in cantharidin. The whole synthesis apparently takes place on one enzyme without formation of an intermediate that can be isolated.

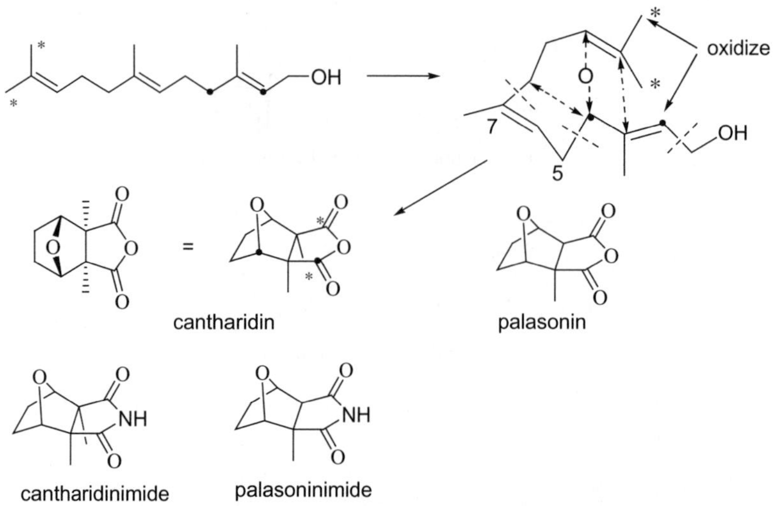

Figure 7.30 The biosynthesis of cantharidin. The asterisks and black dots represent two different experiments with labelled farnesol that showed these particular carbon atoms are retained in cantharidin. The two teminal methyls (asterisked) of farnesol become scrambled during the synthesis. No mechanism has been proposed for this unusual reaction. The excision of atoms 5 to 7, ring closure between C-3 and C-11, insertion of the oxygen bridge, oxidation to acid and anhydride formation all occur without the molecule dissociating from the enzyme. Cantharidin is accompanied by palasonin and the two non-toxic imides. Note that palasonin is chiral but cantharidin has a plane of symmetry.

Cantharidin is accompanied in beetles by smaller amounts of the less toxic palasonin, and by two non-toxic imides, cantharidinimide and palasoninimide (Figure 7.29) in the blister beetle *Hycleus lunata* (Coleoptera: Meloidae) (Dettner *et al.*, 2003). Palasonin was first identified in a plant, and is apparently formed from cantharidin by oxidative demethylation.

There are other insects in several orders, including members of Hemiptera, Coleoptera, Diptera and Hymenoptera, called cantharidiphiles, which seek out and devour blister beetles, dead or alive, to arm themselves with cantharidin.

7.5.3 Lac Insects

A group of homopterous insects (subfamily Lacciferinae), chiefly feeding on various forest trees, produce a dark-red, transparent excretion on the bark of the tree, which eventually covers and protects the insects. The hardened excretion is known as lac. It has been a commercial

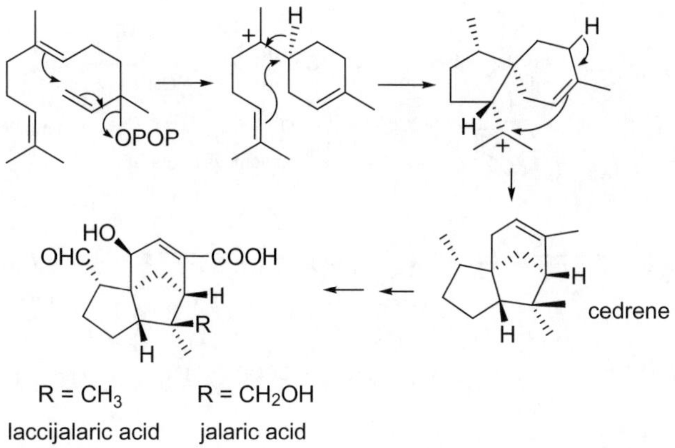

Figure 7.31 The lac produced by *Laccifera* insects consists of a mixture of tricyclic sesquiterpenes derived from cedrene.

product for hundreds of years for making varnish or in a purified form as shellac. As with some other insect products, it is not yet certain whether lac is made by the insects themselves or by symbiotic microorganisms in them. Lac consists of a mixture of hydroxylated palmitic acids and sesquiterpenes. The hydroxypalmitic acids are hydroxylated at C-16, or di-hydroxylated at C-10 and C-11, or trihydroxylated at all three positions.

The principal terpenes are jalaric acid and laccijalaric acid, which are reached by several oxidation steps from cedrene. It is known from plants that cedrenes are formed there from farnesyl diphosphate *via* linalyl diphosphate (Figure 7.31), but there is no information on the insect route. While α-cedrene and cedrol are found in *Juniperus* trees, there is no cedrene in the trees that the lac insects attack, and the lac acids are derived from a different isomer of cedrene from that found in plants. The lac is coloured reddish by the presence of anthraquinone pigments, laccaic acids (Section 9.6.2).

7.6 HOMOTERPENES

While mevalonate is the universal building block of terpenes, insects also make use of homomevalonate, which contains an extra carbon atom derived from a molecule of propionate replacing one of acetate in the initial stage of the synthesis. Labelling experiments by Schooley *et al.* (1973) showed that homomevalonate starts with condensation between one molecule of propionate and one of acetate (Figure 7.32). In some insects it is known that the propionate is derived from isoleucine or

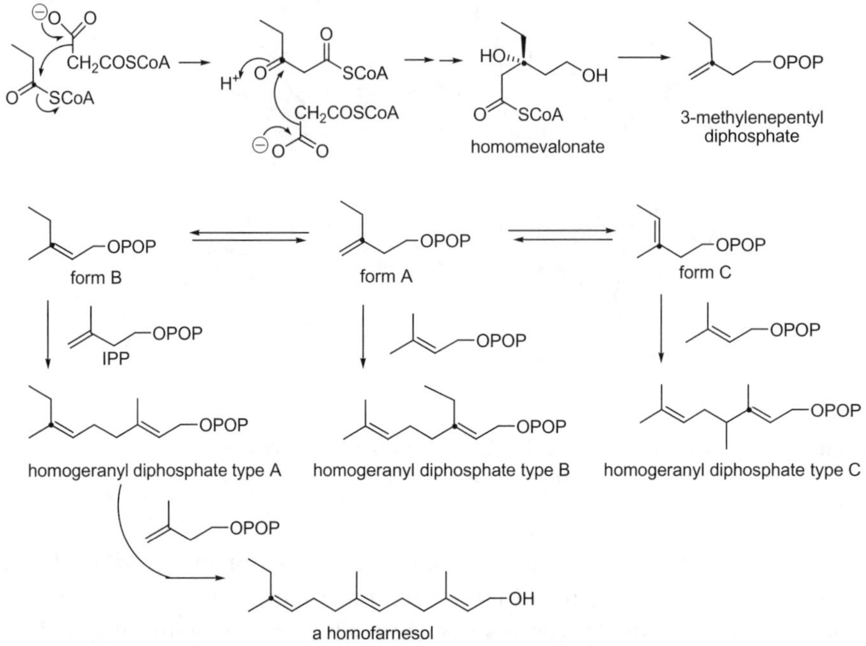

Figure 7.32 The biosynthesis of homomevalonate from acetate (as malonate) and propionate units. The black dot on carbon indicates a labelled carbon of propionate. Isomerization of 3-methylenepentyl diphosphate gives rise to three possible kinds of branching. Form B shown on the left is probably most common and the formation from it of a homo-sesquiterpene is illustrated.

valine only. The intermediate 3-methylenepentyl diphosphate (call it form A) can be isomerized to two other structures (forms B and C). Form A can react with DMAPP to give one form of homogeranyl diphosphate (call it type A). Forms B and C can each react with IPP to give two other types of homogeranyl diphosphate (types B and C) (Figure 7.32). We thus obtain homomonoterpenes with a terminal ethyl branch (type A), an internal ethyl branch (type B) or two vicinal methyl branches (type C). It is also possible for homomevalonate A to condense with homomevalonate B or C to form bishomogeranyl diphosphates A-B or A-C. All possibilities are found among insect compounds. This system of classifying the homoterpene compounds has no general acceptance, but is useful as a shorthand description.

The first homo-monoterpene and bishomo-monoterpene have recently been discovered. The trail pheromone of the ponerine ant *Gnamptogenys striatula* consists of a mixture of three esters, the decyl and dodecyl esters of homogeraniol type C (4S,2E)-3,4,7-trimethyl-2,6-

octadienol and the dodecanoate of bishomogeraniol type A-C (4*S*,2*E*,6*E*)-3,4,7-trimethyl-2,6-nonadienol (Figure 7.33). The dodecanoate of the bishomo-monoterpene was only marginally active as part of the pheromone (Blatrix *et al.*, 2002). A small amount of homo-ocimene accompanies larger amounts of (*E*)-β-ocimene in the Dufour glands of the ant *Labidus praeditor*. The bark beetle *Ips typographus* produces (*E*)-3-methyl-7-methylene-1,3,8-nonatriene, apparently in stressed conditions.

Sesquiterpene homologues are more common, particularly in ants. The three farnesene homologues from the Dufour glands of *Myrmica* ants are an example of simple homosesquiterpenes (Figure 7.34). Their purpose there is unknown. In the Dufour gland of the ant *Manica rubida* nine isomers and homologues of farnesene were found. Their function is again unknown. The major component of the male-produced sex pheromone of the Sunn pest or wheat bug *Eurygaster integriceps*

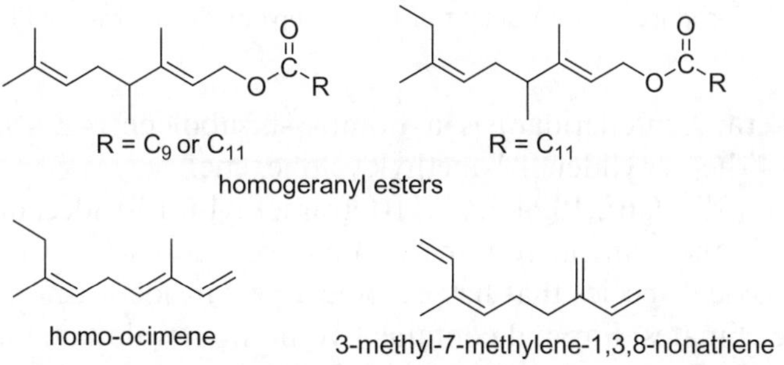

Figure 7.33 Examples of homo-monoterpenes from insects.

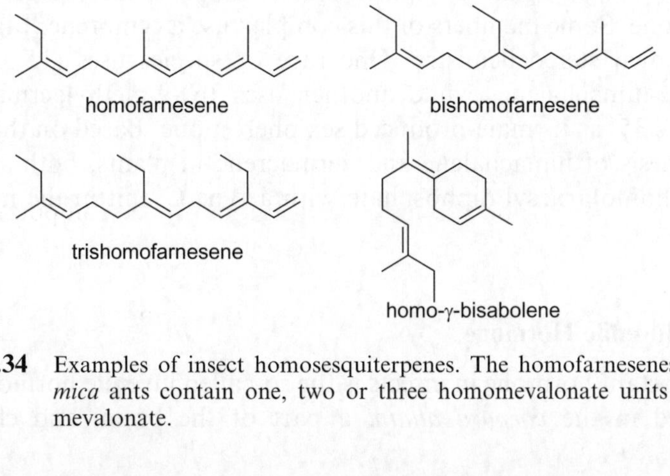

Figure 7.34 Examples of insect homosesquiterpenes. The homofarnesenes of *Myrmica* ants contain one, two or three homomevalonate units replacing mevalonate.

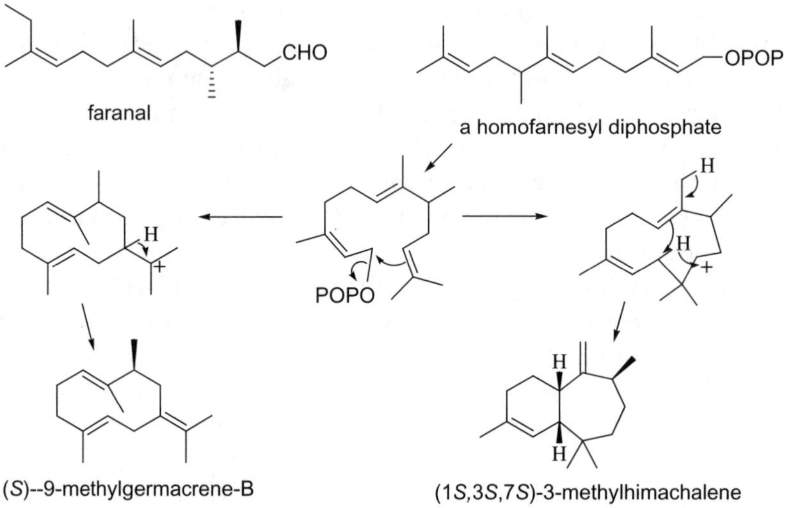

Figure 7.35 The ant trail pheromone faranal, a bishomosesquiterpene. The probable biosynthetic route to homohimachalene and homogermacrene, male-produced sex attractant of two groups of *Lutzomyia* sandflies.

(Heteroptera: Scutelleridae) is a homo-γ-bisabolene (4Z,4′E)-4-(1′,5′-dimethyl-4′-heptenylidene)-1-methylcyclohexene.

Faranal, (3S,4R,6E,10Z)-3,4,7,11-tetramethyl-6,10-tridecadienal (Figure 7.35), is the trail pheromone of Pharaoh's ant *Monomorium pharaonis*, a tropical species that has become a pest inside warm buildings in temperate climates. Faranal contains two homomevalonate units, a type B and a type C, separated by a normal mevalonate.

The sandfly *Lutzomyia longipalpis* (Diptera: Psychodidae), which carries the protozoal disease leishmaniasis in some parts of Brazil, is apparently a complex of species. They have a male-produced sex pheromone. Some members of this complex use a cembrene (Figure 8.3) of unknown stereochemistry. One race or species uses (1S,3S,7R)-3-methyl-α-himachalene, while another uses (S)-9-methylgermacrene-B (Figure 7.35) as its male-produced sex pheromone. Based on the known biosynthesis of himachalene and germacrene in plants, both syntheses require homofarnesyl diphosphate with a type C unit in the middle of the chain.

7.6.1 Juvenile Hormone

An important hormone in insects is the so-called juvenile hormone (JH), produced in the *corpora allata*, a part of the brain, and circulated through the haemolymph. It originally acquired that name because it is required at each moult between immature stages (see Figure 1.4). Juvenile hormone titre falls as the immature stages progress. A reduced JH level at moulting in holometabolous insects leads to pupal formation. When JH is absent at the time of moulting, the insect moults to the adult form. But the hormone is also produced in adult insects and has other functions there. The name therefore is misleading, but has been accepted. Metamorphosis is therefore not so much controlled by a hormone as by the absence of a hormone.

Juvenile hormone also regulates embryogenesis, reproduction and pheromone synthesis, as well as metamorphosis. It was suggested in 1965 that control of pheromone biosynthesis in long-lived insects with multiple reproductive cycles would be by hormones. Beetles, cockroaches and flies fit this description. It has since been found that juvenile hormone controls sex pheromone production in beetles and cockroaches while 20-hydroxyecdysone (moulting hormone, Section 8.5) controls it in flies.

The juvenile hormone first discovered, known as JH I, is a bishomoterpene (Figure 7.36). Later, all the forms from JH I to JH III, corresponding to the farnesene homologues found in the Dufour glands of myrmicine ants, have been found, but only JH III is found in all insect orders. The hormones JH 0, JH I and JH II are found only in Lepidoptera. Notice that the C form of 3-methyl-3-pentenyl diphosphate is added in 4-methyl-JH I (Figure 7.36). The compound JHB$_3$, a bis-epoxide form of JH III, has been found in all the higher flies (Diptera) so far studied, and has been shown to be the only JH produced by larvae and adults in the Australian sheep blowfly *Lucilia cuprina* (Diptera:

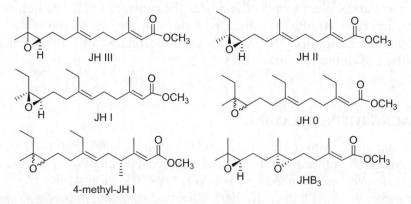

Figure 7.36 The juvenile hormones of insects. Most insect orders use only JH III.

Calliphoridae). Biosynthetic studies have firmly shown that JH III is produced *via* mevalonate, and then through the steps already discussed to farnesyl diphosphate. The catalytic action of diphosphatase releases free farnesol, which is oxidized by a dehydrogenase and NAD^+ to farnesal and then to farnesoic acid. The acid is esterified using S-adenosyl methionine catalysed by a methyl transferase. Finally methyl farnesoate is epoxidized by a cytochrome P450 to juvenile hormone III. These stages have been thoroughly investigated, for example in the yellow fever mosquito *Aedes aegypti* (Diptera: Culicidae) (Borovsky *et al.*, 1992). A specific JH epoxidase has been cloned and characterized from corpora allata of the cockroach *Diploptera punctata* (Blattodea: Blaberidae), which epoxidizes methyl farnesoate with high regio- and stereospecificity (Helvig *et al.*, 2004). Most of the enzymes of the mevalonate pathway in insects seem to act in the cytoplasm.

Methyl farnesoate acts as a juvenile hormone in some Crustacea, and a mixture of methyl farnesoate and JH III is present in the embryos of primitive insects such as cockroaches. There appears to be a kind of evolution of juvenile hormone compounds through appearance of successive epoxidases of the cytochrome P450 family with JHB_3 representing the most advanced form. Some kind of juvenile hormone is present in ticks, spiders and other arthropods.

Juvenile hormone is inactivated either by hydrolysis of the methyl ester by juvenile hormone esterase, followed by opening of the epoxide ring by juvenile hormone epoxide hydrolase to a diol, or sometimes the steps occur in the opposite order. In the yellow fever mosquito *Aedes aegypti* JH III is first hydrolysed to JH acid, then hydrated to the diol acid by JH III epoxide hydrase. The epoxide hydrolase works 17 times faster on the JH III acid than on the JH III itself. Vitellogenesis and male maturation are under JH control. Pheromone release in Lepidoptera and Coleoptera is often under the control of JH. The hormone titre increases steadily in worker bees through the first 15 days of adult life. It does not affect their ovaries but is correlated with the changing pattern of duties of worker bees.

BACKGROUND READING

Bohlmann, J., Meyer-Gauen, G. and Croteau, R. 1998. Plant terpenoid synthases: molecular biology and phylogenetic analysis. *Proceedings of the National Academy of Sciences, USA*, **95**, 4126–4133.

Francke, W. and Dettner, K. 2005. Chemical signaling in beetles. *Topics in Current Chemistry*, **240**, 85–166.

For detailed reaction schemes for terpenes: www.chem.qmul.ac.uk/ lubmb/enzyme/reaction/terp/

REFERENCES

Baker, R., Briner, P. H. and Evans, D. A. 1978. Chemical defence in the termite *Ancistrotermes cavithorax*: ancistrodial and ancistrofuran. *Chemical Communications*, **1978**, 410–411.

Barkawi, L. S., Francke, W., Blomquist, G. J. and Seybold, S. J. 2003. Frontalin: *de novo* biosynthesis of an aggregation pheromone component by *Dendroctonus* spp. bark beetles (Coleoptera: Scolytidae). *Insect Biochemistry and Molecular Biology*, **33**, 773–788.

Bearfield, J. C., Box, C. D., Keeling, C. I., Young, S., Blomquist, G. J. and Tittiger, C. 2008. Isolation, endocrine regulation and transcript distribution of a putative primary JH-responsive gene from the pine engraver, *Ips pini* (Coleoptera: Scolytidae). *Insect Biochemistry and Molecular Biology*, **38**, 256–267.

Blatrix, R., Schulz, C., Jaisson, P., Francke, W. and Hefetz, A. 2002. Trail pheromone of the ponerine ant *Gnamptogenys striatula*: 4-methyl-geranyl esters from Dufour's gland. *Journal of Chemical Ecology*, **28**, 2557–2567.

Borovsky, D., Carlson, D. A. and Ujvary, I. 1992. *In vivo* and *in vitro* biosynthesis and metabolism of methyl farnesoate, juvenile hormone-III and juvenile hormone-III acid in the mosquito *Aedes aegypti*. *Journal of Medical Entomology*, **29**, 619–629.

Brand, J. M., Young, J. C. and Silverstein, R. M. 1979. Insect pheromones: a critical review of recent advances in their chemistry, biology, and application. *Fortschritte der Chemie Organische Naturstoffe – Progress in the Chemistry of Organic Natural Products*, **37**, 1–367.

Burse, A., Frick, S., Schmidt, A., Büchler, R., Kunert, M., Gershenzon, J., Brandt, W. G., and Boland, W. 2008. Implication of HMGR in homeostasis of sequestered and *de novo* produced precursors of the iridoid biosynthesis in leaf beetle larvae. *Insect Biochemistry and Molecular Biology*, **38**, 76–88.

Croteau, R. B., Wheeler, C. J., Cane, D. E., Ebert, R. and Ha, H. J. 1987. Isotopically sensitive branching in the formation of cyclic monoterpenes: proof that (−)-α-pinene and (−)-β-pinene are synthesised by the same monoterpene cyclase via deprotonation of a common intermediate. *Biochemistry*, **26**, 5383–5389.

Dettner, K., Schramm, S., Seidl, V., Klemm, K., Gäde, G., Fietz, O. and Boland, W. 2003. Occurrence of terpene anhydride palasonin and palasoninimide in blister beetle *Hycleus lunata* (Coleoptera: Meloidae). *Biochemical Systematics and Ecology*, **31**, 203–205.

Einhorn, J., Guerrero, A., Ducrot, P. H., Boyer, F. D., Gieselmann, M. and Roelofs, W. 1998. Sex pheromone of the oleander scale *Aspidiotus nerii*, structural characterization and absolute configuration of an unusual functionalized cyclobutane. *Proceedings of the National Academy of Sciences, USA*, **95**, 9867–9872.

Francke, W. and Dettner, K. 2005. Chemical signalling in beetles. *Topics in Current Chemistry*, **240**, 85–166.

Francke, W. and Schulz, S. 1999. Pheromones. In: *Comprehensive Natural Products Chemistry*, vol. 8 (Mori, K., ed.), Elsevier, Oxford, pp. 197–261.

Gilg, A. B., Tittiger, C. and Blomquist, G. J. 2009. Unique animal prenyltransferase with monoterpene synthase activity. *Naturwissenschaften*, **96**, 731–735.

Helvig, C., Koerner, J. F., Unnithan, G. C. and Feyereisen, R. 2004. CYP15A1, the cytochrome P450 that catalyses epoxidation of methyl farnesoate to juvenile hormone in cockroach corpora allata. *Proceedings of the National Academy of Sciences, USA*, **101**, 4024–4029.

Hick, A. J., Luszniak, M. C. and Pickett, J. A. 1999. Volatile isoprenoids that control insect behaviour and development. *Natural Products Reports*, **16**, 39–54.

Honda, K. 1990. GC-MS and ^{13}C-NMR studies on the biosynthesis of terpenoid defensive secretions by the larvae of papilionid butterflies (*Luehdorfia* and *Papilio*). *Insect Biochemistry*, **20**, 245–250.

Kunert, M., Søe, A., Bartram, S., Discher, S., Tolzin-Banasch, K., Nie, L., David, A., Pasteels, J. and Boland, W. 2008. De novo biosynthesis *versus* sequestration: a network of transport systems supports in iridoid-producing leaf beetle lavae both modes of defense. *Insect Biochemistry and Molecular Biology*, **38**, 895–904.

Lanne, B. S., Ivarsson, P., Johnsson, P., Bergström, G. and Wassgren, A.-B. 1989. Biosynthesis of 2-methyl-3-buten-2-ol, a pheromone component of *Ips typographus* (Coleoptera, Scolytidae). *Insect Biochemistry*, **19**, 163–167.

McCormick, J. P. and Carrell, J. E. 1987. Cantharidin biosynthesis and function in meloid beetles. In: *Pheromone Biosynthesis* (Prestwich, G. D. and Blomquist, G. J., ed.), Academic Press, New York and London, pp. 307–350.

Millar, J. G. 2005. Pheromones of true bugs. *Topics in Current Chemistry*, **240**, 37–84.

Oldham, N. J., Veith, M. and Boland, W. 1996. Iridoid monoterpene biosynthesis in insects: evidence for a *de novo* pathway occurring in the defensive glands of *Phaedon armoraciae* (Chrysomelidae) leaf beetle larvae. *Naturwissenschaften*, **83**, 470–473.

Omura, H., Honda, K. and Feeny, P. 2006. From terpenoids to aliphatic acids: further evidence for late-instar switch in osmeterial defense as a characteristic trait of swallowtail butterflies in the tribe Papilionini. *Journal of Chemical Ecology*, **32**, 1999–2012.

Perez, A. L., Gries, R., Gries, G. and Oelschlager, A. C. 1996. Transformation of presumptive precursors to frontalin and *exo*-breviconin by bark beetles and the West Indian sugarcane weevil (Coleoptera). *Bioorganic and Medicinal Chemistry*, **4**, 445–450.

Pickett, J. A., Williams, I. H., Martin, A. P. and Smith, M. C. 1980. Navonov pheromone of the honey bee, *Apis mellifera* L. (Hymenoptera: Apidae). Part I. Chemical characterization. *Journal of Chemical Ecology*, **6**, 425–434.

Quintana, A., Reinhard, J., Faure, R., Uva, P., Bagnères, A.-G., Massiot, G. and Clément, J.-L. 2003. Interspecific variation in terpenoid composition of defensive secretions of European *Reticulitermes* termites. *Journal of Chemical Ecology*, **29**, 639–652.

Raspotnig, G., Kaiser, R., Stabentheiner, E. and Leis, H. J. 2008. Chrysomelidial in the opisthonotal glands of the oribatid mite *Oribotritia berlesei*. *Journal of Chemical Ecology*, **34**, 1081–1088.

Rohmer, M., Seeman, M. and Horbach, S. 1996. Glyceraldehyde 3-phosphate and pyruvate as precursors of isoprenic units in an alternative non-mevalonate pathway for terpenoid biosynthesis. *Journal of the American Chemical Society*, **118**, 2564–2566.

Sandstrom, P., Ginzel, M. D., Bearfield, J. C., Welch, W. H., Blomquist, G. J. and Tittiger, C. 2008. Myrcene hydroxylases do not determine enantiomeric composition of pheromonal ipsdienol in *Ips* spp. *Journal of Chemical Ecology*, **34**, 1584–1592.

Schooley, D. A., Judy, K. J., Bergot, B. J., Hall, M. S. and Siddall, J. B. 1973. Biosynthesis of juvenile hormone of *Manduca sexta* – labelling pattern from mevalonate, propionate and acetate. *Proceedings of the National Academy of Sciences, USA*, **70**, 2921–2925.

Seybold, S. J. and Tittiger, C. 2003. Biochemistry and molecular biology of *de novo* isoprenoid pheromone production in the Scolytidae. *Annual Review of Entomology*, **48**, 425–453.

Seybold, S. J. and Vanderwel, D. 2003. Biosynthesis and endocrine regulation of pheromone production in the Coleoptera. In: *Insect Pheromone Biochemistry and Molecular Biology* (Blomquist, G. J. and Vogt, R. G., ed.), Elsevier, Oxford, pp. 137–200.

Seybold, S. J., Quilici, D. R., Tillman, J. A., Vanderwel, D., Wood, D. L. and Blomquist, G. J. 1995. *De novo* biosynthesis of the aggregation pheromone components ipsenol and ipsdienol by the pine bark beetles *Ips paraconfusus* Lanier and *Ips pini* (Say) (Coleoptera: Scolytidae). *Proceedings of the National Academy of Sciences, USA*, **92**, 8393–8397.

Sillam-Dussès, D., Sémon, E., Lacey, M. J., Robert, A., Lenz, M. and Bordereau, C. 2007. Trail-following pheromones in basal termites, with special reference to *Mastotermes darwiniensis*. *Journal of Chemical Ecology*, **33**, 1960–1977.

Sullivan, B. T., Shepherd, W. P., Pureswaran, D. S., Tashiro, T. and Mori, K. 2007. Evidence that (+)-endo-brevicomin is a male-produced component of the southern pine beetle. *Journal of Chemical Ecology*, **33**, 1510–1527.

Taban, A. H., Fu, J., Blake, J., Awano, A., Tittiger, C. and Blomquist, G. J. 2006. Site of pheromone biosynthesis and isolation of HMG-CoA reductase cDNA in the cotton boll weevil, *Anthonomus grandis*. *Archives of Insect Biochemistry and Physiology*, **62**, 153–163.

Taban, A. H., Tittiger, C., Blomquist, G. J. and Welch, W. H. 2009. Isolation and characterization of farnesyl diphosphate synthase from the cotton boll weevil *Anthonomus grandis*. *Archives of Insect Biochemistry and Physiology*, **71**, 88–104.

Tada, M., Chiba, K. and Hashizume, T. 1982. Formation of dendrolasin, sesquirosefuran, perillen and rosefuran by biomimetic autooxidation. *Agricultural and Biological Chemistry*, **46**, 819–820.

Tillman, J. A., Lu, F., Goddard, L. M., Donaldson, Z. R., Dwinell, S. C., Tittiger, C., Hall, G. M., Storer, A. J., Blomquist, G. J. and Seybold, S. J. 2004. Juvenile hormone regulates *de novo* aggregation pheromone biosynthesis in pine bark beetles, *Ips* spp., through transcriptional control of HMG-CoA reductase. *Journal of Chemical Ecology*, **30**, 2459–2494.

Tschuch, G., Lindemann, P. and Moritz, G. 2008. An unexpected mixture of substances in the defensive secretion of the tubuliferan thrips, *Callococcithrips fuscipennis* (Moulton). *Journal of Chemical Ecology*, **34**, 742–747.

Vanderwel, D. and Oehlschlager, A. C. 1992. Mechanism of brevicomin biosynthesis from (Z)-6-nonen-2-one in a bark beetle. *Journal of the American Chemical Society*, **114**, 5081–5086.

Vanderwel, D., Gries, G., Singh, S. M., Borden, J. H. and Oehlschlager, A. C. 1992a. (*E*)-6-Nonen-2-one and (*Z*)-6-nonen-2-one – biosynthetic precursors of *endo*-brevicomin and *exo*-brevicomin in two bark beetles (Coleoptera, Scolytidae). *Journal of Chemical Ecology*, **18**, 1389–1404.

Vanderwel, D., Johnston, B. and Oehlschlager, A. C. 1992b. Cucujolide biosynthesis in the merchant and rusty grain beetles. *Insect Biochemistry and Molecular Biology*, **22**, 875–883.

Veith, M., Lorenz, M., Boland, W., Simon, H. and Dettner, K. 1994. Biosynthesis of iridoid monoterpenes in insects – defensive secretions from larvae of leaf beetles (Coleoptera, Chrysomelidae). *Tetrahedron*, **50**, 6859–6874.

Weibel, D. B., Oldham, N. J., Feld, B., Glombitza, G., Dettner, K. and Boland, W. 2001. Iridoid biosynthesis in staphylinid rove beetles (Coleoptera: Staphylinidae, Philonhinae). *Insect Biochemistry and Molecular Biology*, **31**, 583–591.

CHAPTER 8
Higher Terpenes and Steroids

8.1 INTRODUCTION

Addition of another unit of isopentenyl diphosphate (IPP) to farnesyl diphosphate in the same way as described for geranyl diphosphate in Section 7.5 gives geranylgeranyl diphosphate (Figure 8.1), the parent of the diterpene compounds. Diterpenes may be acyclic, macrocyclic, or polycyclic and are less volatile than mono- and sesquiterpenes, because of their greater molecular mass. Cyclized ones are generally solids. Insects are adept at synthesizing diterpenes and using them for communication and defence, but the study of biosynthesis of insect diterpenes has hardly begun.

8.2 DITERPENES

A number of simple uncyclized diterpenes are frequently found in insects. Springene, first discovered in facial glands of the South African springbok, has been found in Dufour glands of ants, stingless bees and a parasitic braconid wasp. Simple alcohols such as geranylgeraniol, geranyllinalool and geranylcitronellol and their esters are particularly found in Hymenoptera, but also in termites. Geranylgeraniol has been found in the labial glands of male bumblebees, in the Dufour glands of the stingless bee *Nannotrigona testaceicornis* (Hymenoptera: Apidae), the ant *Ectatomma ruidum* (Hymenoptera: Formcidae) and in the female sex pheromones of click beetles (*Agriotes*, *Sinapus* and *Melanotus*

Biosynthesis in Insects, Advanced Edition
By E. David Morgan
© E. David Morgan 2010
Published by the Royal Society of Chemistry, www.rsc.org

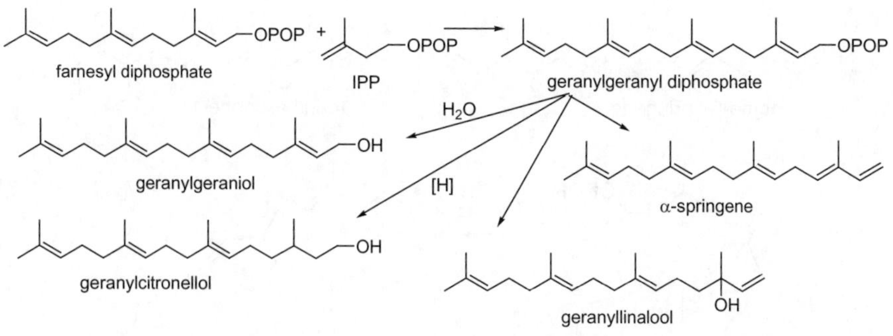

Higher Terpenes and Steroids

Figure 8.1 The formation of geranylgeranyl diphosphate, the parent of the diterpenes and some simple uncyclized diterpenes derived from it that are found in insects.

species, Coleoptera: Elateridae). Male bumblebees may also contain geranylgeraniol, geranylcitronellol, their acetates, and the corresponding aldehydes geranylgeranial and geranylcitronellal. Geranylcitronellol is also the chief component of the defensive secretion of *Reticulitermes* termites (Isoptera: Rhinotermitidae). The replacement of mevalonate by homomevalonate is also observed in some diterpene cases. The first acyclic homoditerpenes, a homospringene and a homogeranyllinalool, have recently been isolated from the parasitic wasp *Habrobracon hebetor* (Hymenoptera: Braconidae) (Howard et al., 2003). Cantharenone, a monocyclic diterpene ketone, has been found in the soldier beetle *Cantharis livida* (Coleoptera: Cantharidae) (Figure 8.2).

Of cyclized diterpenes, the large-ringed cembrene structure (Figure 8.2) seems much used by insects. Direct ring closure from all-*trans*-geranylgeranyl diphosphate gives cembrene A (or neocembrene), which is the trail pheromone of some termites and it, or an isomer of it, is also the sex pheromone of some sandflies, and is the queen pheromone of the ant *Monomorium pharaonis*. Related structures crematofuran and isocrematofuran have been found in the defensive or offensive secretion from the Dufour glands of the Brazilian ant *Crematogaster brevispinosa* (Figure 8.2). These are quite different from the defensive compounds from some European *Crematogaster* (see Figure 5.33), and the first cembrane defensive compounds from ants (Leclercq et al., 2000).

8.2.1 Termites

Termites (Isoptera) appear to use terpenes frequently for sex and trail pheromones and for defensive secretions, and they construct cyclic diterpenes that are not known elsewhere in nature. Females of

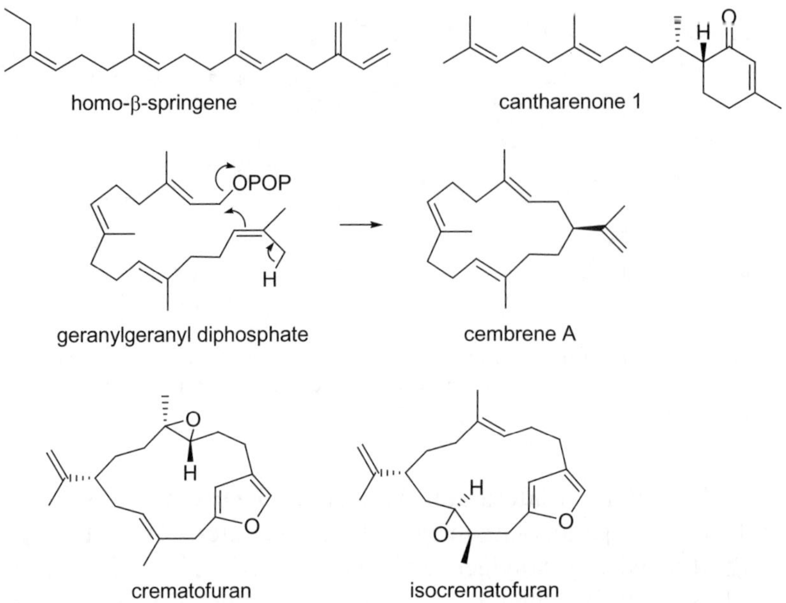

Figure 8.2 Homo-β-springene, an example of a homo-diterpene; cantharenone is an example of a cyclized diterpene from the beetle *Cantharis lividai*; the macrocyclic diterpene cembrene A, and two cembrene derivatives from the defensive secretion of *Crematogaster* ants.

Nasutitermes corniger (given incorrectly as *N. ephratae*) (Isoptera: Termitidae) and a *Nasutitermes* species from Brazil produce a trinervitane-type sexual substance (Budesinsky *et al.*, 2005).

Termites have developed many means of defence against predators, both physical and chemical. The Nasutitermitinae are the largest, most advanced and most widely distributed subfamily of higher termites. The sterile soldier caste has large frontal glands, and heads modified into a turret shape for discharging the frontal gland secretion of terpenes. In addition to the monoterpenes, which act as a "solvent" for the mixture and some sesquiterpenes, there is a mixture of acyclic and cyclic diterpenes in the secretion. The mixture partially evaporates and oxidizes in the air to leave a viscous semi-solid, which traps ants or other invaders. Some 60 different diterpene compounds have been isolated and identified from soldier frontal glands (Budesinsky *et al.*, 2005). They are either acyclic, like geranyllinalool, monocyclic cembranes, bicyclic secotrinervitanes, tricyclic trinervitanes and tetracyclic kempanes, ripperanes or longipanes. Their biosynthetic relationships through cembranes can be recognized in Figure 8.3, though this has not received experimental support. The trinervitanes are most common. Still other

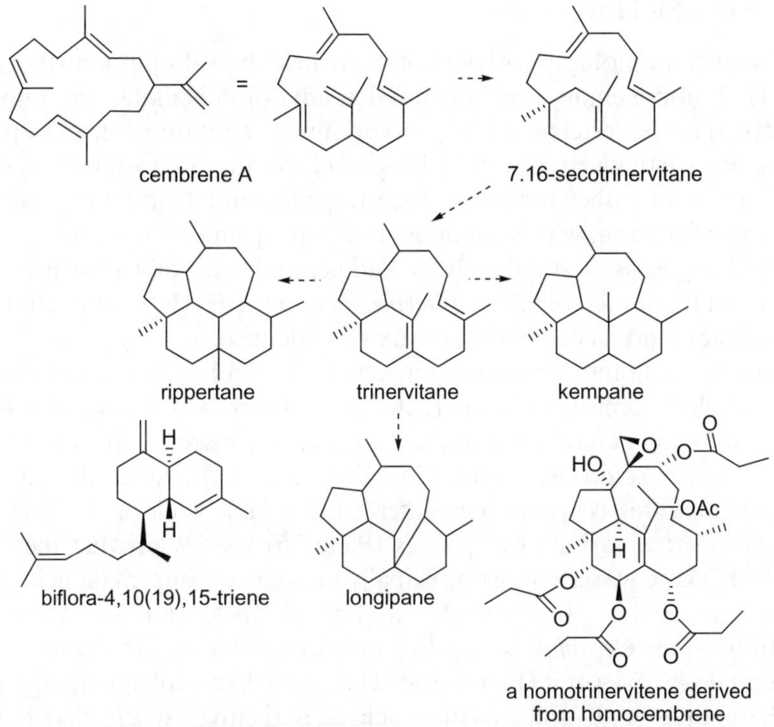

Figure 8.3 Some types of diterpenes from termite defensive secretions and an example of a termite compound derived from homocembrene.

isomeric forms of cembrene are found in *Cubitermes* species (see Figure 8.3) and (−)-biflora-4,10(19),15-triene is a bicyclic diterpene from soldiers of *Cubitermes umbratus* (Isoptera: Termitidae) and other East African species. The tetrapropionate derivative of a trinervitane with an extra methyl group was reported, without comment, from *Hospitalitermes umbrinus* (Isoptera: Termitidae) (Goh et al., 1988). This was the first example of a termite homoditerpene, made with a homomevalonate starter unit.

It is clear these are insect-produced *via* the isoprenoid route. When the soldiers of *Nasutitermes octopilis* were injected with ^{14}C-labelled acetic acid or mevalonolactone, the ^{14}C was incorporated into tri- and tetracyclic diterpenes (Prestwich et al., 1981). A geranylgeranyl diphosphate synthase gene from *Nasutitermes takasagoensis* has been cloned and is highly expressed exclusively in soldier frontal glands (Hojo et al., 2007). Localization experiments on soldier heads showed that the gene was uniformly expressed in epithelial glandular cells.

8.3 SESTERTERPENES

Still higher homologues of terpenes, formed by adding a further isopentenyl diphosphate unit to geranylgeranyl diphosphate, are found in insects. The sesterterpenes, C_{25} compounds, containing five isoprene units, are particularly found in fungi and marine organisms, but some are known in insects. Acyclic sesterterpenes and C_{30} compounds, of unknown function, with structures similar to springene have been found in Dufour glands of stingless bees, and geranylfarnesol was identified in the frontal glands of *Reticulitermes santonensis* (Isoptera: Reticulitermitidae), and as esters in the wax of scale insects.

A group of homopterous scale insects (Coccidae) enclose themselves in wax, which in some cases is an article of commerce. Many scale insects are also important pests. In China, the commercial insect is the China wax scale *Ericerus pela* (Hemiptera: Coccidae), and in India the Indian wax scale *Ceroplastes ceriferus* (Hemiptera: Coccidae), while in Central and South America various *Ceroplastes* (Plate 25) species provide the same product. These waxes consist principally of esters of long-chain acids with long-chain alcohols, but they also contain triterpenes and sesterterpenes. Hydrolysis of *Ceroplastes albolineatus* wax gives geranylfarnesol and cyclized derivatives of it (Figure 8.4). The geranylfarnesol apparently has a *cis*-2,3-double bond, since all the cyclized derivatives where that double bond is still intact have a *cis*-geometry at this point. Many of the products have a C_8 uncyclized tail and a terminal methyl group oxidized to an alcohol or carboxylic acid (Figure 8.5). Though biosynthetic proposals are frequent, no experimental studies have been reported. The mechanism shown is a probable biosynthetic route.

Still longer chains of isoprene units are known. The dolichols, which can be described as farnesyl diphosphate extended by a number of *cis*-isoprene units, and completed by a reduced isoprenol unit (Figure 8.6), are found in all eucaryotic organisms, including insects. Their physiological function is not yet clear although dolichyl phosphate is involved in moving polysaccharides in glycoprotein synthesis (Kornfeld and Kornfeld, 1985). Li *et al.* (1995), bearing in mind the use of homomevalonate by insects, examined the dolichols of the tobacco hornworm *Manduca sexta* (Lepidoptera: Sphingidae) but found no evidence of incorporation of homomevalonate in them. The dolichols of *M. sexta* contained 17, 18, 19 and 20 isoprene units and were quite different from those of its diet.

8.4 TRITERPENES AND STEROIDS

The linking together of two farnesyl diphosphate molecules, head-to-head, gives the C_{30} terpene hydrocarbon squalene, a compound first

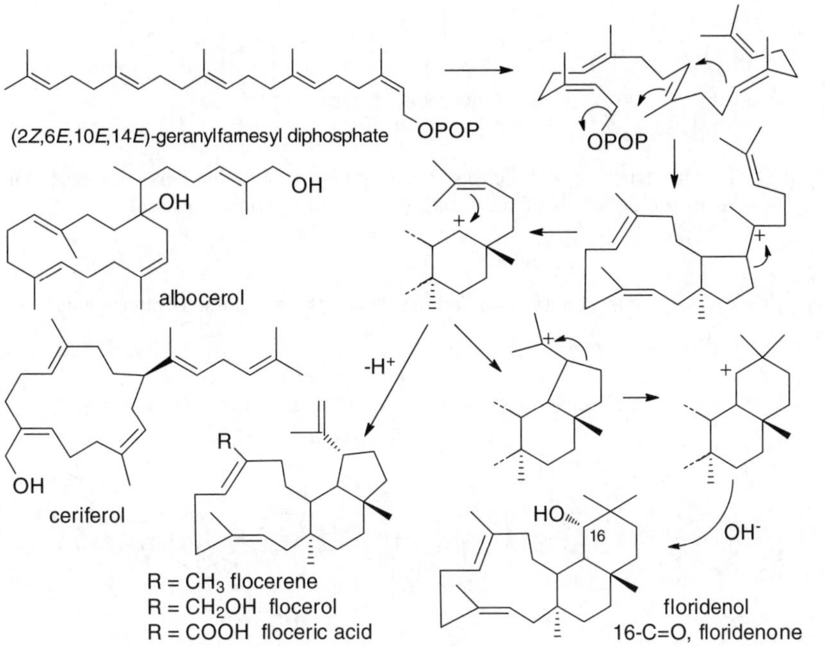

Higher Terpenes and Steroids

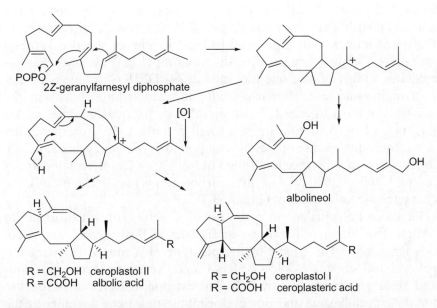

Figure 8.4 Geranylfarnesyl diphosphate with one *cis* double bond, the apparent parent of many sesterterpenes from wax scale insects and some examples of wax scale compounds. Formation of flocerene and floridenol require skeletal rearrangements as indicated.

Figure 8.5 Probable biosynthetic route to some wax scale cyclic sesterterpenes.

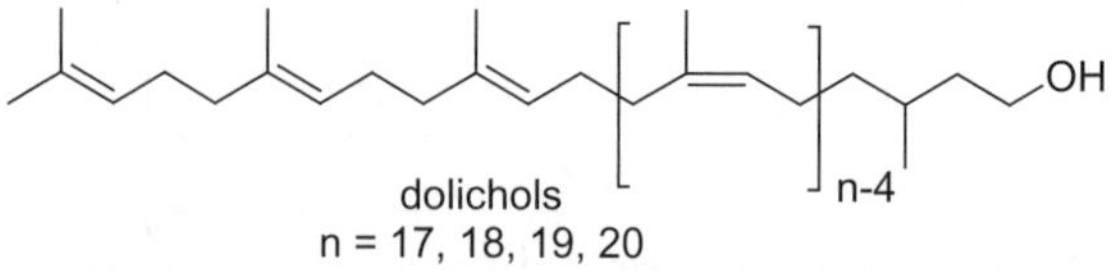

Figure 8.6 The structure of dolichols. The numbers are the numbers of isoprenyl units in the dolichols of the tobacco hornworm *Manduca sexta*.

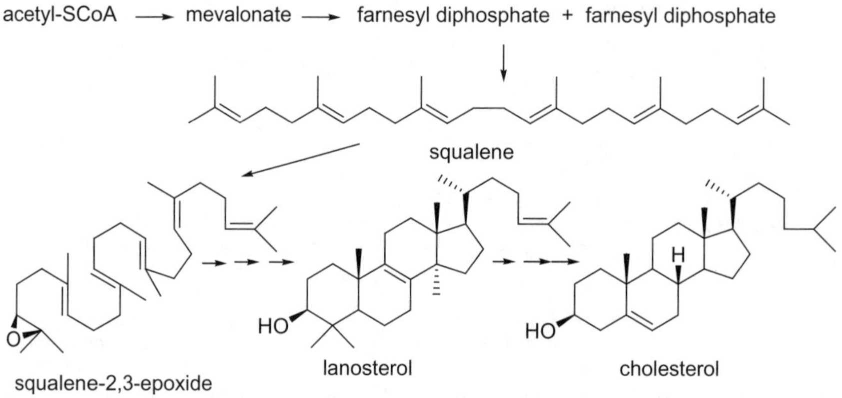

Figure 8.7 A brief summary of the stages of formation of squalene and cholesterol in animals.

found in the liver oil of sharks (*Squalus* species), but is now known to be a widely distributed compound (Figure 8.7). It is found, for example, in the oil of human skin. The head-to-head joining of the two farnesyl groups is a complex reaction with cyclopropyl and cyclobutyl intermediates, and requiring one molecule of NADPH. Squalene is an important intermediate from which all the plant triterpenoids and the steroids are biosynthesized. From mevalonate to squalene there are no less than 14 stereospecific steps, all well described. The triterpenes derived from squalene are polycyclic compounds. Animals cyclize squalene to lanosterol, a substance first found in wool wax. Lanosterol undergoes several reactions, with loss of three carbon atoms to give cholesterol, the characteristic animal sterol (Figure 8.7).

Plants cyclize squalene to cycloartenol, which is converted further in several steps to one of the phytosterols, such as β-sitosterol, campestrol or stigmasterol. The typical higher plant sterol is sitosterol, with campesterol and stigmasterol less widely found. Yeasts produce ergosterol, and algae, fucosterol (Figure 8.8). The essential features of sterols are the three cyclohexane and one cyclopentane ring fused together in the

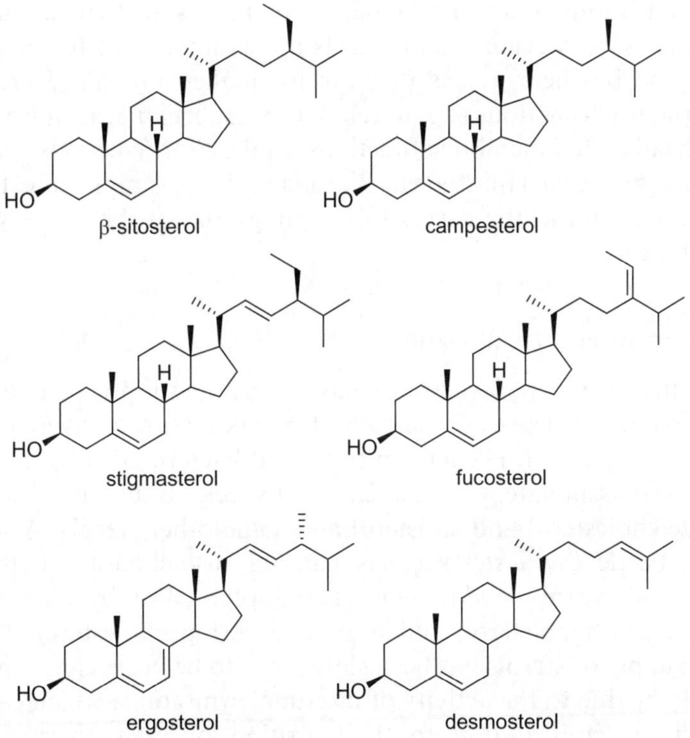

Figure 8.8 Examples of plant, fungal and yeast sterols that are metabolized by insects. Desmosterol is a minor sterol in nature, but an intermediate in the conversion of phytosterols to cholesterol by insects.

way shown. The plant and fungal sterols are distinguished by an extra group at C-24, which is methyl, ethyl or ethynyl, derived from S-adenosyl methionine. Sterols always have an oxygen function at C-3, the position of the hydroxyl in cholesterol.

8.4.1 Sterols in Insects

While it had been known since the 1930s that cholesterol or plant sterols were necessary in nutritional studies of insects, Clark and Bloch (1959) showed in an important paper that the hide beetle *Dermestes vulpinus* (Coleoptera: Dermestidae) is unable to make squalene or sterols. Later the same was demonstrated for other insects, including the blowfly *Calliphora erythrocephala* (*vicina*) (Diptera: Calliphoridae) and the housefly *Musca domestica* (Diptera: Muscidae). Nor can insects convert supplied squalene into sterols. Sterol biosynthesis is almost universal among eukaryotic cells and organisms except for insects. Not enough is known yet about other arthropods to make a statement about them. Nevertheless, it is evident that sterols are as important for insect physiology and biochemistry as they are for higher animals. Sterols have both structural functions (*e.g.* in cellular and subcellular membranes and nerve sheaths) and metabolic functions. In the Florida woods cockroach *Eurycotis floridana* (Blattodea: Blattidae), for example, most of the sterols are found in the nerves ($3.25\,\mu g/mg$) and in the salivary glands ($2.98\,\mu g/mg$).

8.4.2 Phytosterol Dealkylation

Insects that feed on vertebrates have a ready supply of cholesterol. Those feeding on leaves or phloem of plants tend to convert the plant sterols to cholesterol. Feeders on fungi and bacteria also have a variety of sterols to assimilate. Of some 22 insect species tested, most were able to utilize cholesterol and sitosterol and some other sterols. A few, like the hide beetle *Dermestes vulpinus*, can only use cholesterol efficiently, while the silkworm *Bombyx mori* (Lepidoptera: Bombycidae) requires sitosterol or stigmasterol, and is poor at using cholesterol. Where a dietary supply of sterols has been shown not to be necessary, it has been shown to be due to the activity of intestinal symbionts, which can make a significant contribution to the nutrition of many insects. When the beetles *Stegobium paniceum* and *Lasioderma serricorne* (both Coleoptera: Anobiidae) were raised in the absence of yeasts, which they normally have in their gut, the survival of the beetles was much more dependent upon sterols and B vitamins. The dependence could be reversed by re-infecting the beetles with the necessary yeasts (Nasir and Noda, 2003).

Many phytophagous insects are able to dealkylate the C-24 alkyl groups of phytosterols, but some Hymenoptera, Hemiptera, Heteroptera, Coleoptera and Diptera cannot. Included among them are the honeybee *Apis mellifera*, the milkweed bug *Oncopeltus fasciatus* (Hemiptera: Lygaeidae), the cotton stainer *Dysdercus fasciatus* (Heteroptera: Pyrrhocoridae) and other bees and some leaf-cutting ants. The sawflies (Symphyta, a primitive sub-order of Hymenoptera) can dealkylate phytosterols (Schiff and Feldlaufer, 1996). Behmer and Nes (2003) give a table of 60 species of insect listing the composition of their dietary sterols (sometime more than one diet) and their tissue sterols.

There is considerable diversity of steroid metabolism between insect species, which makes it difficult to generalize about the subject. Nevertheless, many insects have developed a very clever dealkylation mechanism that can accommodate ethyl or methyl groups, alkyl or alkenyl,

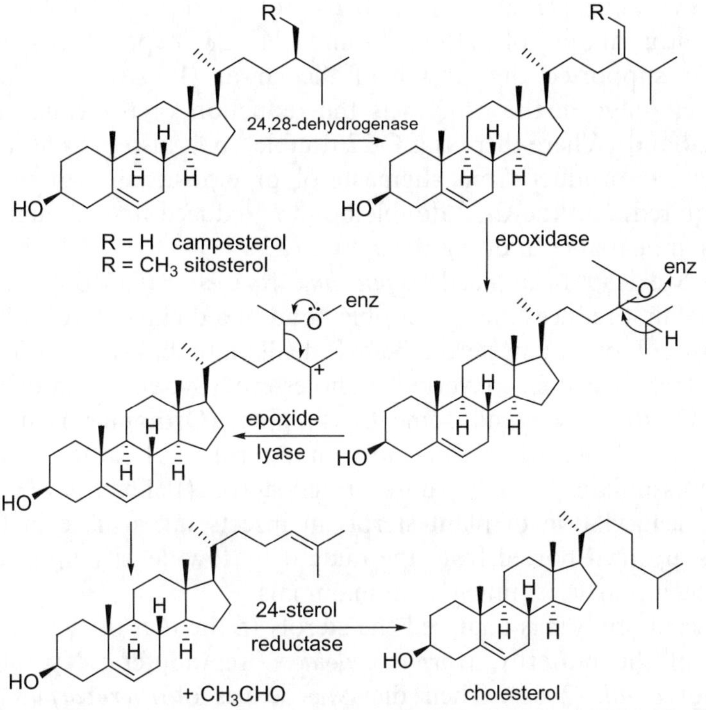

Figure 8.9 Summary of the steps in the de-alkylation of sterols by insects. The mechanism of the epoxide cleavage is not fully known.

both R and S configurations at C-24, and whether or not there is a double bond at C-22. Ikekawa et al. (1993) and Lafont et al. (2005) have summarized the many experiments using radio-labelling and inhibitors of sterol metabolism that have elucidated the process. The C-24 alkyl or alkylidene groups are removed in stages, requiring desaturation, followed by epoxidation, and cleavage of the epoxide (Figure 8.9). A saturated alkyl group, such as those of sitosterol, campesterol or stigmasterol, is first dehydrogenated to an alkylidene group, such as that of fucosterol. The alkylidene group is then epoxidized, and the epoxide cleaved. The mechanism of the cleavage is not fully known, though there has been considerable investigation of it. The discovery that boron trifluoride etherate cleaves the epoxide has given some clue, and the biochemical mechanism is probably as shown in Figure 8.9. The deuterium atom of [25-^{2}H]campesterol migrates to C-24 when dealkylated by the silkworm *Bombyx mori*. Similarly deuterium migration in deuterated 24-methylenecholesterol was found in dealkylation by the yellow mealworm *Tenebrio molitor* (Coleoptera: Tenebrionidae). Both

(24*S*,28*S*)- and (24*R*,28*R*)-epoxyfucosterol are dealkylated by *Bombyx mori*, but a mixture of (24*S*,28*R*)- and (24*R*,28*S*)-epoxyfucosterol only partially supported the growth of the larvae (Ikekawa *et al.*, 1993). The final stage in the process is the reduction of the resulting C-24 double bond. Where there is a C-22 double bond also present, as in the dealkylated product from stigmasterol or ergosterol, two reductases are required, and the C-22 double bond is reduced first, though not all insects are capable of doing this.

The Mexican bean beetle *Epilachna varivestis* (Coleoptera: Coccinellidae) first reduces the C-5 double bond of a dietary sterol, whether it is sitosterol or stigmasterol, before dealkylation, giving cholestanol, some of which is then converted to cholesterol (Svoboda and Feldlaufer, 1991). On the other hand, some grasshoppers (Orthoptera) can convert 24-alkylsterols like sitosterol and campesterol to cholesterol, but they cannot assimilate Δ^7 or Δ^{22}-unsaturated sterols (Behmer *et al.*, 1999).

The dealkylation of plant sterols in insects takes place in the gut. Sterols are transported from the midgut by low-density lipoprotein in lipoprotein particles, much as in mammals.

A recent study has analysed the sterols in the diet of the larvae and adults of the butterfly *Morpho peleides* (Lepidoptera: Nymphalidae) (Connor *et al.*, 2006). Their diet was leaves of *Pterocarpus bayessii* containing chiefly sitosterol, but with some stigmasterol and a little cholesterol. The tissues of the larvae contained mostly cholesterol with less sitosterol and still less stigmasterol. In the adults the proportion of cholesterol was still greater and sitosterol was a minor component.

It is a little surprising to learn that nine species of astigmatid mites (Acari: Acaridae) contain crinosterol (Figure 8.10), a 24-alkylsterol, first identified from marine animals, plants and micro-organisms (Murakami *et al.*, 2007). When one of these mites *Rhizoglyphus robini* (Figure 8.11) was fed on an artificial diet containing [^{2}H$_3$]methionine, [^{2}H$_2$]crinosterol was detected, indicating that a [^{2}H$_3$]methyl group was incorporated into

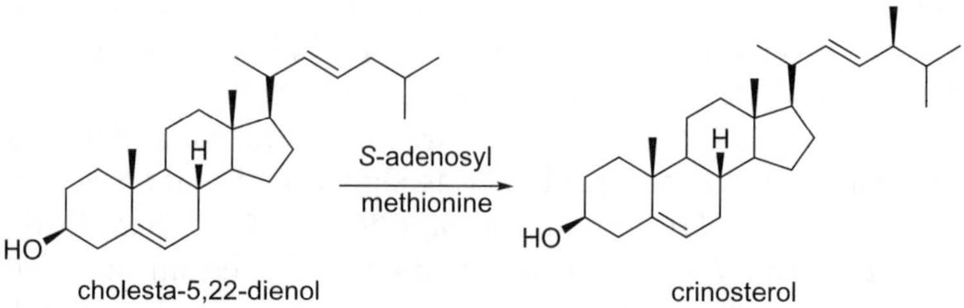

Figure 8.10 The probable way in which crinosterol is formed by mites.

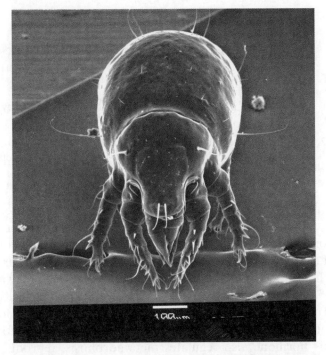

Figure 8.11 A scanning electron micrograph of the mite *Rhyzoglyphus robini*. Courtesy of Dr N. J. Fashing, College of William and Mary.

the side-chain of the sterol, so that mites can alkylate sterols like plants, the opposite of what insects do.

8.5 INSECT MOULTING HORMONE – ECDYSTEROIDS

An essential stage in the growth and development of insects is moulting or ecdysis. Growth is a discontinuous process, initiated at each stage by the moulting hormone. It was apparent from the 1920s that moulting is regulated by a hormone, and isolation studies began in Germany in the 1940s, but it was not until 1954 that a pure crystalline compound named ecdysone (Figure 8.12) was obtained (Butenandt and Karlson, 1954). That, however, turned out to be a pre-hormone, rapidly converted in the haemolymph to the true hormone, 20-hydroxyecdysone (Figure 8.12), in many insects. This compound contains five hydroxyl groups, which render it water-soluble. The group of polyhydroxysterols that serve as moulting hormone in all insects, which are called ecdysteroids, all have a 3β-hydroxyl group, characteristic of sterols, and several other hydroxyl groups, plus a *cis*-fusion of rings A and B, a 7-ene-6-one, and a *trans*-

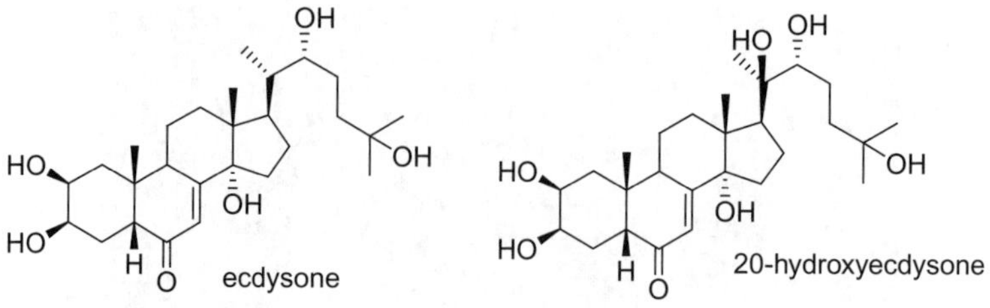

Figure 8.12 The structures of ecdysone, the first moulting hormone isolated, and 20-hydroxyecdysone, the true hormone of the majority of insects.

fusion of rings C and D. Lafont et al. (2005) list 70 ecdysteroids (including biosynthetic intermediates) found in insects together with the species from which they were first isolated. Almost simultaneously with the report of the structure of ecdysone, the first ecdysteroid was found in a plant. Their function in plants is still debatable, though they appear in at least 5% of all higher plants and ferns, concentrated in leaves and flowers. Today over 400 ecdysteroids from all sources are known with the list still growing; they are catalogued on a website (Lafont et al., 2002). This hormone also regulates metamorphosis, reproduction and diapause.

8.5.1 Ecdysteroid Biosynthesis

The study of insect moulting hormone has absorbed the attention of many investigators over half a century, and yet some of the early stages of the process are still unknown. Most of the synthesis takes place in the prothoracic gland. In higher Diptera the prothoracic gland is fused with the corpus allatum (site of juvenile hormone synthesis) and other structures into the ring gland. Cholesterol is the source of 20-hydroxyecdysone, the moulting hormone of the majority of insects. A protein prothoracotropic hormone (PTTH) from the brain stimulates the prothoracic gland or ring gland to begin the synthesis that results in the formation of 20-hydroxyecydsone or its equivalent. Cholesterol is first converted by microsomal cytochrome P450 in the prothoracic gland to 7-dehydrocholesterol (Figure 8.13), by removal of 7-H_β (a *cis* elimination). 7-Dehydrocholesterol is then transferred to the mitochondria. By three more steps not yet clear, and generally referred to as "the black box", an α-hydroxyl group is introduced at C-14, an unsaturated ketone at C-6 and the 3β-OH is oxidized to a ketone, giving a possible, but unconfirmed, hydroxydiketone (**A** in Figure 8.13). Moving then from

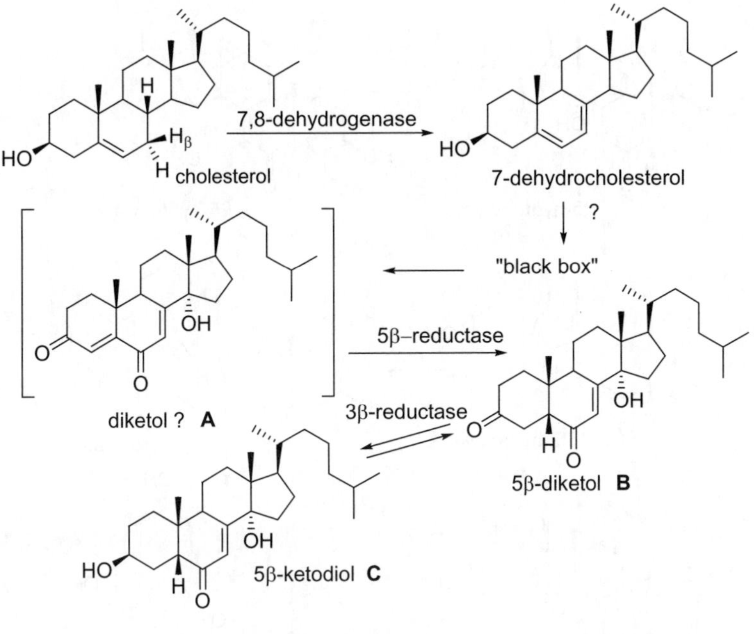

Figure 8.13 The first stages in the biosynthesis of ecdysteroids from cholesterol, including the still-unknown intermediate stages described as "the black box". These unknown steps occur in the mitochondria. **A**, **B** and **C** are identified in the text.

mitochondria to cytosol, a 5β-reductase catalyses the formation of the first well-established intermediates, 14α-hydroxy-5β-cholest-7-en-3, 6-dione (**B** in Figure 8.13) and 3β,14α-dihydroxycholest-7-en-6-one (**C** in Figure 8.13).

From this stage it seems that two sequences run in parallel, with either a ketone or an alcohol at C-3. The relative importance of ketone and alcohol vary with species. A 3β-hydroxysteroid dehydrogenase has been isolated from the crab *Carcinus maenas* (Decapoda: Portunidae) which, in the presence of NADH, reversibly catalyses the reduction of **B** to **C** in Figure 8.14 and in the two following stages. Certainly the next step for each sequence is hydroxylation at C-25, catalysed by a microsomal cytochrome P450 25-hydroxylase (Figure 8.14). Both intermediates are then transported back to the mitochondria where the introduction of the C-22 hydroxyl is catalysed by another cytochrome P450 hydroxylase. Both compounds again are hydroxylated at C-2, catalysed by another mitochondrial P450 and released into the cytosol where the 3-dehydroecdysone is converted to ecdysone. All of this occurs in the prothoracic gland. The end product in the prothoracic gland

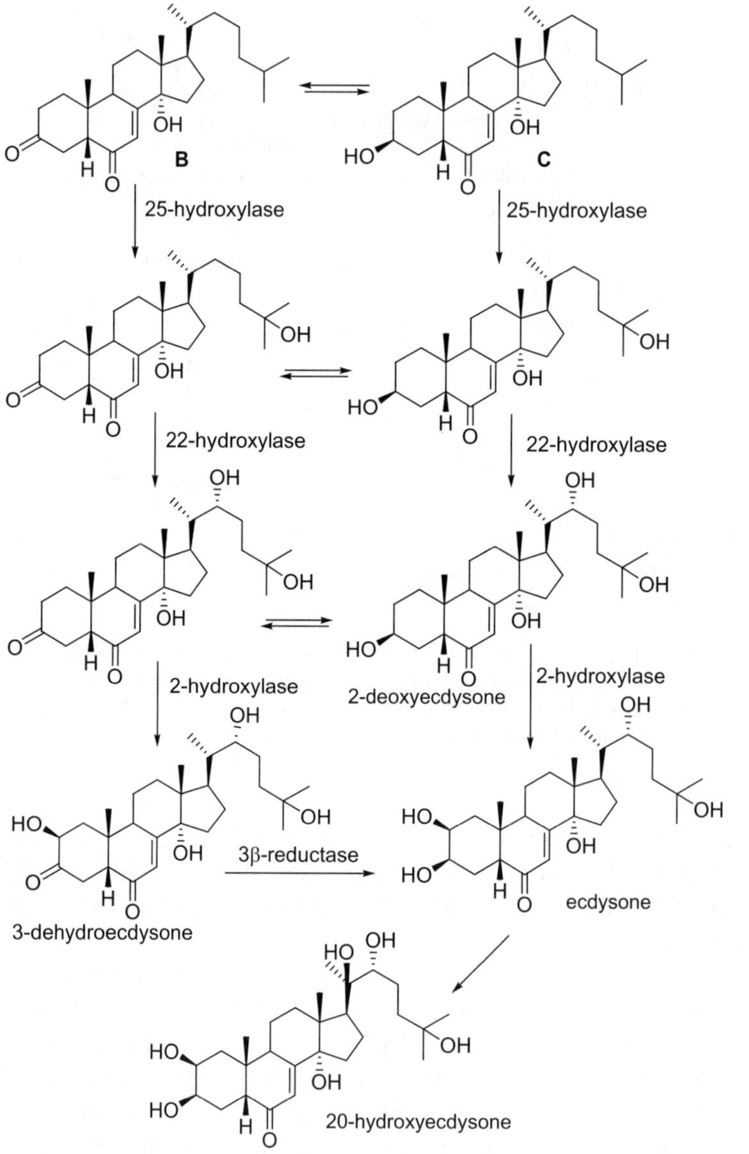

Figure 8.14 The later stages of synthesis of ecdysone and 20-hydroxyedysone. Whether there is a ketone or alcohol at C-2 varies with species.

is either ecdysone, the first compound that was isolated, and thought for a time to be the hormone, or 3-deoxyecdysone. It is not known how the intermediate sterols are carried between the subcellular organelles.

Later it was found that hydroxylation of ecdysone at (20S) was necessary to give 20-hydroxyecdysone, the true hormone. The final stage occurs principally in the midgut, the Malpighian tubules, epidermis and fat body. The enzyme catalysing the last step, in the peripheral tissues, ecdysone 20-mono-oxygenase (and NADPH) has been studied in great detail (Lafont *et al.*, 2005); it has been found in both microsomal and mitochondrial fractions, depending upon the species or tissue examined. Through study of the genetics of *Drosophila melanogaster* a group of mutant genes called Haloween (because they cause unusual marking of the epidermis) have been identified. Among these genes are those coding for the enzymes catalysing the last four steps in the ecdysteroid biosynthesis sequence (the C-25-, 22-, 2- and 20-hydroxylases) (Huang *et al.*, 2008). All these hydroxylases are cytochrome P450 enzymes. The ecdysone 20-mono-oxygenase has been identified in *Drosophila melanogaster* Malpighian tubules and fat body, but also in ovariole nurse and follicle cells (Section 8.5.2).

Some insects of the Diptera, Coleoptera, Hemiptera and Hymenoptera (and the tobacco hornworm *Manduca sexta*) that are unable to remove the 24-alkyl groups of plant sterols, use makisterone A (20-hydroxy-24α-methylecdysone, Figure 8.15) as their moulting hormone (Svoboda and Feldlaufer, 1991), Radio-labelling experiments have shown that the honeybee *Apis mellifera* transforms plant campesterol to

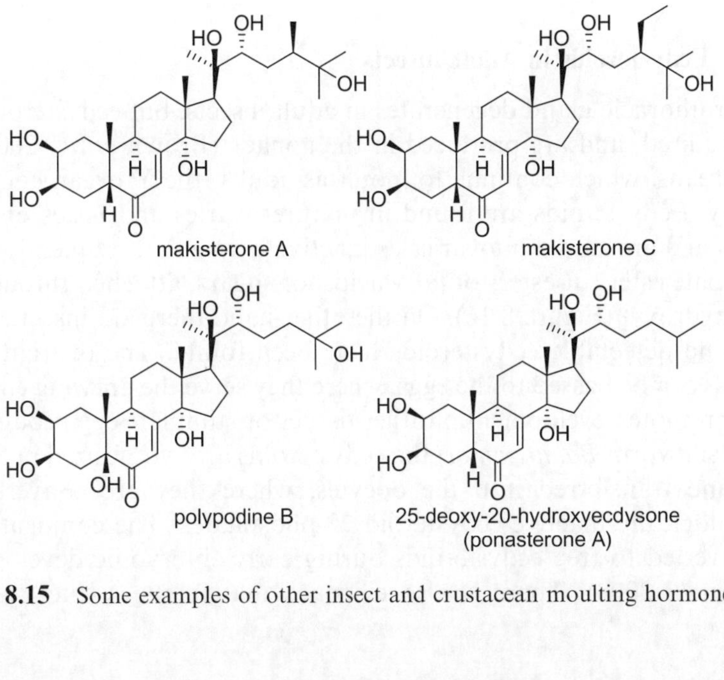

Figure 8.15 Some examples of other insect and crustacean moulting hormones.

makisterone A. The cotton stainer bug *Dysdercus fasciatus* (Heteroptera: Pyrrhocoridae) uses a mixture of makisterone A and makisterone C (Figure 8.15). The inability of leaf-cutting ants to de-alkylate phytosterols has been noted (Section 8.4.2). *Acromyrmex octospinosus* feed mainly upon fungal sterols like ergosterol, which has a (24R)-methyl group (Figure 8.8) and consequently produce 24-*epi*-makisterone A. The isolation of many other ecdysteroids from plants has provided many variations of the ecdysone theme. These have helped to understand the essential structure-activity relationships. The 5β-fusion of the A and B rings is essential for activity, and polypodine B (a plant ecdysteroid, Figure 8.15) with a 5β-OH is about 500 times more active than 20-hydroxyecdysone *in vitro*.

Ecdysteroids have been found in many more primitive phyla (Coelenterata, Platyhelminthes, Annelida) and higher ones (Mollusca and Echinodermata) but their functions there are not known. In Crustacea, moulting hormone is produced in the Y-organ, which usually releases ecdysone, which is activated to 20-hydroxyecdysone in peripheral tissues, as in insects. In the crabs *Carcinus maenas* and *Menippe mercenaria* (Decapoda: Menippidae) the Y-organs produce 3-desoxyecdysone and 25-desoxyecdysone, the latter is oxidized to the active compound ponasterone A (Figure 8.15), which is also used by some other crustaceans. Ecdysteroids function as moulting hormone in ticks (Acari), but no secretory gland has been identified (Rees, 2004).

8.5.2 Ecdysteroids in Adult Insects

The prothoracic gland degenerates in adult insects, but ecdysteroids are still required, and are produced in the gonads (Brown *et al.*, 2009). In crustaceans, which continue to moult as adults, the Y-organ continues activity. Ecdysteroids are found in mature ovaries and testes of many insects and are stored in ovaries as inactive *conjugates*, frequently either phosphate esters or esters of fatty acids, or sugars, attached through the C-22 hydroxyl (Figure 8.16). On the other hand there are insect eggs in which no detectable ecdysteroids have been found. The ovarian ecdysteroids can be passed to the eggs, where they serve the growing embryo, or to promote development of other tissues or stimulate egg production. In the silkworm *Bombyx mori* the ecdysteroids are synthesized in follicle cells and transferred into the oocytes, where they are converted to physiologically inactive ecdysteroid 22-phosphates. The conjugates are re-converted to free ecdysteroids during early embryonic development (Ito *et al.*, 2008). There is also evidence that in the cricket *Achaeta*

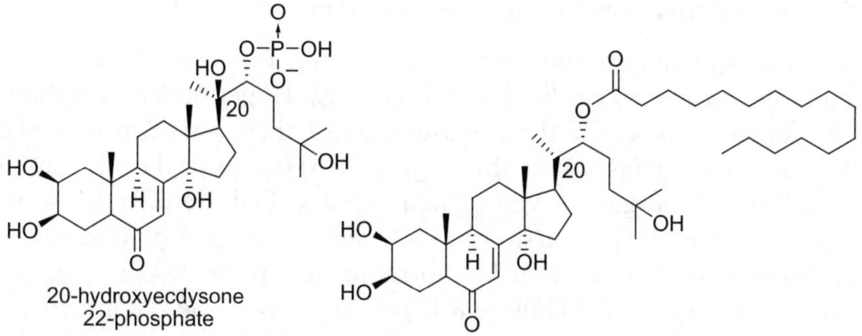

Figure 8.16 Structures of the most commonly found ecdysteroid conjugates.

domestica (Orthoptera: Gryllidae) and the cockroach *Periplaneta americana* (Blattodea: Blattidae) the maternal ecdysteroid fatty-acid esters are hydrolysed in the early stages of embryogenesis. The amounts of ecdysteroids found in males have been generally small, but the males of the malaria mosquito *Anopheles gambiae* (Diptera: Culicidae) produce large amounts of 20-hydroxyecdysone in their accessory glands, store them there and deliver them to the female at mating (Pondeville *et al.*, 2008). The authors suggest that the ecdysteroid may not be therefore a true male hormone but rather an "allohormone" (which might be defined as a hormone of another) (see nuptial gifts, Section 11.1.1).

Conjugated ecdysteroids are also found in the ovaries of crustaceans, myriapods and ticks. The centipede *Lithobius forficatus* (Chilopoda, of the superclass Myriapoda) contains and excretes ecdysone and 20-hydroxyecdysone from ovaries and testes.

Fatty-acid esters of ecdysteroids were first identified in the ovaries of the southern cattle tick *Rhipicephalus* (*Boophilus*) *microplus* (Acari: Ixodidae), but the eggs of some ticks contain no conjugates, but appreciable free hormone. The greatest diversity of compounds and their conjugates is found in eggs and embryos of the desert locust *Schistocerca gregaria* (Orthoptera: Acrididae) and the lepidopteran *Bombyx mori*. Fifteen ecdysteroids and phosphates have been recognized in *S. gregaria*, and eleven conjugates were found in *Bombyx mori*.

The major conjugate in newly laid eggs of the tobacco hornworm *Manduca sexta* is 26-hydroxyecdysone 26-phosphate During embryogenesis the 26-hydroxyecdysone 26-phosphate is hydrolysed to 26-hydroxyecdysone and converted to nine other compounds, including a phosphate, a glucoside, epimers and carboxylic acids.

8.5.3 Ecdysteroid Inactivation and Excretion

Inactivation of ecdysteroids can be temporary, for re-use at an appropriate time, or in another life stage. It can also be permanent, combined with excretion. As seen in the previous section, there is no consistency of behaviour among insects in this regard. 20-Hydroxyecdysone can be excreted intact *via* gut or Malpighian tubules. Ecdysteroids are sometimes inactivated by conversion to C-22 sulfates or phosphates, since a free 22-hydroxyl is important for hormone action, or they may be oxidized to ecdysonoic acid (Figure 8.17) or to a lesser extent converted to glycosides. In the Egyptian cotton leafworm *Spodoptera littoralis* (Lepidoptera: Noctuidae) 20-hydroxyedysone is converted first to 20-26-dihydroxyecdysone, which is then oxidized further to ecdysonioic acid through the aldehyde (Figure 8.17). In the tobacco hornworm *Manduca sexta* (Lepidoptera: Sphingidae) during pupal-adult development, the major metabolites are 3-*epi*-20-hydroxyecdysonoic acid and a phosphate of 3-*epi*-20-hydroxyecdysone. 3-*Epi*-ecdysone formation is particularly found in Lepidoptera. These studies may be made by analysis of endogenous material, or by feeding with ecdysteroids, or by injection of labelled ecdysteroids. Some insects are able to feed on plants containing phytoecdysteroids without being affected by them. When [^{3}H]ecdysone was injected into adult females of the cricket *Gryllus bimaculatus* (Orthoptera: Gryllidae), it was converted into at least 24 metabolites, most of them in the gut and faeces, with 14-deoxyecdysone the major

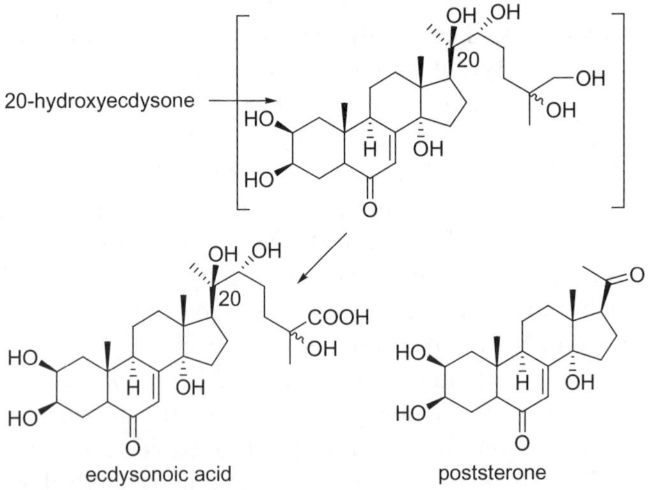

Figure 8.17 The formation of the inactivation product ecdysonoic acid, and the side-chain degraded product posterone.

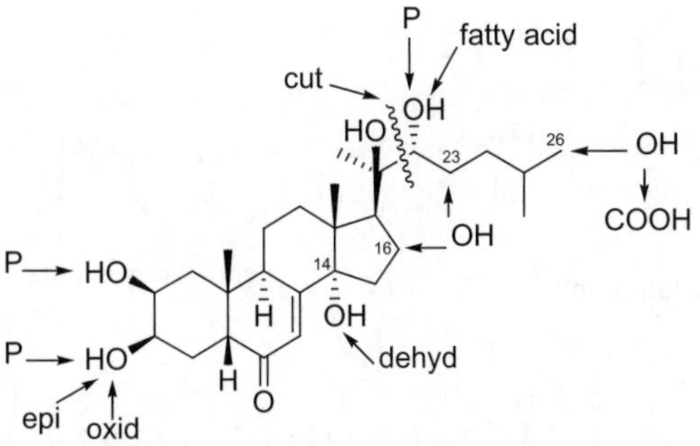

Figure 8.18 A summary of the routes to temporary or permanent inactivation of 20-hydroxyecdysone. P indicates phosphorylation; alternatively, sometimes an acetate is formed at one of these points. Fatty acid indicates formation of an ester, epi means epimerization, oxid is oxidation, dehyd indicates removal of the 14-hydroxyl, a minor route. Hydroxylation at C-26 is common with further oxidation to the carboxylic acid. Hydroxylation at C-16 or C-23 is rare. Cleavage of six atoms from the side-chain is of unknown importance.

metabolite. Generally, labelled ecdysteroids have been labelled in the side-chain, so that if the side-chain has been removed in metabolism (as in the example of posterone, Figure 8.17), such metabolites have not been recognized. Figure 8.18 summarizes the commonest of the known metabolizing routes from ecdysone.

8.6 STEROL PHEROMONES

The high molecular mass and low volatility of sterols make them unlikely candidates for pheromones. 3-Cholestanone is used as a trail pheromone by tent caterpillars (*Malacosoma* species; Family Lasiocampidae) (Figure 8.19), where it is probably detected by contact with the substrate. Two sterol glucosides, blattellostanosides A and B (Figure 8.19), are aggregation pheromones of the cockroach *Blattella germanica*. These not only retain the side-chain ethyl group of sitosterol but have been chlorinated as well. They are emitted with the frass and are quite involatile and must be sensed by contact. Both these compounds and the cholestanone have the 5β-configurstion of the ecdysteroids and unlike the more planar 5α-configuration of most mammalian hormones. The mounting pheromone of the hard tick *Dermacentor variabilis* (Acari: Ixodoidae) is cholesteryl oleate. It is excreted by the mature female, and accumulates on its cuticle. It too probably requires contact for perception. Cholesteryl oleate and other cholesteryl esters are attractant pheromones for some other ticks.

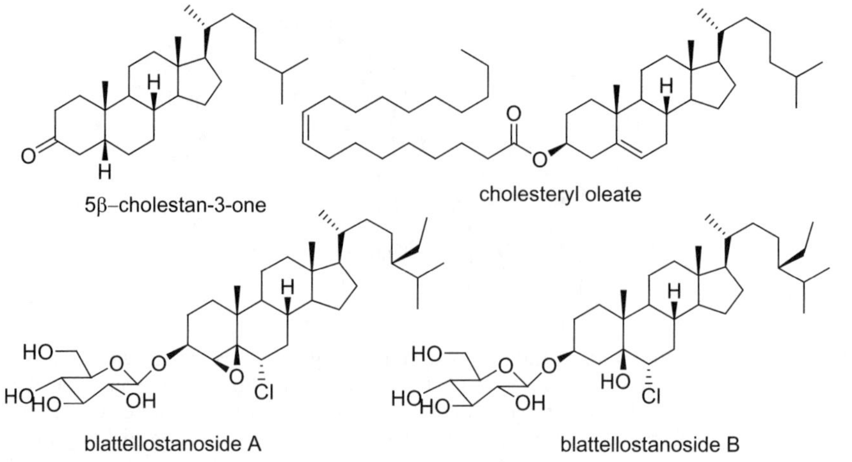

Figure 8.19 Some sterol pheromones. 5β-Cholestan-3-one is a trail pheromone of tent caterpillars; cholesteryl oleate is the mounting pheromone of the tick *Dermacentor variabilis*. Of the two blattellostanoside aggregation pheromones of the German cockroach, blattellostanoside A is much the more active.

8.7 STEROID DEFENSIVE SUBSTANCES

There are a number of examples where sterols have been adapted for defence by insects, but there is limited information about their biosynthesis. Some examples can illustrate the types of structures that insects, particularly beetles, can produce from sterols. Leaf beetles (Chrysomelidae) are notable for their originality in producing defensive secretions. The subtribe Chrysolinina tend to produce steroids (Pasteels and Daloze, 1977), the Chrysomelina produce isoxazolinones (Section 10.3) and the Doryphorina use triterpenoid saponins. The chrysomelid beetle *Chrysolina carnifex* produces in its body, and stores in its elytra and pronotal glands, a large amount of 20-hydroxyecdysone 22-acetate (Laurent *et al.*, 2003). The concentration is estimated to be about 0.15 M. While the 22-hydroxyl group is protected by the acetate it is hormonally inactive, but active circulating ecdysteroids in insects are normally about 10^{-6} M.

Other members of the Chrysolinina subtribe use cardiac glycosides or cardenolides. Adults release them from the pronotum and elytra. Van Oycke *et al.* (1987) showed that they were made by the beetles and not taken directly from plants. When *Chrysolina coerulans* were fed leaves coated with [4-^{14}C]cholesterol the labelled cholesterol was incorporated into six cardenolides. The major compounds are the cardenolide

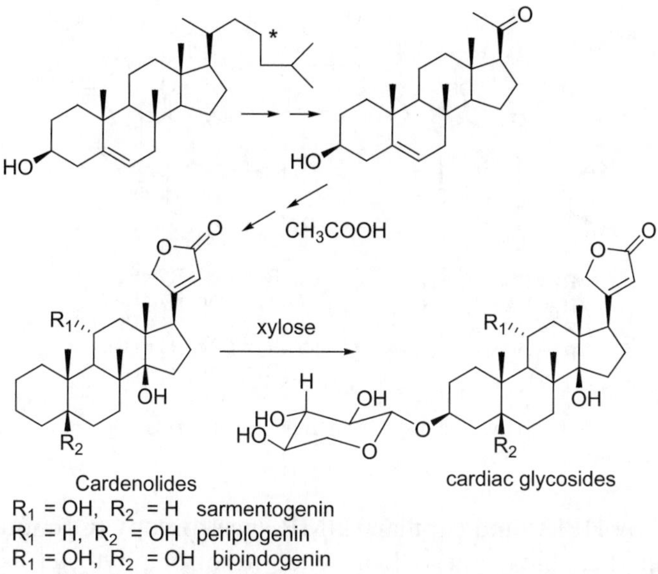

Higher Terpenes and Steroids

Figure 8.20 Summary of the formation of three cardenolides found in the beetle *Chrysolina coerulans*. Both the cardenolides and their xylosides (or cardiac glycosides) are present in the secretion. The asterisk indicates a ^{14}C-label which is lost in the synthesis.

sarmentogenin, and the cardiac glycosides sarmentogenin xyloside, periplogenin xyloside and bipindogenin xyloside. In plants the sterol side-chain is degraded to two carbon atoms, and then an acetate unit is added to give the lactone. To see if the same route was used by the beetles [1,2-^{3}H,23-^{14}C]cholesterol was added to the diet. The label at C-23 was lost in the cardenolides, showing that insects created them in the same way as plants (Figure 8.20). With sugar molecules attached they are called cardiac glycosides. In some *Chrysolina* species the compounds are also found in the eggs, but it is not clear if they come from parents, or are synthesized by the embryo in the egg. Other chrysomelid beetles attach 2-deoxyhexoses to cardenolides and others make the steroids without the sugars. The cardiac glycosides from plants, like digitoxin, originally used in arrow poisons, are strongly toxic in mammals, and cause vomiting and affect their nervous systems.

Another group of beetles, the fireflies (Coleoptera: Lampyridae), of the genus *Photinus*, known to be repellent to predators, contain steroidal pyrones, closely related to the toad poisons bufodienolides, called lucibufagins (Goetz *et al.*, 1981). Many of these compounds have been identified (Figure 8.21); in one species *Lucidida atra* 13 lucibufagins were

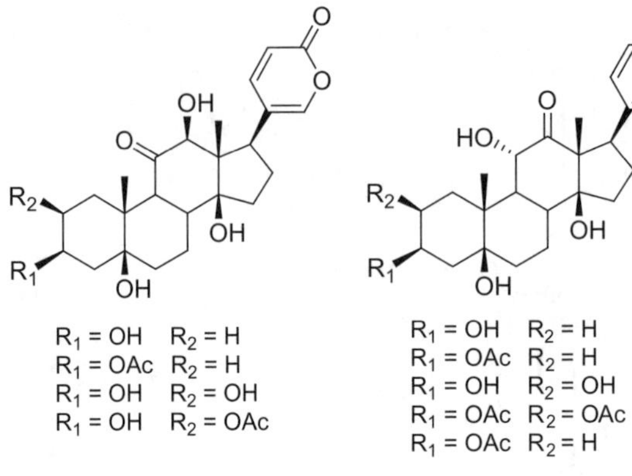

Figure 8.21 Lucibufagins, defensive and repellent steroidal pyrones from *Photinus* fireflies.

identified by HPLC and capillary NMR spectroscopy (Gronquist *et al.*, 2005), but their biosynthesis from cholesterol or phytosterols has not been studied. Females of *Photurus* fireflies which copy the *Photinus* flashes, lure and catch male *Photinus* and sequester their lucibufagins and add them to their own defences, partially altered to other compounds (Gonzalez *et al.*, 1999).

8.7.1 Saponins from Triterpenes

Saponins are plant triterpenoid glycosides, close relatives of the cardiac glycosides. They make a strong foam when shaken with water. They are bitter tasting and generally toxic to lower forms of life. They have been much used by primitive people for killing fish.

The defensive glands of adult beetles of *Platyphora*, *Leptinotarsa*, *Labidomera* and *Desmogramma* (Chrysomelidae) have been found to contain six identified triterpene saponins, with one, two or three sugar residues. Since insects cannot make triterpenes, and the plants the beetles were feeding on did not contain saponins, biosynthetic studies were made to find how the insects acquired them. First the beetles were fed on sweet potato (*Ipomoea batatas*) leaves, which contain α- and β-amyrin (widely distributed plant triterpenoids) to confirm they could produce the saponins (Laurent *et al.*, 2003). Feeding *Platyphora kollari* with [2,2,3-^{2}H$_3$]β-amyrin produced up to 40% of labelled saponin (Figure 8.22). The enzymes required to convert β-amyrin are a cytochrome P450 to oxidize the C-28 methyl group to COOH and to introduce a hydroxyl group at C-23 and a glycosylase to couple the product with a sugar; these are generally present in all living organisms. The first step is the conversion of β-amyrin to

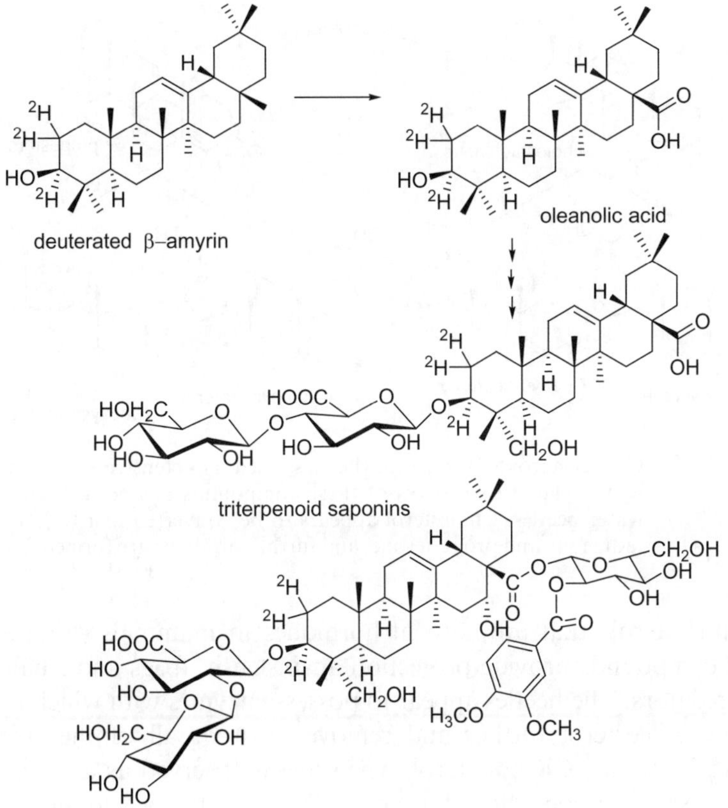

Higher Terpenes and Steroids

Figure 8.22 Two examples of triterpenoid saponins, synthesized from the triterpene β-amyrin by chrysomelid beetles, *via* oleanolic acid. A gluconic acid is attached at C-3 in each compound identified.

oleanolic acid (Ghostin *et al.*, 2007). Although both α- and β-amyrin are present in the food of the beetle in comparable amounts, almost exclusively the β-amyrin was metabolized. α-Amyrin would have given ursane glycosides. A later survey of about 30 species of neotropical Chrysomelid species by LC-MS showed that oleanane glycosides were widely distributed among them. The *Oreina* genus of chrysomelid beetles also make saponins but, in addition, sequester pyrrolizidine alkaloids from their food plants (Section 11.3.2) (Van Oycke *et al.*, 1987).

8.8 MAMMALIAN HORMONES IN INSECTS

Some water beetles of the Dytiscidae (predacious diving beetles) and Hygrobiidae (squeak beetles) store in their paired thoracic defensive glands

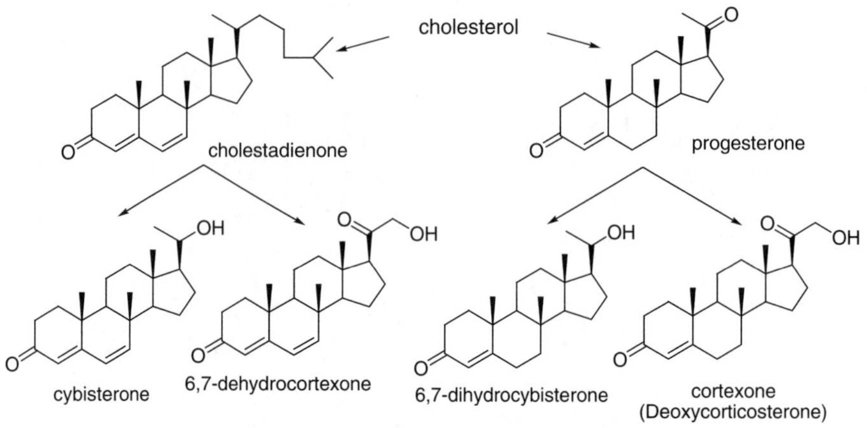

Figure 8.23 Corticosteroids from water beetles, used as defensive secretions. Other hydroxylated derivatives of these compounds have also been found in water beetles. Cholesterol appears to be converted first to both cholestadienone and progesterone, and further products are formed from these.

the same sterols that are adrenal hormones in mammals (Figure 8.23). These compounds provide protection against fish, frogs and small mammal predators. The beetles appear to possess enzymes with which they can degrade cholesterol further and remove most or all of the side-chain (Figure 8.23). [4-^{14}C]Cholesterol, when injected, served as a precursor of all the corticosterones (Schildknecht, 1970). Insects injected with labelled mevalonolactone, as expected, produced no labelled sterols. The pathway to dienones and enones presumably diverges at an early stage since there was a greater incorporation of [4-^{14}C]cholestadienone into the two dienones (Figure 8.23), while [4-^{14}C]progesterone was better incorporated into the two enones. Specific incorporation of labelled cholesterol varied from 0.51% for cortexone to 1.20% for dihydrocybisterone. Labelling experiments with cholesterol showed that the 4β and 7β hydrogen atoms were eliminated in forming the diene (Schildknecht, 1970).

There is about 1 mg of deoxycorticosterone (also known as cortexone or systematically as 4-pregnene-21-ol-3,20-dione) in the prothoracic defensive glands of each beetle of *Cybister limbatus*. In *Ilybius fenestratus* (Plate 26) there is testosterone, oestradiol and three compounds related to cortexone. *Agabus guttatus* also has testosterone, oestradiol and nine oxygenated pregnane derivatives (Francke and Dettner, 2005). *Agabus affinis* contains a pregnanedione (Figure 8.24). However, strains of bacillus isolated from the foregut of *A. affinis* were able to convert pregnenolone into 7α-hydroxypregnenolone, and androst-4-en-3,7-dione into a large number of other products (Francke and Dettner, 2005).

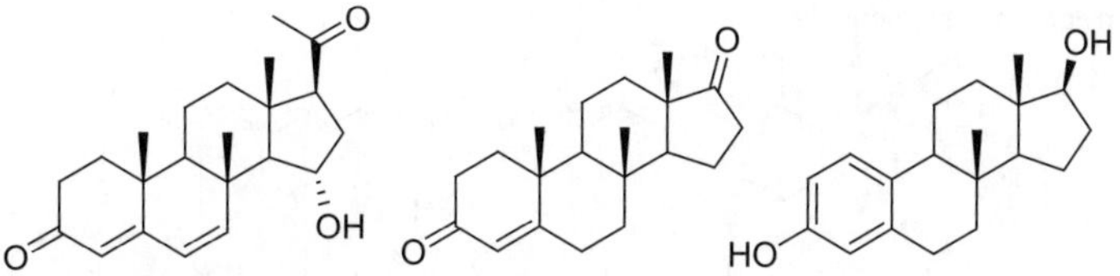

Figure 8.24 Steroid hormones used for defence by water beetles.

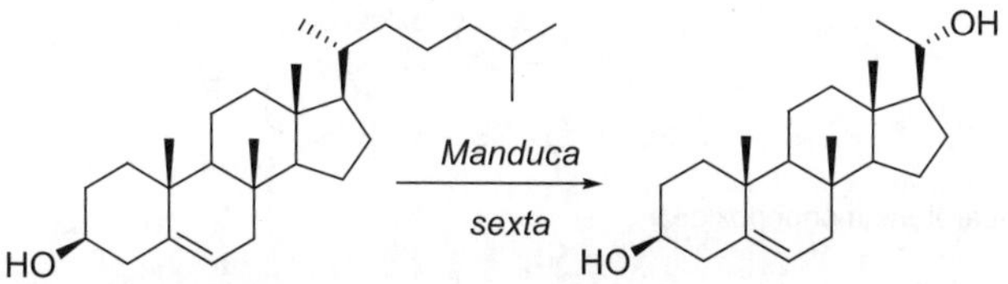

15α-hydroxypregna-4,6-dien-3,20-dione
Agabus affinis

testosterone

oestradiol

Figure 8.25 The major metabolite of labelled cholesterol in pupae of the tobacco hornworm *Manduca sexta* is pregn-5-en-3,20-diol, of no known function.

The carrion beetle *Silpha novaboracensis* (Coleoptera: Silphidae) also makes pregnane derivatives (Meinwald *et al.*, 1987).

When studying the formation of ecdysteroids in ovaries and eggs of the tobacco hornworm *Manduca sexta* (Section 8.5.2), [4-^{14}C]cholesterol was injected into female pupae. Some of the cholesterol was converted in adult females to 26-hydroxyecdysone 26-phosphate, but the major metabolite in the eggs was a conjugate of a C_{21} sterol, identified as the triglucoside of 3β,20(*R*)-pregn-5-en-3,20-diol (Figure 8.25). No function for the conjugate was discovered. It remained with the egg shell after hatching.

Some mammalian hormones have already been identified among the defensive secretions (Section 8.7). At least 12 sterols of the pregnane, androstane and oestrane types have been identified in insects and crustaceans. Except when used as defensive secretions, no function has been identified for them. Oestradiol was detected in the ovaries of the silkworm *Bombyx mori*, but it has no significant effect on development.

8.9 TETRATERPENES

Tetraterpenes are represented by only one group of compounds, the carotenes, the most widely distributed pigments in nature, with about

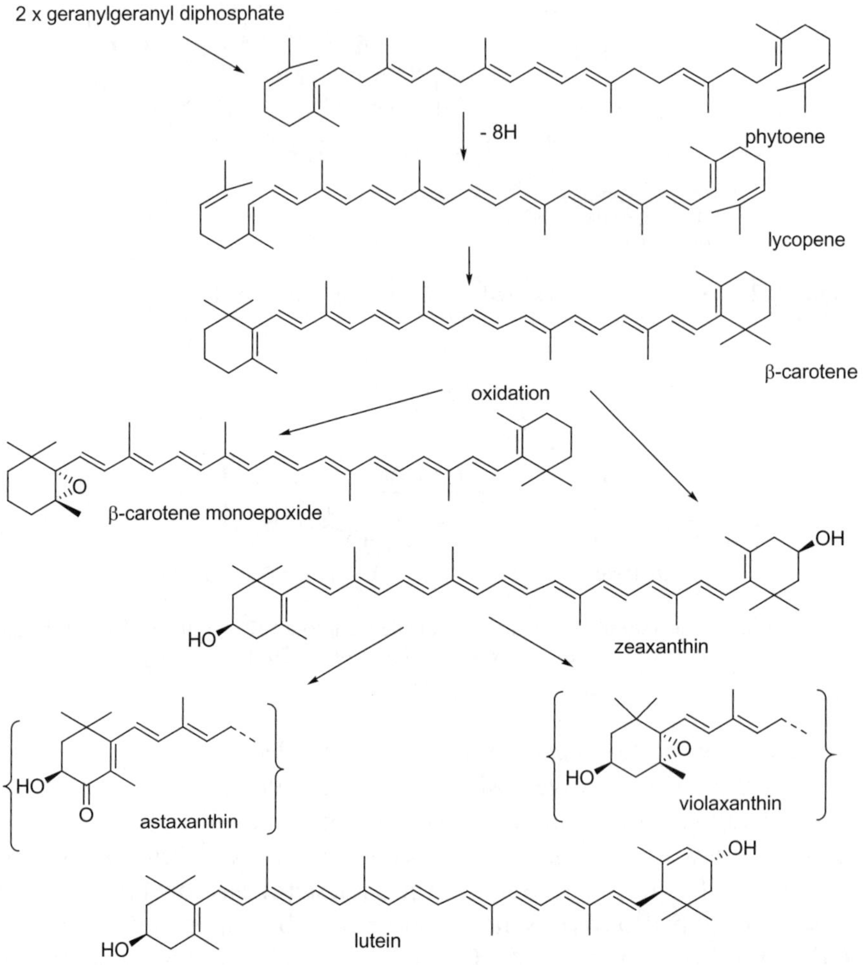

Figure 8.26 The formation of tetraterpenes by condensation of two molecules of geranylgeranyl pyrophosphate, and conversion to carotenes, with some examples of carotenes found in insects.

700 known. They are constructed in a manner similar to squalene, by head-to-head condensation of two molecules of geranylgeranyl diphosphate, also through a cyclopropane intermediate, but NAPDPH is not required as for the biosynthesis of squalene, rather a double bond is formed at the point of fusion. The initial product is phytoene (Figure 8.26). It is desaturated in stages to lycopene, the pigment of tomatoes. Carotenes have long conjugated systems of double bonds, usually all-*trans*, and are strongly coloured. They are synthesized in bacteria, certain fungi and the plastids of plants *via* the methylerythritol

phosphate process (Section 7.22). It is thought that not only insects, but mammals, birds and amphibians were unable to synthesize them, but there is an unconfirmed report that bovine corpus luteum can synthesize carotenes. Carotenes are highly unsaturated and are very unstable to oxygen, heat and light, making them difficult to isolate and purify, nevertheless they are found in many places, from butter to flamingos.

Carotenes give colour to many fruit and vegetables, besides lycopene in tomatoes and paprika; zeaxanthin is the yellow pigment of maize (*Zea mais*); astaxanthin is a further oxidation product of carotene and gives the pink colour of shrimps, boiled lobsters, salmon and the pink bollworm *Pectinophora gossypiella* (Lepidoptera: Gelechiidae). Violaxanthin is a common plant pigment, also widely found in Lepidoptera. The pigment of many insects is due to the presence of carotenes and carotene-protein complexes. Carotene-protein compounds give green, blue-green, blue and red pigments in insect integument and haemolymph. Lycopene, β-carotene, zeaxanthin, violaxanthin, astaxanthin, xanthophyll and β-carotene monoepoxide and at least eight other carotenes have been found in insects. The conspicuous colours of the elytra of chrysomelid beetles are due to carotenes. The red colour of the firebug *Pyrrhocoris apterus* (Heteroptera: Pyrrhocoridae) is due to lycopene. The chrysomelid Colorado potato beetle *Leptinotarsa decemlineata* is coloured by carotenes from its food, potato leaves; and the two-spotted stink bug *Perillus bioculatus* (Hemiptera: Pentatomidae), which preys upon the Colorado beetle and its eggs, acquires the same colours. An early paper claimed that the green mantid *Shodromantis bioculata* (Mantodea: Mantidae) contains carotene, and retained its colour even if it is raised from colourless eggs on a diet lacking in carotenes. This is the only evidence found suggesting insects could make carotenes. Otherwise, all the evidence suggests that carotenes come from insect food, though insects seem capable of altering them. The only carotene in newly laid locust eggs (both *Schistocerca gregaria* and *Locusta migratoria*, (both Orthoptera: Acrididae) is β-carotene, but during incubation this slowly disappears and is replaced by astaxanthin. Ladybirds get their yellow or orange pigments from aphids, on which they feed. The aphids in turn get them from their plant food. In an examination of 114 species of lepidoptera, forty-one carotenes were recorded (Czeczuga, 1986). Lutein, also known as xanthophyll, is found in almost all Lepidoptera examined (the name xanthophyll is also used for all oxygenated carotenes).

The difference between the well-known white and yellow strains of cocoons and silk of the silkworm *Bombyx mori* has been traced to a genetic defect in the protein that carries carotenoids (Tsuchida *et al.*, 2004). The white mutant inadequately absorbs carotenes, particularly lutein.

The carotenes of one aphid *Macrosiphum liriodendri* (Homoptera: Aphididae), which occurs in two colour variants, have been studied in detail. This aphid, living on tulip trees (*Liriodendron tulipifera*), exists in green and pink forms, which appear to differ in their level of cyclizing enzymes (Andrews et al., 1971). The green form have all cyclized carotenes, while the pink form have two partly cyclized and two uncyclized carotenes (Figure 8.27). It must be presumed the aphids have the necessary enzymes to cyclize lycopene. Many other insects, particularly aphids, contain carotenes along with other pigments.

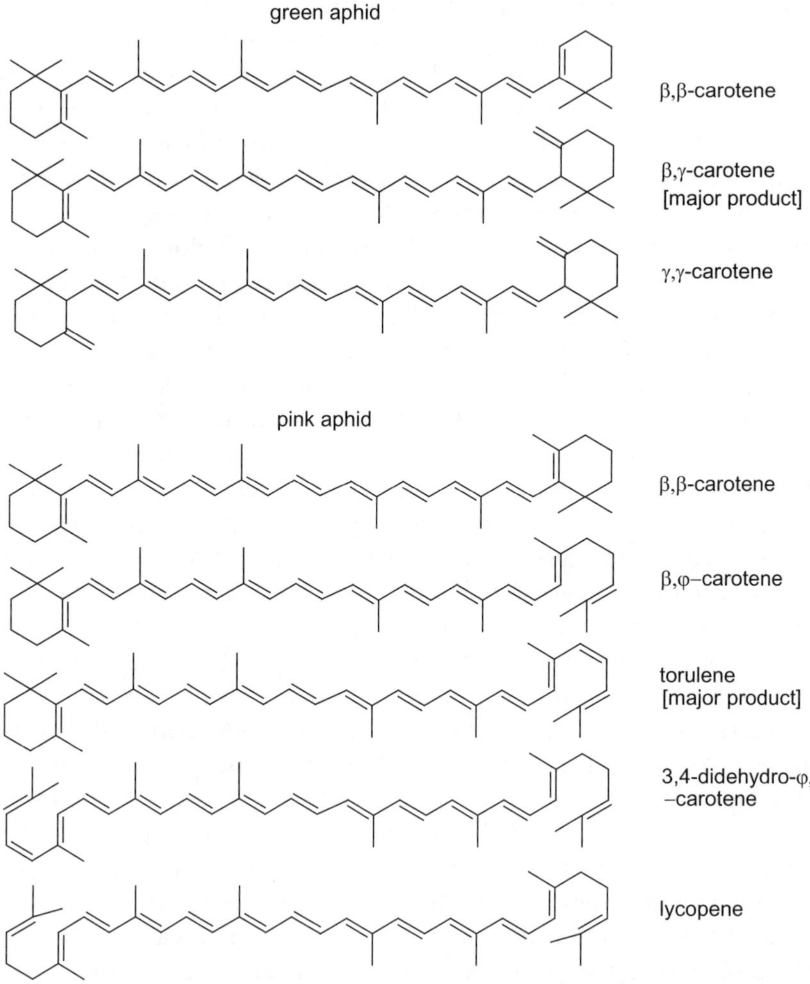

Figure 8.27 The carotene pigments in the two colour forms of *Macrosiphum liriodendri* aphids. The systematic names are used here. β,β-Carotene is more commonly known as β-carotene.

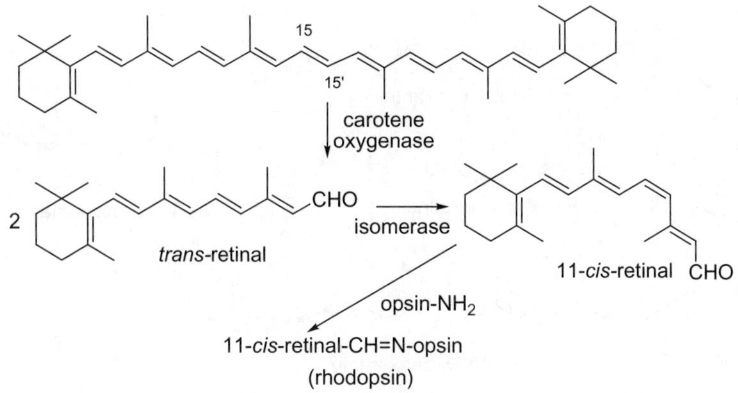

Figure 8.28 The formation of retinal and derivatives from carotenoids. The functions of carotene oxygenase and retinal isomerase are contained in the one polypeptide. After formation of 11-*cis*-retinal, combination of it with the protein opsin through formation of a Schiff's base gives the visual pigment rhodopsin.

8.9.1 Insect Vision

It is emerging that the visual process is highly conserved among animals. Cleavage of β-carotene by oxidation gives two molecules of vitamin A aldehyde or retinal (Figure 8.28). Vitamin A is needed for vision in the entire animal kingdom. Retinal is isomerized to 11-*cis*-retinal which combines with the protein opsin through a Schiff's base to form the purple visual pigment rhodopsin. Opsin is embedded in the photoreceptor membrane of the eye. Depending upon the structure of the opsin protein, rhodopsins are formed which have different absorption maxima, in humans, for red (570 nm), green (530 nm) and blue (440 nm) light, giving colour vision. Insects lack the red-absorption pigment, but have a UV-absorbing rhodopsin with maximum at 350 nm, as well as blue at 450 nm and green at 550 nm.

The study of blind mutants of *Drosophila* called ninaB has recently advanced the whole science of animal vision. We now know that carotene is cleaved by a non-haem oxygenase, carotene-15,15′-oxygenase to give retinal (von Lintig and Vogt, 2000). This has led to the cloning of the human enzyme, which has also shown that the enzyme is well conserved among animals. All-*trans*-retinal has to be isomerized to 11-*cis*-retinal to produce the active visual pigment (Figure 8.28). By cloning the gene in the greater wax moth *Galleria mellonella* (Lepidoptera: Pyralidae), it has been shown that one polypeptide incorporates the oxygenase and the isomerase functions (isomero-oxygenase), and this isomerase operates in the dark (Oberhauser *et al.*, 2008).

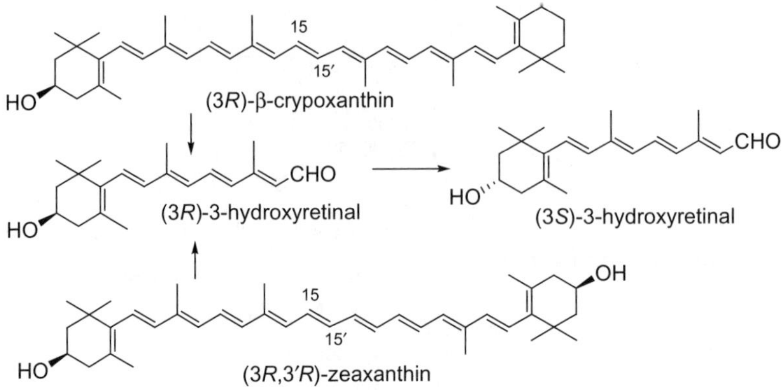

Figure 8.29 Later-evolving insects have used 3-hydroxyretinal in forming the visual pigment rhodopsin, while only higher Diptera isomerize (3R)-hydroxyretinal to (3S)-hydroxyretinal and use that in rhodopsin. A molecule of β-cryptoxanthin provides only one molecule of 3-hydroxyretinal but zeaxanthin gives two molecules of 3-hydroxyretinal.

Retinal, formed by cleavage of β-carotene was the original insect component of rhodopsin, but around the end of the carboniferous period (see Figure 1.1) some insects acquired the ability to use (3R)-3-hydroxyretinal, which can be obtained by cleaving (3R)-β-cryptoxanthin or (3R,3R')-zeaxanthin (Figure 8.29). The orders Odonata, Hemiptera, Neuroptera, Coleoptera, Lepidoptera and lower sub-orders of Diptera, which evolved about this time, use (3R)-3-hydroxyretinal (Seki and Vogt, 1998). The higher sub-order of Diptera, the Cyclorrhapha, which evolved during the Jurassic period, contain both (3R)-3-hydroxyretinal and (3S)-3-hydroxyretinal, but only incorporate (3S)-3-hydroxyretinal into their rhodopsin (Figure 8.29). Since there are no plant carotenoids containing a (3S)-3-hydroxyl group, they must have acquired the ability to isomerize the (3S)-hydroxyl in a yet unknown way.

BACKGROUND READING

Britton, G. 1983. *The Biochemistry of Natural Pigments*, Cambridge University Press, Cambridge, Chapter 2, Carotenes, 366 pp.

Dewick, P. M. 2009. *Medicinal Natural Products: A Biosynthetic Approach*. John Wiley, Chichester, Chapter 5, The mevalonate and methylerythritol phosphate pathways. Terpenoids and steroids, pp. 187–310.

Mann, J. 1987. *Secondary Metabolism*, Oxford University Press, Oxford, Chapter 3, Metabolites derived from mevalonate: isoprenoids, pp. 95–171.

Gilbert, L. I., Rybczynski, R. and Warren, J. T. 2002. Control and biochemical nature of the ecdysteroidogenic pathway. *Annual Review of Entomology*, **47**, 883–916.

Mann, J. 1994. *Chemical Aspects of Biosynthesis*, Oxford University Press, Oxford, Chapter 4, Terpenoids and steroids-the isoprenoids, pp. 31–52.

Needham, A. E. 1978. Insect biochromes: their chemistry and role. In: *Biochemistry of Insects* (Rockstein, M., ed.), Academic Press, New York and London, pp. 233–305.

Rees, H. H. 1977. *Insect Biochemistry*, Chapman and Hall, London, 64 pp.

Torssell, K. B. G. 1997. *Natural Product Chemistry*, Swedish Pharmaceutical Society, Stockholm, Chapter 5, The mevalonic acid pathway. The terpenes, pp. 167–225.

REFERENCES

Andrewes, A. G., Kjosen, H., Liaaen-Jensen, S., Weisgraber, K. H., Lousberg, R. J. and Weiss, U. 1971. Animal carotenoids. 7. Carotenes of two colour variants of the aphid *Macrosiphum liriodendri* – identification of natural γ,γ-carotene. *Acta Chemica Scandinavica*, **25**, 3878–3880.

Behmer, S. T. and Nes, W. D. 2003. Insect sterol nutrition and physiology: a global view. *Advances in Insect Physiology*, **31**, 1–72.

Behmer, S. T., Elias, D. O. and Grebenok, R. J. 1999. Phytosterol metabolism and absorption in the generalist grasshopper *Schistocerca americana* (Orthoptera: Acrididae). *Archives of Insect Biochemistry and Physiology*, **42**, 13–25.

Brown, M. R., Sieglaff, D. H. and Rees, H. H. 2009. Gonadal ecdysteroidogenesis in Arthropoda: occurrence and regulation. *Annual Review of Entomology*, **54**, 105–125.

Budesinsky, M., Valterová, I., Sémon, E., Cancello, E. and Bordereau, C. 2005. NMR structure determination of $(11E)$-trinervita-1(14),2,11-triene, a new diterpene from sexual glands of termites. *Tetrahedron*, **61**, 10699–10704.

Butenandt, A. and Karlson, P. 1954. Über die Isolierung eines Metamorphosehormons der Insekten in kristallisierter form. *Zeitschrift für Naturforschung*, **9b**, 389–391.

Clark, A. J. and Bloch, K. 1959. The absence of sterol synthesis in insects. *Journal of Biological Chemistry*, **234**, 2578–2582.

Connor, W. E., Wang, Y., Green, M. and Lin, D. S. 2006. Effect of diet and metamorphosis upon the sterol composition of the butterfly *Morpho peleides*. *Journal of lipid Research*, **47**, 1444–1448.

Czeczuga, B. 1986. The presence of carotenes in various species of Lepidoptera. *Biochemical Systematics and Ecology*, **14**, 345–351.

Francke, W. and Dettner, K. 2005. Chemical signalling in beetles. *Topics in Current Chemistry*, **240**, 85–166.

Ghostin, J., Habib-Jiwan, J.-L., Rozenberg, R., Daloze, D., Pasteels, J. M. and Braekman, J.-C. 2007. Triterpenoid saponin hemibiosynthesis of a leaf beetle's (*Platyphora kollari*) defensive secretion. *Naturwissenschaften*, **94**, 601–605.

Goetz, M., Meinwald, J. and Eisner, T. 1981. Lucibufagins. 4. New defensive steroids and a pterin from the firefly *Photinus pyralis* (Coleoptera, Lampyridae). *Experientia*, **37**, 679–680.

Goh, S. H., Chuah, C. H., Beloeil, J. C. and Morellet, N. 1988. High molecular weight diterpenes and a new C-17 methylated trinervitene skeleton from a Malaysian termite *Hospitalitermes umbrinus*. *Tetrahedron Letters*, **29**, 113–116.

Gonzalez, A., Schroeder, F. C., Attygalle, A. B., Svatoš, A., Meinwald, J. and Eisner, T. 1999. Metabolic transformation of acquired lucibufagins by firefly "femmes fatales". *Chemoecology*, **9**, 105–112.

Gronquist, M., Meinwalt, J., Eisner, T. and Schroeder, F. C. 2005. Exploring uncharted terrain in nature's structure space using capillary NMR spectroscopy; 13 steroids from 50 fireflies. *Journal of the American Chemical Society*, **127**, 10810–10811.

Hojo, M., Matsumoto, T. and Miura T. 2007. Cloning and expression of a geranylgeranyl diphosphate synthase gene: insights into the synthesis of termite defence secretion. *Insect Molecular Biology*, **16**, 121–131.

Howard, R. W., Baker, J. E. and Morgan, E. D. 2003. Novel diterpenoids and hydrocarbons in the Dufour gland of the ectoparasitoid *Habrobracon hebetor* (Say). *Archives of Insect Biochemistry and Physiology*, **54**, 95–109.

Huang, X., Warren, J. T. and Gilbert, L. I. 2008. New players in the regulation of ecdysone biosynthesis. *Journal of Genetics and Genomics*, **35**, 1–10.

Ikekawa, N., Morisaki, M. and Fujimoto, Y. 1993. Sterol metabolism in insects: dealkylation of phytosterols to cholesterol. *Accounts of Chemical Research*, **26**, 139–146.

Ito, Y., Yasuda, A. and Sonobe, H. 2008. Synthesis and phosphorylation of ecdysteroids during ovarian development in the silkworm *Bombyx mori*. *Zoological Science*, **25**, 721–727.

Kornfeld, R. and Kornfeld, S. 1985. Assembly of asparagine-linked oligosaccharides. *Annual Review of Biochemistry*, **54**, 631–664.

Lafont, R., Harmatha, J., Marian-Poll, F., Dinan, L. and Wilson, I. D. 2002. *Ecdybase*. http://ecdybase.org.

Lafont, R., Dauphin-Villemant, C., Warren, J. T. and Rees, H. H. 2005. Ecdysteroid chemistry and biochemistry. In: *Comprehensive Molecular Insect Science* (Gilbert, L. I., Gill, S. and Iatrou, K., ed.), Elsevier, Amsterdam, vol. 3, pp. 125–195.

Laurent, P., Dooms, C., Braekman, J.-C., Daloze, D., Habib-Jiwan, J.-L., Rozenberg, R., Termonia, A. and Pasteels, J. M. 2003. Recycling plant wax constituents for chemical defense: hemi-biosynthesis of triterpene saponins from β-amyrin in a leaf beetle. *Naturwissenschaften*, **90**, 524–527.

Leclercq, S., de Biseau, J. C., Braekman, J. C., Daloze, D., Quinet, Y., Luhmer, M. Sundin, A. and Pasteels, J. M. 2000. Furanocembranoid diterpenes as defensive compounds in the Dufour gland of the ant *Crematogaster brevispinosa rochai*. *Tetrahedron*, **56**, 2037–2042.

Li, H., Wang, H., Hackett, M. and Schooley, D. A. 1995. The structure of dolichols isolated from *Manduca sexta* larvae. *Insect Biochemistry and Molecular Biology*, **25**, 1019–1026.

von Lintig, J. and Vogt, K. 2000. Filling the gap in vitamin A research. Molecular identification of an enzyme cleaving β-carotene to retinal. *Journal of Biological Chemistry*, **275**, 11915–11920.

Meinwald, J., Roach, B. and Eisner, T. 1987. Defence mechanisms of arthropods. 81. Defensive steroids from a carrion beetle (*Silpha novaboraensis*). *Journal of Chemical Ecology*, **13**, 35–38.

Murakami, K., Watanabe, B., Nishida, R., Mori, N. and Kuwahara, Y. 2007. Identification of crinosterol from astigamatid mites. *Insect Biochemistry and Molecular Biology*, **37**, 506–511.

Nasir, H. and Noda, H. 2003. Yeast-like symbiotes as a sterol source in anobiid beetles (Coleoptera, Anobiidae): Possible metabolic pathways from fungal sterols to 7-dehydrocholesterol. *Archives of Insect Biochemistry and Physiology*, **52**, 175–182.

Oberhauser, V., Voolstra, O., Bangert, A., von Lintig, J. and Vogt, K. 2008. NinaB combines carotenoid oxygenase and retinoid isomerase activity in a single polypeptide. *Proceedings of the National Academy of Sciences, USA*, **105**, 19000–19005.

Pasteels, J. M. and Daloze, D. 1977. Cardiac glycosides in the defensive secretion of chrysomelid beetles: evidence for their production by the insects. *Science*, **197**, 70–72.

Pondeville, E., Maria, A., Jacques, J. C., Bourgouin C. and Dauphin-Villemant, C. 2008. *Anopheles gambiae* males produce and transfer the vitellogenic steroid hormone 20-hydroxyecdysone to females during mating. *Proceedings of the National Academy of Sciences, USA*, **105**, 19631–19636.

Prestwich, G. D., Jones, R. W. and Collins, M. S. 1981. Terpene biosynthesis by nasute termite soldiers (Isoptera, Nasutitermitinae). *Insect Biochemistry*, **11**, 331–336.

Rees, H. H. 2004. Hormonal control of tick development and reproduction. *Parasitology*, **129**, Supplement S127–S143.

Schiff, N. M. and Feldlaufer, M. F. 1996. Neutral sterols of sawflies (Symphyta): their relationship to other Hymenoptera. *Lipids*, **31**, 441–443.

Schildknecht, H. 1970. The defensive chemistry of land and water beetles. *Angewandte Chemie, International Edition in English.* **9**, 1–9.

Seki, T. and Vogt, K. 1998. Evolutionary aspects of the diversity of visual pigment chromophores in the class Insecta. *Comparative Biochemistry and Physiology B*, **119**, 53–64.

Svoboda, J. A. and Feldlaufer, M. F. 1991. Neutral sterol metabolism in insects. *Lipids*, **26**, 614–618.

Tsuchida, K., Katagiri, C., Tanaka, Y., Tabunoki, H., Sato, R., Maekawa, H., Takada, N., Banno, Y., Fujii, H., Wells, M. A. and Zouni, Z. E. 2004. The basis for colorless hemolymph and cocoons in the Y-gene recessive *Bombyx mori* mutants: a defect in the cellular uptake of carotenoids. *Journal of Insect Physiology*, **50**, 975–983.

Van Oycke, S., Braekman, J. C., Daloze, D. and Pasteels, J. M. 1987. Cardenolide biosynthesis in chrysomelid beetles. *Experientia*, **43**, 460–462.

CHAPTER 9
Aromatic Compounds

9.1 AROMATIC COMPOUNDS IN NATURE

Plants and micro-organisms have an exclusive route to create benzene-ring compounds, not shared by animals. The great majority of aromatic compounds in nature are produced by plants and micro-organisms through the shikimic acid pathway. All animals are dependent upon plants for basic aromatic compounds, either directly in food, or indirectly. A small number of aromatic compounds are made by micro-organisms and possibly by insects from polyketides, some of them described in Chapter 6. Aromatic compounds from polyketides usually have a number of oxygen functional groups, pointing to their polyketide origin. Simple benzenoid compounds, such as the amino acids phenylalanine and tryptophan, must be obtained by animals, including insects, in their food. When a mutant strain of *E. coli* bacterium, produced by ionizing radiation, was unable to make the three aromatic amino acids, as well as *p*-aminobenzoic acid and *p*-hydroxybenzoic acid, it was discovered that shikimic acid could restore to the bacterium the ability to make the amino acids. The synthesis route to these benzenoid compounds was therefore called the shikimic acid pathway. While the shikimic acid steps are unavailable to insects, the metabolism of all animals is vitally dependent upon them.

9.2 THE SHIKIMIC ACID PATHWAY

The shikimic acid pathway requires the C_4 sugar erythrose-4-phosphate and phosphoenol pyruvic acid (a derivative of pyruvic acid, locked in its enol form, see Figures 1.1 and 2.19) as starting materials. A total of ten carbon atoms are required, four from erythrose and six from two molecules of pyruvate, with one carbon atom later lost as CO_2. The final product therefore is a C_9 compound, so that such products are referred to as phenyl-C_3 compounds. Many plant substances have this phenyl-C_3 structure. An outline of the shikimic acid pathway is shown in Figure 9.1. Greater detail of the mechanisms of the reactions and the enzymes required can be found on the Queen Mary College enzymes site (www.chem.qmul.ac.uk/iubmb/enzyme/reaction/misc/shikim.html).

In the first step, a molecule of phosphoenol pyruvic acid is attached to a carboxyl group on the enzyme. The resulting ester loses phosphate to give an enolpyruvate ester, which condenses readily with the aldehyde of erythrose-4-phosphate, to give 3-deoxy-D-arabino-heptulosonate-7-phosphate (DAHP), catalysed by DAHP synthetase. A second loss of phosphate and a similar condensation through the nucleophilic addition of the enol to the α-keto-group derived from pyruvic acid, gives the six-membered ring that will become a phenyl group. This first cyclized product is dehydroquinic acid; the ring closure catalysed by dehydroquinate synthase. By dehydration of a β-hydroxyketone catalysed by 3-dehydroquinate dehydrase, followed by reduction of the ketone (through a Schiff's base formed with an NH_2 group on the enzyme), shikimic acid is reached, the key compound in the series.

Shikimic acid is phosphorylated (with adenosine triphosphate catalysed by shikimate kinase) before addition of the second molecule of phosphoenol pyruvic acid. Loss of another phosphate gives an enol ether of shikimic acid phosphate and pyruvic acid. Loss of phosphate for the fourth time gives a second double bond, and another key intermediate called chorismic acid. This compound undergoes a Claisen rearrangement to prephenic acid, followed by a concerted loss of CO_2 and H_2O to give the third double bond and the first aromatic phenyl-C_3 compound, phenylpyruvic acid (Figure 9.1).

9.3 PHENYL-C_3 COMPOUNDS

One of the first important derivatives of phenylpyruvic acid is the essential amino acid phenylalanine, produced with the aid of pyridoxamine (Figure 2.17), coupled with the deamination of glutamic acid to 2-oxoglutaric acid. Phenylalanine is described as an essential amino acid because all animals must have it in their diet. Tyrosine (*p*-hydroxyphenylalanine) is made in plants directly from prephenic acid (Figure 9.2) by oxidation, concurrently with decarboxylation. Mammals can make tyrosine by oxidation of phenylalanine, but insects cannot,

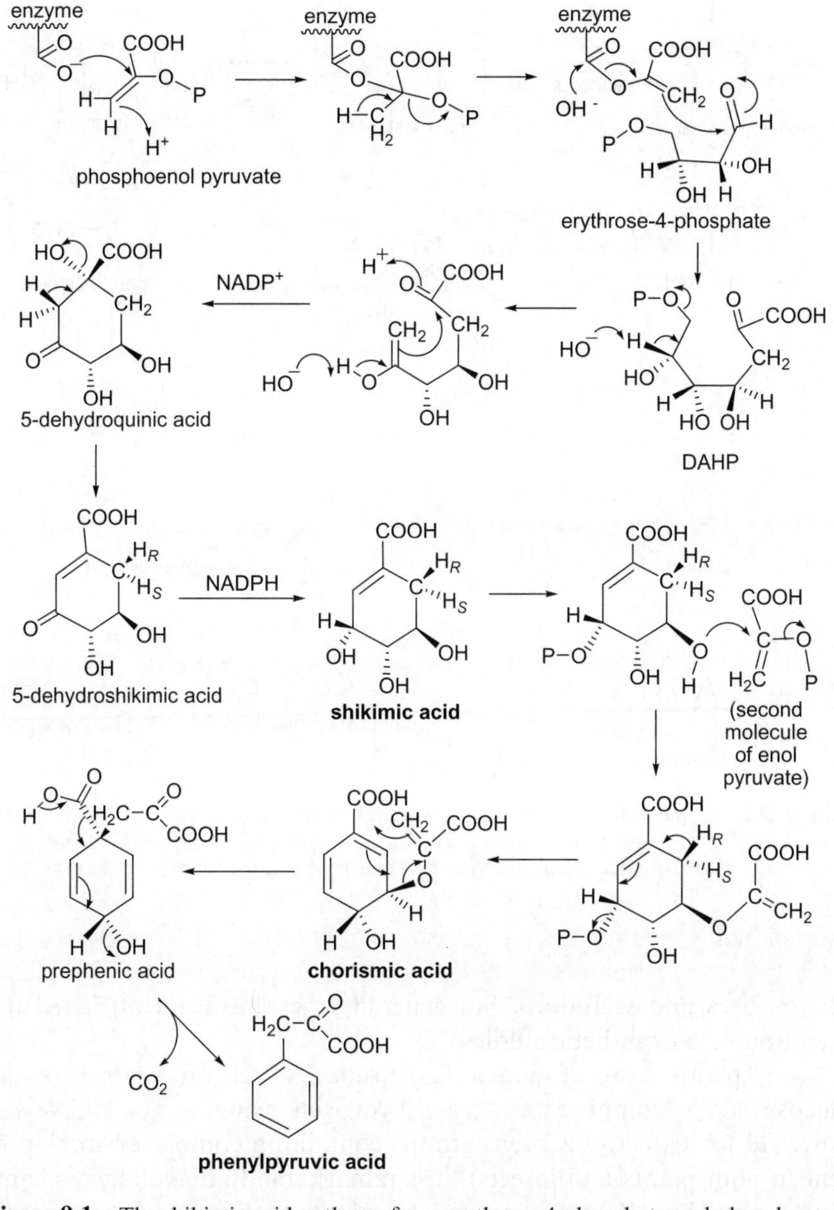

Aromatic Compounds

Figure 9.1 The shikimic acid pathway from erythrose-4-phosphate and phosphoenolpyruvic acid to phenylpyruvic acid.

and must obtain it in food. Cinnamic acid is formed from phenylalanine, and not, as might have been expected, from phenylpyruvic acid. Some other simple phenyl-C_3 compounds and derivatives are shown in Figure 9.2. Many such compounds have been found in insect

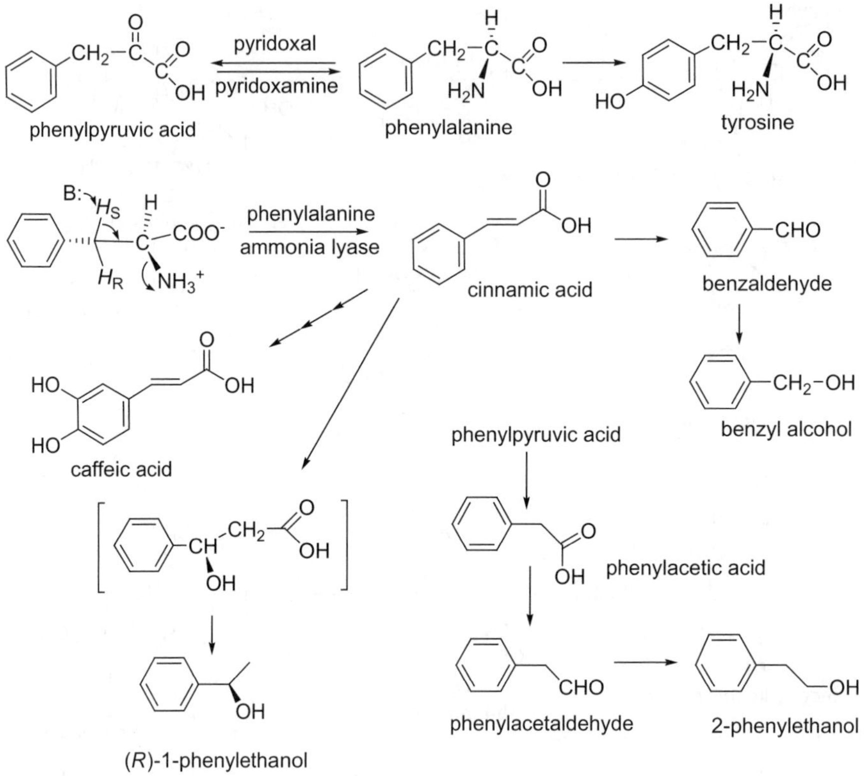

Figure 9.2 Some simple phenyl-C_3 derivatives from phenylpyruvic acid, including some insect pheromones. The formation of 1-phenylethanol has not been investigated. Benzaldehyde can be formed in more than one way.

pheromones and secretions, but generally they have not attracted the attention of biosynthetic studies.

Many plants store cyanogenic glucosides, which on removal of the glucose can decompose to release hydrogen cyanide. As HCN is a powerful toxicant to all haem groups containing complexed iron (present in both plants and insects), it is remarkable that cyanogenic compounds can be safely sequestered by plants or insects. Some insects sequester cyanogens from food plants (Section 11.4), others can make them (Zagrobelny *et al.*, 2008). One of these cyanogens is mandelonitrile (Figure 9.3).

The flat-backed millipede *Oxidus gracilis* (Diplopoda: Paradoxosomatidae) has been known for over 100 years to produce benzaldehyde and HCN when disturbed. It was shown to convert racemic

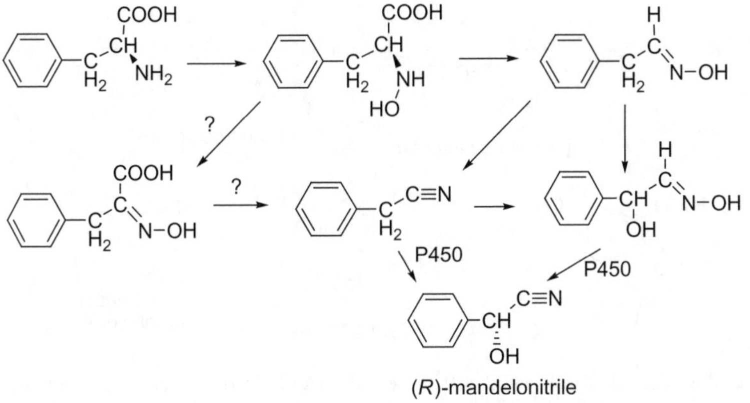

Figure 9.3 The investigated biosynthesis of mandelonitrile from phenylalanine in the millipede *Harpaphe haydeniana*. The studies have not settled which of the intermediate steps is followed, but those with a question mark are less likely.

[2-^{14}C]phenylalanine and [1(ring)-^{14}C]phenylalanine to benzaldehyde, while only [2-^{14}C]phenylalanine gave labelled hydrogen cyanide (Towers *et al.*, 1972). Neither labelled benzaldehyde nor hydrogen cyanide was formed from [2-^{14}C]tyrosine. In the millipede *Apheloria corrugata* (Diplopoda: Polydesmidae) the secretory gland consists of two chambers. The mandelonitrile is stored in the inner chamber and lytic enzymes in the outer chamber, with a muscle to close the opening between them (Eisner and Meinwald, 1966). The valve only opens when the secretion is discharged (compare with the bombardier beetle, Section 9.5.1).

The steps from phenylalanine to mandelonitrile were studied in the yellow spotted millipede *Harpaphe haydeniana* (Diplopoda: Polydesmidae), by feeding 21 different potential precursors, labelled with either ^{14}C or tritium. The compound most efficiently incorporated was phenylacetonitrile. From the total results the biosynthetic scheme shown in Figure 9.3 was constructed (Duffey *et al.*, 1974). Phenylalanine was much less well incorporated into mandelonitrile than N-hydroxyphenylalanine, but that may be because there are many competing uses for phenylalanine. Phenylpyruvic acid oxime was also incorporated, but not as well as phenylacetaldehyde oxime, which left two possible routes open in the middle of the reaction sequence. It does not seem that the uncertainty has yet been resolved. Mandelonitrile is often accompanied by benzoyl cyanide in millipede secretions (Mori *et al.*, 1995).

Many species of tiger beetles (Coleoptera: Cicindelidae) also store mandelonitrile, releasing benzaldehyde and hydrogen cyanide. They

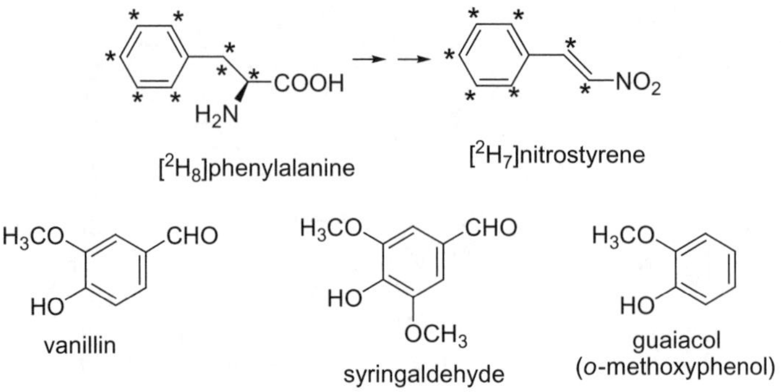

Figure 9.4 Biosynthesis of 2-nitroethenylbenzene (nitrostyrene), a defensive compound from a millipede. Asterisks indicate deuterium labelling. Some simple aromatic compounds found in abdominal glands of adult male leaf-footed beetles, which are probably recognition pheromones.

probably produce the mandelonitrile from phenylalanine. Several species contain benzoic acid, thiobenzoic acid, phenylacetic acid or methyl salicylate (Francke and Dettner, 2005). Benzaldehyde has been found widely in Coleoptera and in some Hymenoptera. Benzoic acid (in Coleoptera) and phenylacetic acid (in ants) are also widely found, as well as a number of simple esters of phenylacetic acid.

9.3.1 Aromatic Pheromones

A number of compounds easily recognized as derivatives of the phenyl-C_3 series have been identified as pheromones (Figure 9.4). Adult males of a number of Lepidoptera use simple phenyl-C_1, and -C_2 compounds like benzyl alcohol, benzaldehyde, *p*-hydroxybenzaldehyde, phenylacetaldehyde and 2-phenylethyl acetate as copulating pheromones. For example, ethyl cinnamate has been found as a component of the male sex pheromone of the oriental fruit fly *Grapholitha* (*Laspeyresia*) *molesta* (Lepidoptera: Baculoviridae). 2-Phenylethanol has been found in the mandibular glands of the ant *Camponotus clarithorax*, and provides the male sex pheromone of the bertha armyworm *Mamestra configurata* (Lepidoptera: Noctuidae). In the latter case labelling showed that the 2-phenylethanol was derived from phenylalanine, and the route probably proceeds *via* cinnamic acid, rather than phenylpyruvic acid, since the pheromone was also produced from [3-^{14}C]cinnamic acid (Weatherstone and Percy, 1976). It was suggested that phenyllactic acid was also an intermediate and the final precursor to phenylethanol was its glucoside.

(*R*)-1-Phenylethanol is the trail pheromone of the ant *Aphaenogaster cockerelli*. Coniferyl alcohol is an attractant for females of the oriental fruit fly, but its immediate precursors are obtained from flowers (Section 11.6). Adult males of leaf-footed bugs (*Leptoglossus* species, Hemiptera: Coreidae) use a variety of simple aromatic compounds in abdominal glands, apparently for species recognition. Among these are guaiacol, vanillin, cinnamyl alcohol and syringaldehyde (Figure 9.4). The biosynthesis of phenol and guaiacol in *Leptoglossus phyllopus* (plate 27) has been traced to tyrosine (Section 9.4).

The curious example of (*E*)-2-nitroethenylbenzene (2-nitrostyrene) (Figure 9.4) has been found in the defensive secretion of the millipedes *Eucondylodesmus elegans* (Diplopoda: Doratodesmidae) and *Thelodesmus armatus* (Diplopoda: Pyrgodesmidae) and shown to be repellent to ants. When [^{2}H$_8$]phenylalanine was fed to the millipedes, [^{2}H$_7$]-2-nitroethenylbenzene was detected by mass spectrometry (Kuwahara *et al.*, 2002). Nitrostyrene has long been known for its fungistatic, insecticidal and bactericidal effects, but this was the first time it had been found as a natural product.

9.3.2 Compounds from Chorismic Acid

Various other aromatic compounds are obtained by plants and microorganisms from the intermediate chorismic acid in Figure 9.1, including the other "essential" amino acid tryptophan. In the presence of water, catalysed by isochorismate synthase, chorismic acid gives isochorismic acid, which loses pyruvic acid to give salicylic acid. The action of L-glutamine (with pyridoxal phosphate, providing ammonia) on chorismic acid gives 2-amino-2-desoxyisochorismic acid, which eliminates water and pyruvic acid to give anthranilic acid (Figure 9.5). Addition of ribose phosphate to anthranilic acid gives after several more stages tryptophan (Figure 9.5), which again can only be made by plants and microorganisms. From tryptophan are derived indole and skatole, evil-smelling compounds found as trail pheromones in ants and defensive secretion of caddis flies (Trichoptera). In the presence of *p*-aminobenzoate synthase, chorismic acid and ammonia (again from glutamine) give *p*-aminobenzoic acid (Figure 9.5). At our present state of knowledge we do not know which, if any, of these syntheses can be accomplished by insects, or whether the compounds are made by symbiotic bacteria harboured by insects.

Methyl anthranilate has been found in the mandibular gland of males of the ant *Camponotus nearticus*, and is one component of the trail pheromone (with methyl nicotinate) of an *Aenictus* species of army ant. Glomerine and homoglomerine, from anthranilic acid, form the defensive secretion of the myriapod *Glomeris marginata* (Diplopoda:

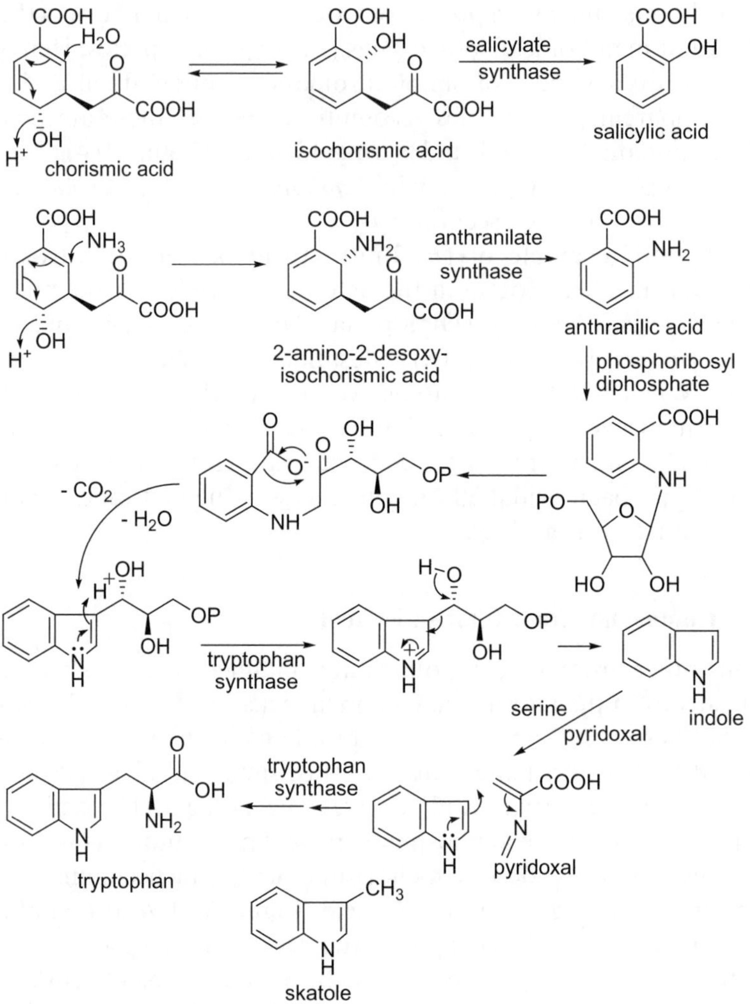

Figure 9.5 Aromatic compounds from chorismic acid. The route from chorismic acid to tryptophan is found only in plants and micro-organisms. The indole formed as an intermediate is not released. Indole and skatole biosynthesis in insects has not been studied, but they are probably formed from tryptophan *via* indoleacetic acid.

Glomeridae) that is toxic to mice and causes paralysis in spiders. It has been shown that anthranilic acid labelled with ^{14}C in the carboxyl group, injected or fed to *G. marginata*, was incorporated into homoglomerine (Schildknecht and Wenneis, 1967). Alkaline hydrolysis of labelled homoglomerine gave N-methylanthranilic acid with 97% of the activity of homoglomerine, showing that all the radioactivity was still located in

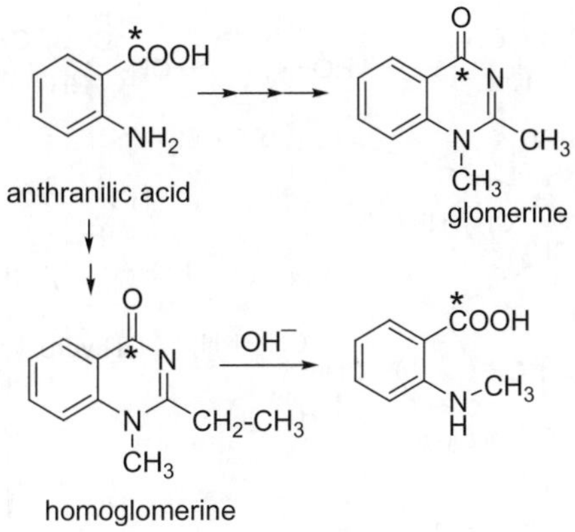

Aromatic Compounds

Figure 9.6 Anthranilic acid labelled with ^{14}C in the carboxyl group gives labelled glomerine and homoglomerine. The label is still found in the carboxyl group of N-methylanthranilic acid produced by degradation of homoglomerine.

the carboxyl group (Figure 9.6). The compounds were not effective against ants or beetles.

9.3.3 Aromatic Amines

By a sequence of important reactions the essential amino acids phenylalanine and tryptophan are converted into amines which have a physiological role in animals as neurotransmitters, neuromodulators or neurohormones. A rather non-specific aromatic amino acid decarboxylase catalyses the decarboxylation by pyridoxal phosphate (Section 2.2.6). The pharmacological importance of these amines is beyond the range of this book. The aromatic amines are just a few of the neurotransmitters known. Those used by insects overlap with those used by mammals, but our knowledge of them is incomplete (Blenau and Baumann, 2001). Tyramine, dopamine, serotonin (5-hydroxytryptamine) and octopamine are known to be important insect neurotransmitters. Octopamine is important for rapid action, such as flying or jumping; tyramine and serotonin have octopamine-opposing effects.

9.3.3.1 Adrenaline Group. Some of the essential amino acid tyrosine is hydroxylated by animals, including insects, to dihydroxyphenylalanine (DOPA) (Figure 9.7). DOPA is decarboxylated to dopamine, which in turn gives noradrenaline (norepinephrine). Noradrenaline is methylated to adrenaline (in mammals, a hormone from the adrenal glands). Both dopamine and noradrenaline are found in the venom of wasps and

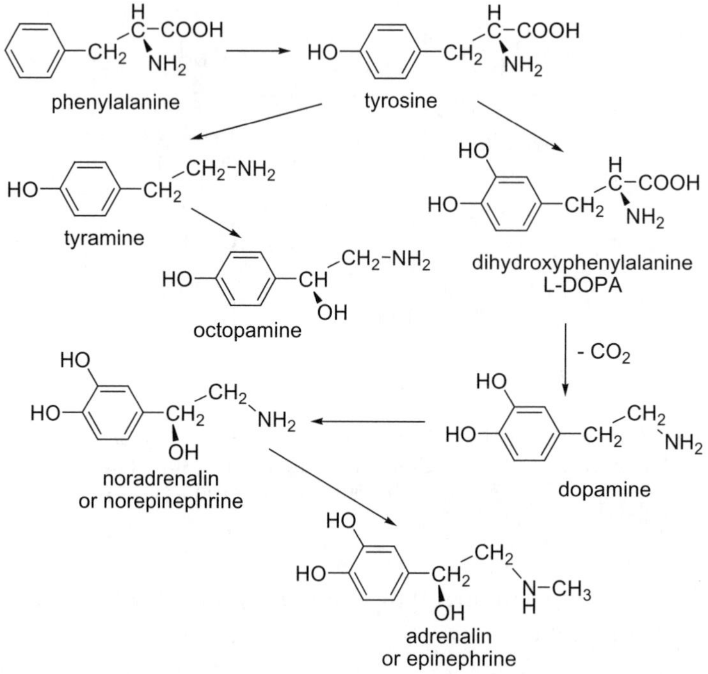

Figure 9.7 Neurotransmitters from phenylalanine.

the honeybee; adrenaline is a minor component of wasp venom. There is about 1 mg of dopamine per gram of bee venom. These substances appear to be present in the nervous systems of insects and their relatives too. Dopamine is found in high concentration in the bee brain, and dopamine, or something like it, stimulates production of pheromone in female ticks after they have taken a blood meal. Tyramine is formed by decarboxylation of tyrosine by pyridoxal phosphate. Hydroxylation of tyramine gives octopamine (Figure 9.7).

9.3.3.2 Serotonin Group. A relationship similar to that between the amino acid tyrosine and dopamine exists between the amino acid tryptophan and serotonin or 5-hydroxytryptamine (Figure 9.8). Serotonin also is present in high concentration in the bee brain. Tryptamine is known in the venom of scorpions, while serotonin is found widely in the venom of honeybees, centipedes and two spiders. It has not been found in the venom of ants or solitary wasps but social wasps have it in quantity, as much as 1 µg per insect. It is present in the irritating hairs of the larvae of the Tiger moth *Arctia caja* (Lepidoptera: Arctiidae) and a saturnid butterfly larva, but in a study of the hairs of twelve species from

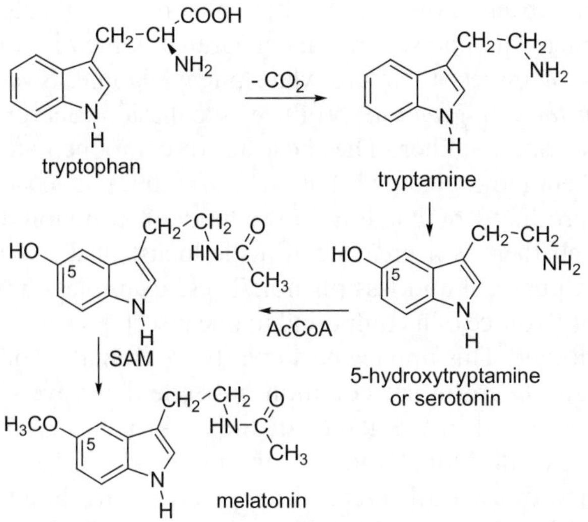

Aromatic Compounds 265

Figure 9.8 Neurotransmitters and the hormone melatonin from tryptophan.

four lepidopteran families no serotonin was found. Melatonin, formed by the action of *S*-adenosyl methionine (Section 2.2.8) on N-acetylserotonin, is a hormone controlling daily (circadian) rhythm of sleep in mammals. It has been shown to be present in the brain of the desert locust *Schistocerca gregaria* (Orthoptera: Acrididae) and the silkworm *Bombyx mori* (Lepidoptera: Bombyciae). Itoh *et al.* (1995) found the synthesis and release of melatonin in the silkworm occurs in a circadian rhythm and suggested it was related to day length.

9.4 PHENOLS

Phenols are found widely among insects, but chiefly in Coleoptera, Orthoptera (crickets and locusts), Isoptera (termites) and Blattodea (cockroaches). Phenol itself is widely scattered, from the defensive secretion of millipedes (diplopods) and opilionids (daddy longlegs or harvestmen) to grasshoppers and beetles. Phenols provide protection upwards against predators and downwards against micro-organisms. They can be formed through a variety of biosynthetic routes. Some phenols have already been encountered among acetogenins (Chapter 6). They can also be formed from tyrosine or cinnamic acid. Phenol and guaiacol (*o*-methoxyphenol) from the ventral abdominal glands of the Florida leaf-footed bug *Leptoglossus phyllopus* (Hemiptera: Coreidae) (Plate 27) are formed from tyrosine (Duffey *et al.*, 1977). Three species of millipede have been shown to produce phenol and guaiacol from tyrosine, but phenylalanine was not incorporated, which is consistent with the inability of insects to hydroxylate phenylalanine. Swarming desert locusts *Schistocerca gregaria* produce a cohesion pheromone, which keeps the swarms together. The most active components of the pheromone are phenol and guaiacol. Dillon *et al.* (2002) have shown that the guaiacol is produced by bacteria in the hindgut, probably from vanillic acid, of which there is a great deal in the locust faeces. Axenic locusts produced no guaiacol and less phenol. These examples show how demonstration of a source of a compound in one insect does not mean it is the source in another. This applies particularly to aromatic compounds.

o-Cresol and *m*-cresol are common in beetle defensive glands and *p*-ethylphenol is found in the glands of the cockroach *Periplaneta americana* (Blattodea: Blattidae) and is probably responsible for that insect's characteristic odour. Both *o*-cresol and *p*-cresol are known in the defensive secretion of the eastern lubber grasshopper *Romalea microptera* (Orthoptera: Acrididae). Other alkylphenols (*e.g.* 2,3-dimethylphenol (guaiacol or methylcatechol) and 3-ethyl-6-methylphenol (of unknown origin) are present in some opilionids.

p-Hydroxybenzoic acid can be formed in any of several ways. For example, benzoic acid can be obtained by β-oxidation of cinnamic acid, or by oxidation of benzaldehyde, in turn obtained by a reverse aldol condensation from cinnamic acid. *p*-Hydroxybenzoic acid can arise from 3-dehydroshikimic acid or chorismic acid, as well as from tyrosine or cinnamic acid. Further hydroxylation is introduced by flavin adenine dinucleotide (FAD Section 2.2.4), catalysed by a phenolic hydroxylase. FAD is reduced to $FADH_2$ by NADPH; reaction of $FADH_2$ with oxygen gives a hydroperoxide, which reacts at the activated *ortho*-position to the phenol (Figure 9.9). Protocatechuic acid can also be formed from 3-dehydroshikimic acid by dehydration and enolization. Which, if any, of these routes is available to insects is still not known.

The predaceous diving beetles (Coleoptera: Dytiscidae) have a pair of pygidial glands which provide deterrents. They frequently contain benzoic acid, phenylacetic acid or *p*-hydroxybenzaldehyde. The pygidial glands of the great diving beetle *Dytiscus marginalis* contain a yellow pigment, marginalin (Figure 9.10). The gland also contains the clues to the biosynthesis of marginalin, *p*-hydroxybenzaldehyde and homogentisic acid (Figure 9.12). *In vitro* base-catalysed condensation of these latter two gives marginalin.

Nymphs and adults of lace bugs (Hemiptera: Tingidae) secrete from abdominal glands phenolic defensive compounds which may have an acetogenin origin. The azalea lace bug *Stephanitis pyrioides* (Plate 28)

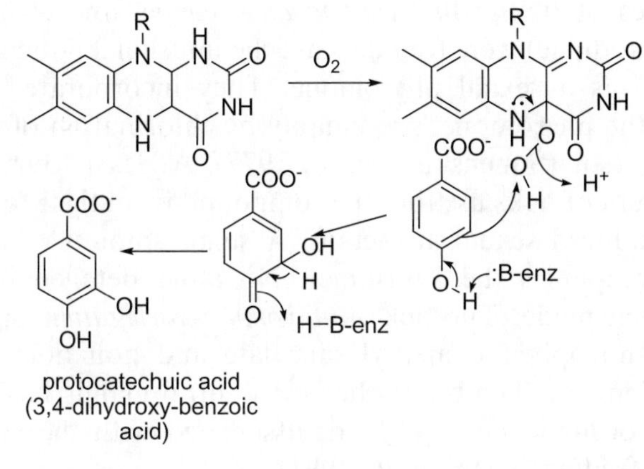

Aromatic Compounds

Figure 9.9 The oxidation of *p*-hydroxybenzoic acid by oxygen and FAD to protocatechuic acid.

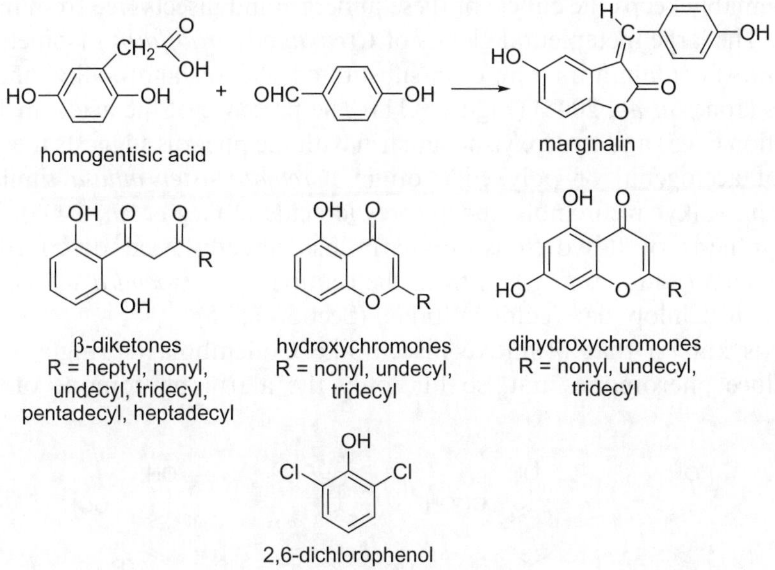

Figure 9.10 The probable route to marginalin, defensive secretion of *Dytiscus marginalis* and defensive compounds from *Stephanitis* lace bugs. 2,6-Dichlorophenol is the sexual pheromone of a number of species of ticks.

secretes 2,6-dihydroxyacetophenone, a β-diketone where R is nonyl (Figure 9.10) and a chromone where R is nonyl. The rhododendron lace bug *Stephanitis rhododendri* secretes a clear fluid from abdominal glands which contain a number of phenolic β-diketones (Figure 9.10) where R is odd numbered from heptyl to heptadecyl and chromones where R is undecyl or tridecyl (Oliver *et al.*, 1987).

Two species of ticks (*Amblyomma americanum* and *A. maculatum*, both Acari: Ixodidae) were found to use the unusual 2,6-dichlorophenol (Figure 9.10) as a sexual pheromone. They incorporated ^{36}Cl from Na^{36}Cl into the pheromone, presumably by chlorination of tyrosine or another precursor (Sonenshine *et al.*, 1977). At least fourteen species from five genera of ticks use this one compound from their foveal glands as female-produced sexual attractant. A small amount of 2,4-dichlorophenol accompanies the 2,6-isomer. No more detailed biosynthetic study has been made. The tick *Amblyomma variegatum* on the other hand uses *o*-nitrophenol, methyl salicylate and nonanoic acid (ratio 2:1:8 µg per female). The biosynthesis in ventral glands has been demonstrated to occur in dermal glands associated with the ventrolateral cuticle, after feeding (Diehl *et al.*, 1991).

The paired metapleural glands on the thoraces of ants have been shown in several examples to secrete an antibiotic or antifungal mixture, that presumably keeps the cuticle of these underground insects free from infection. The large metapleural glands of *Crematogaster difformis* (sometimes *deformis*) contain a mixture of *m*-substituted phenols and 6-alkylsalicylic acids (Jones *et al.*, 2005) (Figure 9.11). The presence of the acids, mellein (Section 6.3.5) and hydroxyisocoumarin with the phenols suggests they are all of acetogenin or polyketide origin. *Crematogaster inflata* similarly contains alkyl resorcinols and resorcylic acids (Jones *et al.*, 2005). The biosynthesis of 8-hydroxyisocoumarin has already been described as centipedin (Kim *et al.*, 1998) from the centipede *Scolopendra subspinipes mutilans* (Chilopoda: Scolopendridae) (Section 6.3.5).

It is known that primitive insects like Collembola (springtails) do produce pheromones but, so far, only the alarm pheromone of one

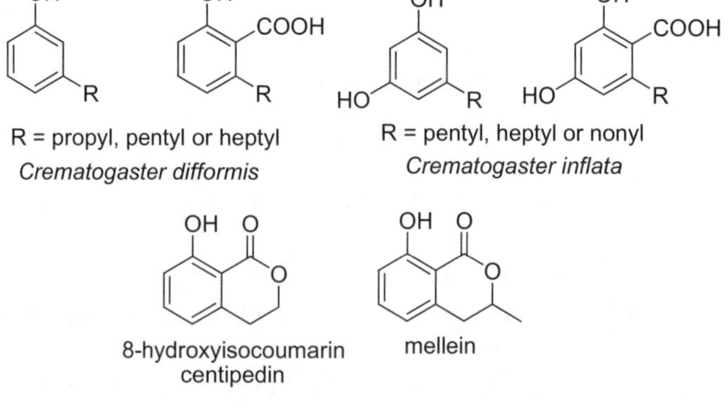

Figure 9.11 Protective phenolic compounds from the metapleural glands of *Crematogaster* ants.

species *Neanura muscorum* (Neanuridae) (Plate 29) has been identified as 1,3-dimethoxybenzene (Messer *et al.*, 1999). Its origin is unknown. Other simple aromatic compounds such as 2,4-dimethoxyaniline, phenol and 2-aminophenol were found in whole body extracts.

9.5 QUINONES

Quinones are distributed widely from opilionids and millipedes to grasshoppers, cockroaches and caddis flies (Trichoptera), but are most frequently found in beetles (Coleoptera).

There are diverse routes to the biosynthesis of quinones, and it is said that their structure is poor guidance to their origins. More complex ones often have polyketide origins (Chapter 6), and simpler ones are derived from shikimic acid. Quinones can be formed from tyrosine through oxidation to homogentisic acid (Figure 9.12). This uses a complex sequence of reactions involving hydroxylation and a 1,2-shift of the side chain and decarboxylation, catalysed by 4-hydroxyphenylpyruvate dioxygenase, an enzyme present in most organisms. Labelling experiments have shown that both oxygen atoms of dioxygen are incorporated into homogentisic acid. If homogentisic acid undergoes another decarboxylation, methylhydroquinone is produced.

Phenols have been shown to be oxidized to quinones in both millipedes and beetles. The tenebrionoid beetle *Zophobas rugipes* secretes phenol, *m-cresol* and *m*-ethylphenol from paired prothoracic glands, and benzoquinone, toluquinone and ethylbenzoquinone from paired abdominal glands (Tschinkel, 1969). It was shown some time ago in the beetle

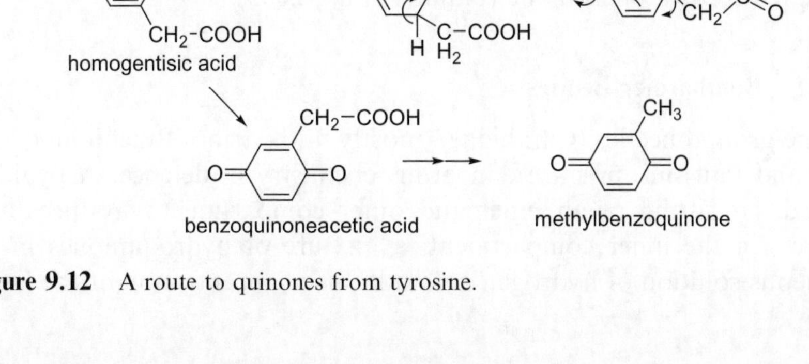

Figure 9.12 A route to quinones from tyrosine.

Eleodes longicollis (Tenebrionidae) that quinones can arise by two independent pathways. Labelled tyrosine, acetic, propionic and malonic acids were all used. Benzoquinone itself was preferentially made from labelled tyrosine, while simple alkylquinones were produced by the acetate pathway (Eisner and Meinwald, 1966; Eisner *et al.*, 1977). Propionic acid was only incorporated well into ethylbenzoquinone. The confusion left by early biosynthetic studies on quinones has not yet been clarified. It has been said that opilionids and millipedes make quinones from pre-existing aromatic substances while insects can make them from acetates. On the other hand, the benzoquinone and methylbenzoquinone in the secretion of the millipede *Narceus gordanus* (Diplopoda: Spirobolidae) were shown to be made from ^{14}C-labelled 6-methylsalicylic acid (Figure 4.2), indicating a polyketide pathway (Monro *et al.*, 1962). The opilionid *Vonones sayi* (Cosmetidae) secretes a mixture of two quinones, 2,3-dimethylbenzoquinone and trimethylbenzoquinone, both of which are solids, but the melting point of the mixture is lowered below room temperature. The quinones are stored as a pure liquid, since quinones are unstable in water, and mixed with fluid from the mouth only on ejection (Eisner *et al.*, 1971).

Benzoquinone, hydroquinone, methyl-, methoxy-, dimethyl- and ethyl-benzoquinones have all been identified in beetles, and some of them in cockroaches, grasshoppers and an earwig *Doru taeniatum* (Dermaptera: Forficulidae). An examination of 147 species of tenebrionid beetles from 55 genera showed they all contain methylbenzoquinone and ethylbenzoquinone, but only occasionally benzoquinone and then only in traces (Tschinkel, 1975). Three species of tenebrionid beetle *Argoporus alutacea*, *A. bicolor* and *A. rufipes* all produce a viscous, orange defensive secretion consisting of four naphthoquinones (Figure 9.13) as well as benzoquinone and its methyl and ethyl derivatives (Tschinkel, 1975). The female-produced sex pheromone of the German cockroach *Blattella germanica* (Blattellidae) has been identified as gentisylquinone isovalerate and given the trivial name blattellaquinone (Figure 9.13). The authors suggest it may be a defensive compound adapted to be a pheromone (Nojima *et al.*, 2005).

9.5.1 Bombardier Beetles

Some ground beetles (Carabidae), mostly in the tribes Brachinini (Plate 30) and Paussini, practise dangerous chemistry in defence. A pygidial gland, consisting of an inner and outer compartment, produces and stores, in the inner compartment, a mixture of hydroquinones in an aqueous solution of hydrogen peroxide. The concentration of the latter

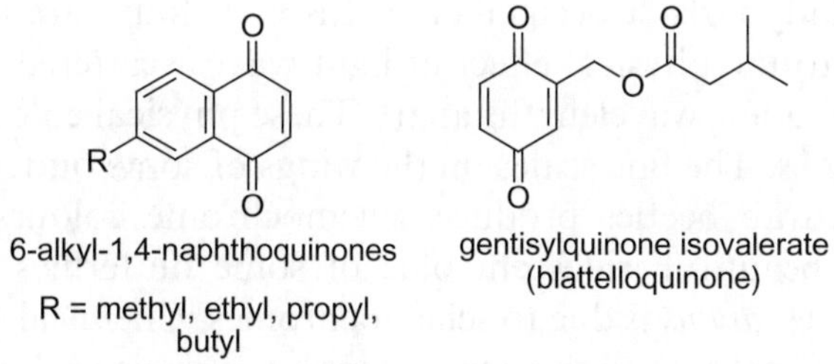

Figure 9.13 Alkylnaphthoquinones are defensive secretions of *Argoporus* beetles. Blatelloquinone is the female sex pheromone of the German cockroach *Blattella germanica*.

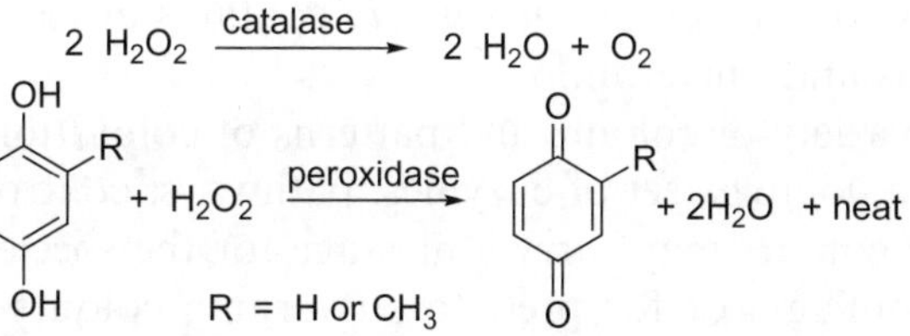

Figure 9.14 The reactions in the abdominal gland of bombardier beetles.

can be up to 28%. When the beetle is disturbed or attacked, the mixture is discharged into the outer thick-walled chamber, which is supplied with many small glands secreting a mixture of four catalases and three peroxidases (Schildknecht, 1971; Eisner and Aneshansley, 1999). The resulting reactions are given in Figure 9.14. Quinones, gas and heat are evolved in a very rapid reaction, with an audible "plop", taking the temperature of the exploded mixture to 100 °C. It is noteworthy that the temperature optimum of the catalase is between 70 and 80 °C (Aneshansley *et al.*, 1969). The discharge occurs at the rate of about 500 times per second, for about 30 discharges. The beetles of Brachinini can turn their abdominal tips through an arc of 270° to direct the explosion at any predator.

9.6 INSECT PIGMENTS

The colours of insects are as varied as those of a fashion designer' dress show; from the satiny black of the elytra of some beetles and cuticle of black wood ants, through the gaudy colours of some butterflies, to the pinstripes of the Colorado potato beetle (*Leptinotarsa*) and the polka dots of the ladybirds (Coccinellidae). Insect colours are due either to pigments or to the physical effect of light waves scattered from regular lines or ridges a few wavelengths apart. These physical colours are called schemochromes. The fine scales on the wings of some butterflies and the surface of scarab beetles produce schemochrome colours (Chapman, 1998a). The beautiful iridescent blue of some butterflies like *Morpho rhetenor* and *M. didius* is due to schemochromes. Chemical substances in the cuticle, on scales or in the haemolymph that cause colours are called chemochromes. Insects do not confine themselves to one kind of pigment and, it is becoming clear, many mix both schemochromes and chemochromes in the same structure. For example, the orange sulfur butterfly *Colias eurytheme* (Pieridae) (Plate 31) mixes pterin pigments with ultraviolet iridescence (Rutowski *et al.*, 2005). The haemolymph of the migratory locust *Locusta migratoria* contains at least melanin, carotenes, pterins and biliverdin.

The great variety of colours and patterns of coloration of insects are governed by a complex set of enzymes, pathways, control elements and genetics. The colours may carry a message for the species (recognition, mating, camouflage) or for predators (warning colours of aposematic species). However, it must be remembered that the spectra of colours seen by vertebrates and insects are different; that of insects is shifted about 100–150 nm towards the ultraviolet (Section 8.9.1).

Carotenes, the most widely distributed natural pigments, have already been considered in Chapter 8. Flavonoids are sequestered by insects from plants, probably the only animal group that do so.

9.6.1 Melanin

Melanin is an insoluble, unreactive and irregular polymer of indole rimgs, derived from tyrosine. It is the major pigment of insects, as well as of vertebrates, of human skin and animal hair, and the sepia of squid. It has been the subject of a great volume of research, but only the first stages of the formation of melanin are well understood. While vertebrate melanin is essentially a polymer of DOPA, insect melanin can be a polymer of either dopamine or DOPA, depending upon its purpose. The melanin used in encapsulation of invading micro-organisms and wound healing is formed from DOPA (Section 4.2.3.1). Reactive quinone intermediates in the biosynthetic pathway have antibiotic properties. Melanin in cuticles can be either in the form of dense granules, or it can be diffusely spread in regions of the cuticle. Melanin granules are an important component of the cuticle and help to strengthen it (Riley, 1997), and also act as a protectant from UV damage. The melanin in granules is formed from DOPA, the diffuse melanin appears to be a polymer of dopamine.

The initial and critical step in melanin formation is the oxidation of tyrosine to DOPA (dihydroxyphenylalanine), catalysed by phenoloxidase (Section 9.3.3.1), and the further oxidation of DOPA to dopaquinone by the same enzyme (Figure 9.15). A non-enzymic reaction

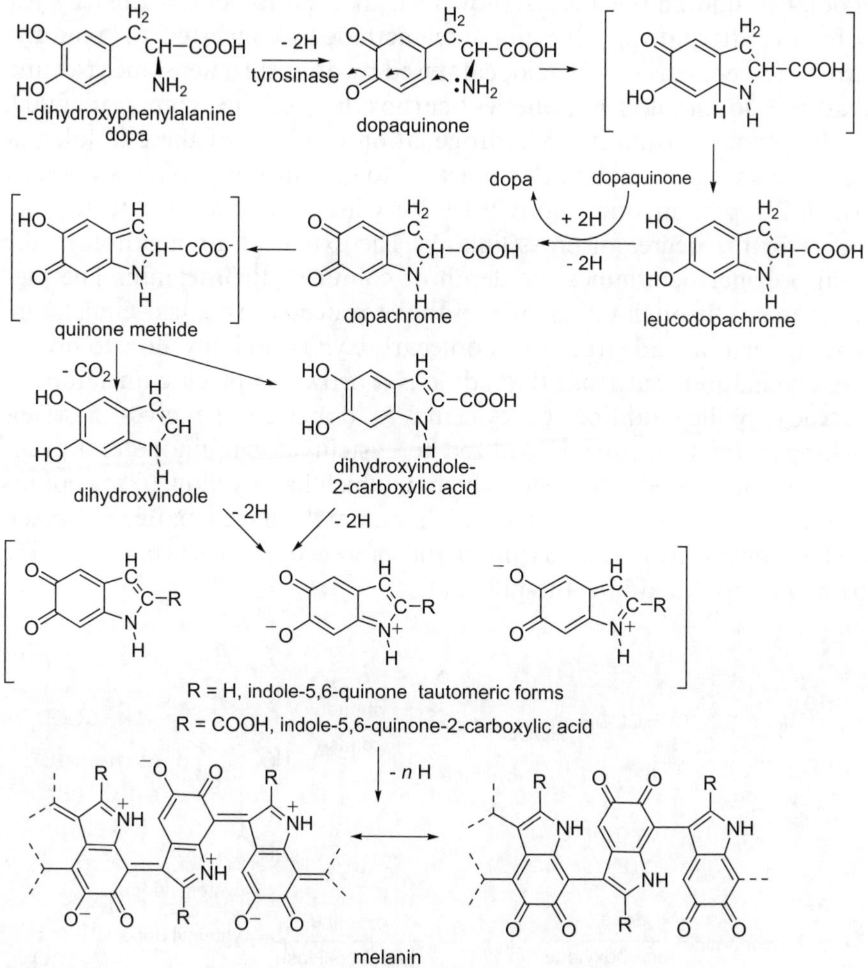

Figure 9.15 The formation of melanin from DOPA, showing a fragment of the melanin structure with two of the resonance forms that contribute to the conjugated system of double bonds that gives it its strong light absorption. Three tautomeric forms of indole-5,6-quinone are shown that all contribute to the structure of the polymer.

cyclizes dopaquinone to leucodopachrome, which then undergoes a spontaneous exchange with dopaquinone. One molecule of leucodopachrome is oxidized to dopachrome and one molecule of dopaquinone is reduced back to DOPA. Dopachrome, which has a red colour, rearranges through a quinone methide in the presence of dopachrome isomerase to quinone methide. Here two processes divide. The quinone methide is either decarboxylated to dihydroxyindole with unknown enzymic assistance and oxidized with the aid of phenoloxidase again to indole-5,6-quinone, which is shown in three tautomeric forms (Figure 9.15) or, alternatively, the quinone methide is converted to 5,6-dihydroxyindole-2-carboxylic acid, catalysed by dopachrome isomerase, and oxidized to indole-5,6-quinone-2-carboxylic acid (Sugumaran, 2002; 2009). Radical oxidative dehydrogenation of either of these indolequinones links the aromatic units together to give the polymer, a section of which is represented in Figure 9.15. Very little is known about the final stages or the degree of cross-linking. The extent of conjugation in the final polymer determines the depth of colour of the melanin. The pigment made through either of the above processes is called eumelanin. The eumelanin made from the indolecarboxylic acid is a yellow to brown eumelanin, and that from the indole is a brown to black eumelanin.

Nucleophilic addition of cysteine to dopaquinone gives 5-cysteinyldopa, which in turn is oxidized to cysteinyldopaquinone and eventually incorporated into phaeomelanin, which has a yellow to red colour (Figure 9.16). Many insects (and higher animals) have a pattern of black and yellow, which depends upon the presence or absence of cysteine being incorporated into the pigment.

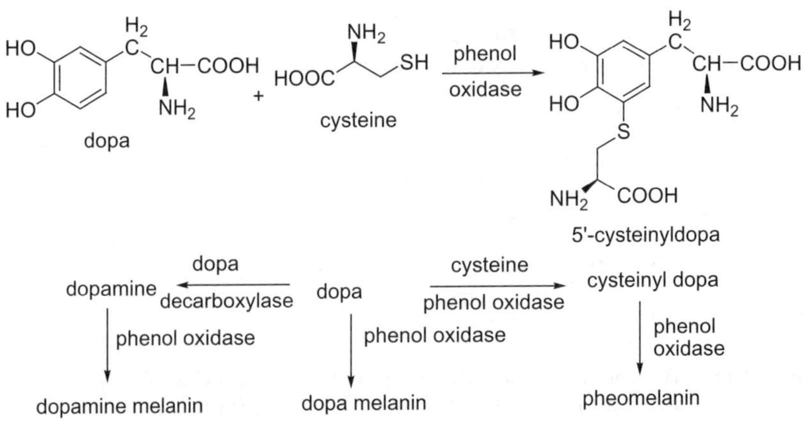

Figure 9.16 Cysteinyldopa and the different kinds of melanin pigments formed through dopa.

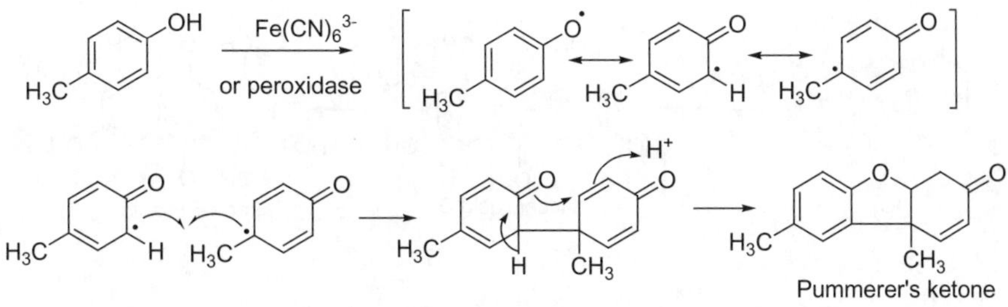

Aromatic Compounds

Figure 9.17 An *in vitro* reaction that is a possible model for the oxidative coupling of phenols that leads to the formation of melanin and other polymers.

The oxidative dehydrogenation of phenols and quinones to give dimers and polymers (as here in melanin and in the aphins, see Section 9.6.3) is a common occurrence in animals, comparable to the production of lignin and tannins in plants. The reaction is usually seen as a radical reaction. Free radicals remain in melanin, and can be detected by electron spin resonance spectroscopy (Kayser and Palivan, 2006). Something of a chemical model of the radical linking is the *in vitro* coupling of *p*-cresol to give the compound known as Pummerer' ketone, using ferricyanide, peroxidase or phenol oxidase (Figure 9.17).

9.6.2 Naphthoquinones and Anthraquinones

The homopteran sub-order of Hemiptera generally contains pterin pigments, but two groups have shown great originality in producing other pigments. These are the superfamilies Coccoidea (scale and lac insects and mealy bugs) and Aphidoidea (the aphids or plant lice). Both groups feed on phloem sap of plants. The Coccoidea contain anthraquinones, and the Aphidoidea have complex naphthoquinones. All pigments of the scale insects are polyketide anthraquinones. It is not known whether the pigments are the product of the insect or microbial symbionts (Section 6.2). These very interesting coloured compounds have not received any biosynthetic attention.

The artist' colour, Venetian red, is produced by the scale insect *Kermes* (*Coccus*) *ilicis* (Hemiptera: Kermesidae) feeding on an oak, *Quercus coccifera*. The pigment is kermesic acid (Figure 9.18). It is probably the oldest used insect pigment. Cochineal, the food colouring, is obtained from dried females of *Dactylopius coccus* (formerly *Coccus cacti*, Hemiptera: Dactylopiidae) (Plate 32), a bug feeding on *Opuntia* cactus (prickly pear). There are far fewer males and they contain less pigment. The pigment, carminic acid, has the same structure as kermesic acid but

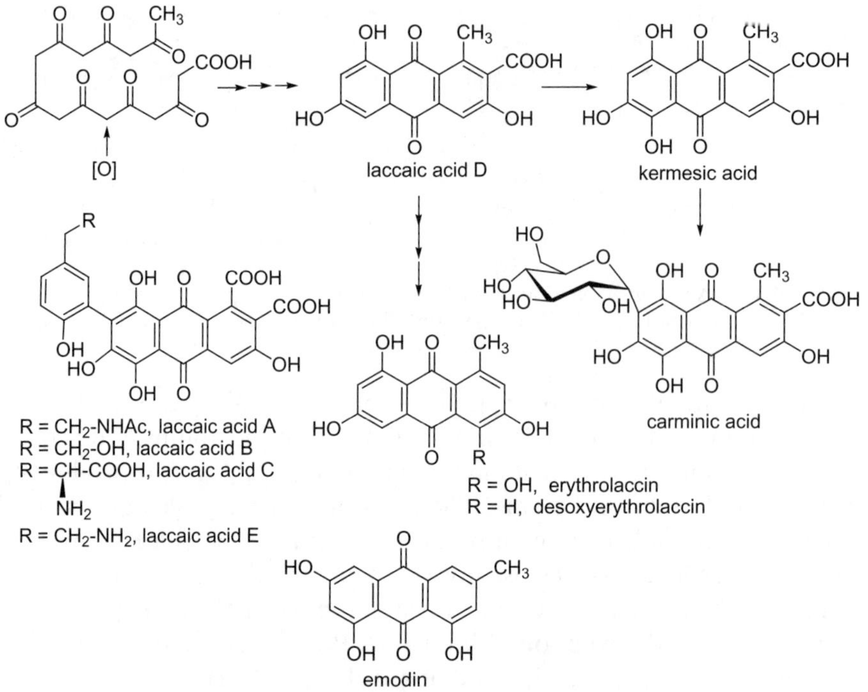

Figure 9.18 Some anthroquinone pigments of scale and lac insects, with a presumed parent polyketide. Kermesic acid is the artist' pigment Venetian red and carminic acid is the food colouring cochineal. The laccaic acids A, B, C and E are derivatives of kermesic acid with tyrosine or a derivative of it attached. Emodin is from an Australian *Eriococcus* species.

with a C-glucoside attached. Carminic acid is a potent deterrent to ants. It has been suggested that it evolved as a chemical weapon against predation (Eisner *et al.*, 1980). These pigments are present in cells of the fat body and in eggs as potassium salts.

Lac insects (Section 7.5.3) also produce pigments, consisting of more polar laccaic acids and fewer polar anthraquinones. The most commonly encountered examples are given in Figure 9.18. The additional benzene ring in laccaic acids is from a tyrosine molecule linked to the naphthoquinone by oxidative dehydrogenation. The anthraquinones can comprise up to 50% of the female body weight, but most of what is produced is secreted externally in the lac.

9.6.3 Aphins

Aphids (plant lice or greenfly (Hemiptera: Aphidoidea)) make pigments, called aphins and aphinins, not found in any other insects. The complex chemistry of aphid pigments was studied by Lord Todd and his students in the 1950s and 1960s. Up to 1975, 79 species had been examined (Brown, 1975). Aphins are dimeric naphthaquinones. The two most important are protoaphin-*fb* (isolated first from the common bean aphid (*Aphis fabae*) and protoaphin-*sl* (first isolated from the brown willow aphid *Tuberolachnus salignus* and only differing at one chiral centre) (Figure 9.19). There is restricted rotation about the bond joining the two parts of the molecule. The protoaphins are found in the haemolymph. The isolated material is brownish-yellow, but they are substantially ionized at physiological pH, and then give a deep purple colour. Different species of aphid vary in colour through shades of green, brown, red to almost black. Aphins are characteristic of the darker species. The green pigment is aphinin. Some species, *e.g.* greenfly, *Macrosiphium rosae* (Plate 33), contain only aphinin. The extraordinary situation exists that the structure of this pigment has been left incompletely solved for 40 years. It is probably as given in Figure 9.19. It was for long presumed that these pigments were made by symbiont micro-organisms but, surprisingly, it has been shown that aphids treated with antibiotics and having no bacteria in them still produced aphins (Walters *et al.*, 1994).

The subject of aphid pigments has been taken up again recently, but still without biosynthetic attention. Horikawa *et al.* (2004) isolated furanaphin, a yellow pigment from the yellow aphid *Aphis spiraecola*, feeding on Japanese knotweed *Fallopia japonica* (syn. *Polygonum cuspidatum*) together with the known compound 6-hydroxymusizin (Figure 9.20). Subsequently this group isolated two red pigments uroleuconaphins A_1 and B_1 (confusingly called rhododactynaphins–*jc*-1 and –*jc*-2 in a paper on apoptosis) from *Uroleucon nigrotuberculatum* (Figure 9.20) (Horikawa *et al.*, 2006). Later they identified four more yellowish pigments from the same aphid (Horikawa *et al.*, 2008).

The aphid *Aphis neri* feeding on the toxic shrub *Nerium oleander* is bright orange, possibly as a warning colour. It contains neriaphin and a glucoside of musizin (Figure 9.21) and some other related naphthalene derivatives. The quinone methide **a** is probably formed by condensation of neriaphin (before the glucoside was attached) with biacetyl and elimination of water, a rare example of biacetyl taking part in the formation of a natural product. The condensation could be repeated *in vitro* in weakly alkaline conditions (Brown *et al.*, 1969).

9.6.4 Pterins

The white and yellow pigments of butterflies were first identified by Heinrich Wieland in the 1930s. They belonged to a whole new class of

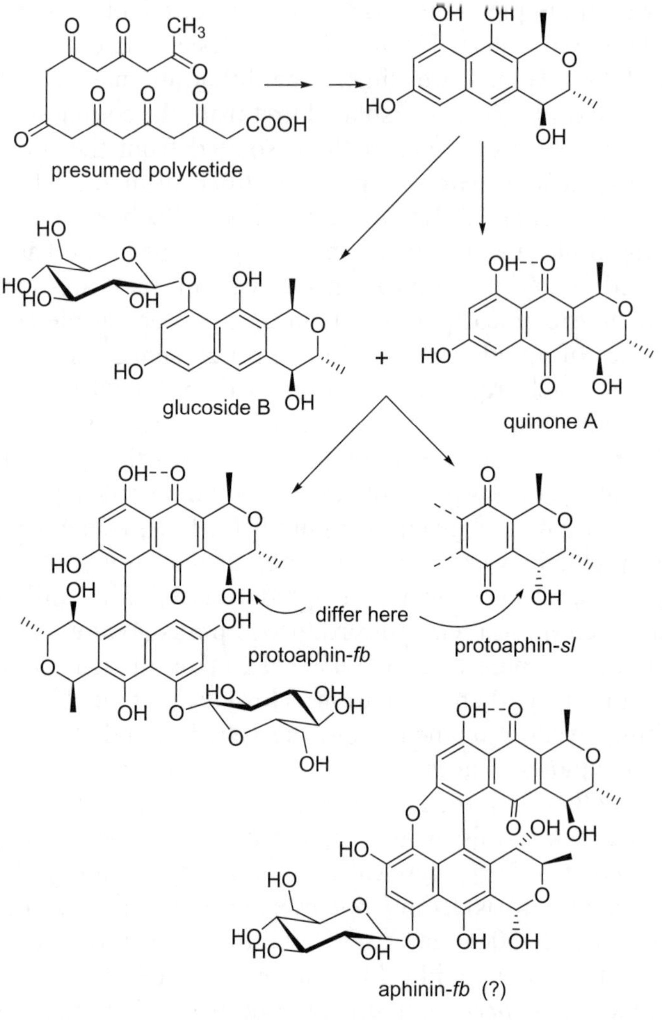

Figure 9.19 The formation of the aphins, pigments of aphids. The naphthalene derivative first formed is converted into quinone A and glucoside B, which linked together give the aphins. Note that aphin-*fb* and aphin-*sl* differ by only one chiral centre. There is restricted rotation about the central bond in both compounds. The structure of aphinin-*fb* is probably correct, but the study has never been completed.

nitrogen heterocycles, which he called pteridines (Greek *pteros* = wing). Leucopterin, the white pigment of the cabbage white butterfly *Pieris brassicae* and the small white *P. rapae*, and xanthopterin, the yellow pigment from the brimstone butterfly *Gonepteryx rhamni* (all Lepidoptera: Pieridae) (Plate 34), were the first elucidated (Figure 9.22). Later it was found that compounds with the pteridine structure occur widely in

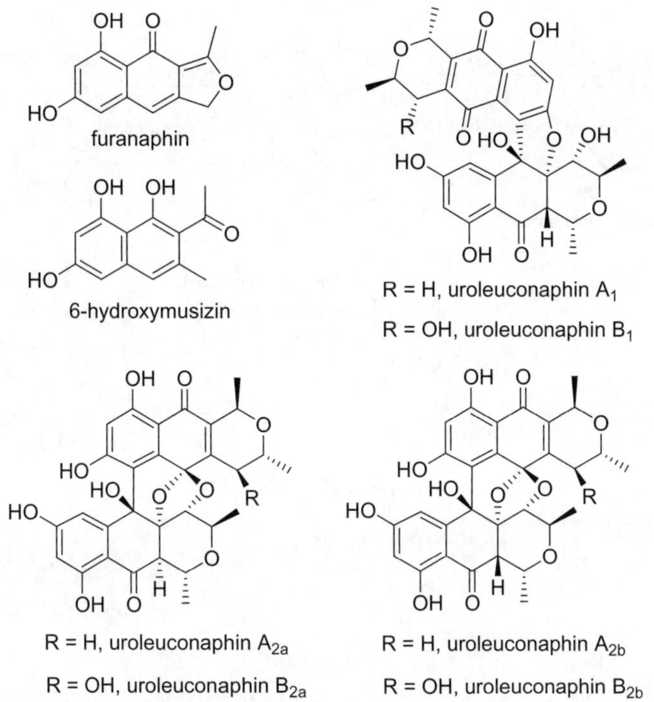

Figure 9.20 Furanaphin and 8-hydroxymusizin from *Aphis spiraecola*, and a new series of aphins from *Uroleucon nigrotuberculatum*.

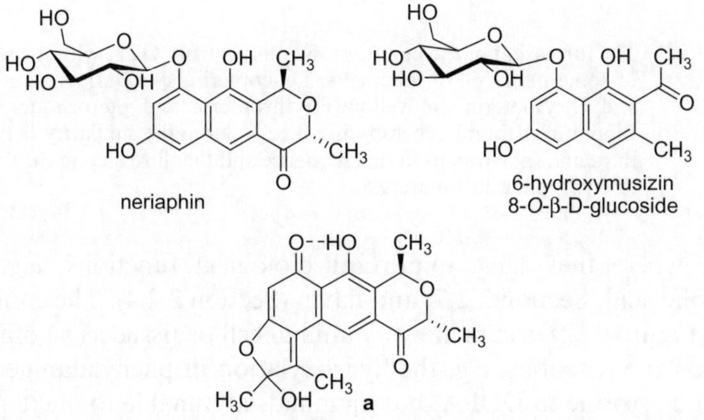

Figure 9.21 Some of the pigments from *Aphis neri*. Compound **a** can be made by reaction between biacetyl and a ketone derivative of the first intermediate in Figure 9.19.

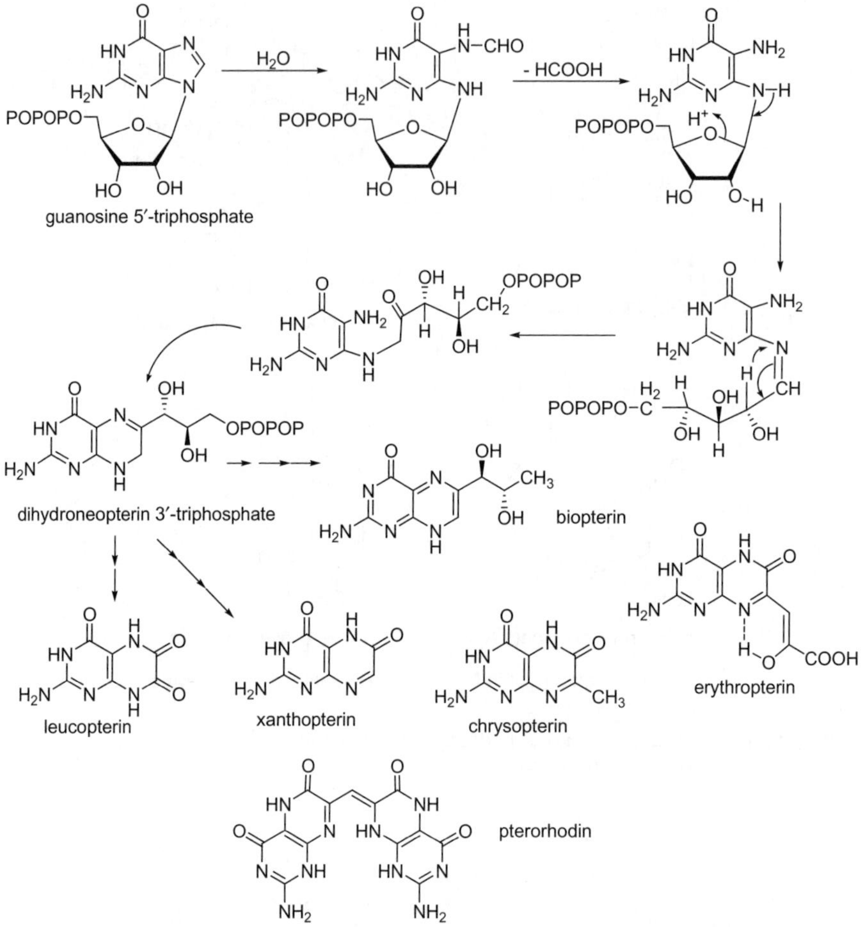

Figure 9.22 The formation of the parent of the pterins from GTP, with the structures of some insect pterin pigments. Leucopterin is colourless, xanthopterin and chrysopterin are yellow, erythropterin and pterorhodin are red. Biopterin is found generally in all cells. Note the similarity between the third and fourth steps in this sequence and the third step in the formation of tryptophan in Figure 9.5.

nature, where they have important biological functions, *e.g.* tetrahydrofolic acid (Section 2.2.7) and flavin (Section 2.2.4). The compound biotin (Figure 9.22) occurs in every animal cell or tissue as a cofactor of some enzyme reactions, *e.g.* the hydroxylation of phenylalanine to tyrosine and tyrosine to DOPA, but mammals are unable to make pterins. Biotin is thought to be a growth factor for some insects.

Pterin biosynthesis begins with opening of the imidazole ring of guanosine 5′-triphosphate, loss of formic acid, to give a derivative of a diaminopyrimidine, an open-chain keto-sugar, which cyclizes again with the amine group to give dihydroneopterin triphosphate (Figure 9.22). The enzyme GTP cyclohydrolase, which surprisingly catalyses all these steps, is present in all types of organisms.

Pterins are found as body pigments of Lepidoptera and Hymenoptera. The pterins of wing scales of Lepidoptera are crystalline solids. Xanthopterin, as well as being present in the wings of *Gonepteryx rhamni*, is widely distributed in insects and other animals. It provides the yellow colour of common wasps (*Vespa vulgaris*, *V. germanica* and *V. crabro*, the hornet) (Hymenoptera: Vespidae). There is red erythropterin in the forewings of the orange tip butterfly *Anthocharis* (*Euchloë*) *cardamines* (Pieridae) (Plate 35). The black and orange warning colour of the milkweed bug *Oncopeltus fasciatus* (Hemiptera: Lygaeidae) (Plate 36) is due to part opaque black melanin and part transparent cuticle that allows the underlying pterins to show through. Five different pterins were identified in the red heteropteran fire bug *Pyrrhocoris apterus* (Hemiptera: Pyrrhocridae), with erythropterin the most abundant (Bel *et al.*, 1997).

Pterins are also found in compound eyes of insects, and there tend to be associated with ommochromes (Section 9.6.6).

9.6.5 Tetrapyrroles

The porphyrins are macrocyclic tetrapyrroles which, by ring-opening, give the bilins or bilanes, pigments widely found from cyanobacteria and plants to mammals (where they are called bile pigments). Both porphyrins and bilins tend to be associated with proteins. They range in colour across the visible spectrum from red, orange, yellow or brown to blue or green.

Biosynthesis of the tetrapyrroles begins with attachment of glycine to pyridoxal phosphate (Section 2.2.6), followed by a Claisen condensation between the attached glycine and succinyl CoA, to give, after cleavage from the pyridoxal, 5-aminolaevulinic acid. Condensation of two of these molecules, catalysed by aminolaevulinate dehydratase, gives the first pyrrole, porphobilinogen (Figure 9.23). Porphobilinogen molecules are coupled by deamination to give a dimer, a trimer and a tetramer, of which the last can be cyclized to give a variety of tetrapyrroles, the important one here being protoporphyrin IX. Porphyrin biosynthesis has been shown to take place in the wings of adult *Pieris brassicae* (Rilkvan Gessel and Kayser, 2007).

When an iron ion is chelated at the centre of the ring, a haem group is formed, an essential part of cytochrome proteins, found in all insects, but present in such small quantities that the colour is not seen. Only a few insects that live in conditions of low oxygen pressure produce

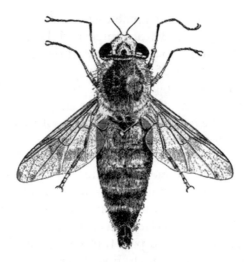

Figure 9.23 The botfly *Gasterophilus intestinalis*, one of the few insects to have haemoglobin in its haemolymph. (CSIRO, Australia.)

haemoglobin. Examples are the midge *Chironomus* (Diptera: Chironomidae), whose larvae live in holes in mud, the botfly *Gasterophilus* (Diptera: Gasterophilidae) (Figure 9.23), a parasite in the stomach of horses, and the diving bugs or backswimmers *Anisops* and *Buenoa* (Hemiptera: Notonectidae) that live in stagnant or slow moving water. The blood-sucking bug *Rhodnius prolixus* (Hemiptera: Reduviidae) absorbs haemoglobin from its food, converts it to biliverdin and stores it in pericardial cells.

Oxidative ring opening at the α-*meso*-position of protoporphyrin IX (between rings A and B) gives the blue-green pigment biliverdin. Reduction of the central –CH= gives the orange-coloured bilirubin (Figure 9.24). Bilins are found in Phasmida, Mantida, Orthoptera and Lepidoptera. There is biliverdin in the haemolymph of the tobacco hornworm *Manduca sexta* (Sphingidae) (Plate 6), and in the wings of many moths and butterflies. The green colour of many grasshoppers and lepidopteran larvae is due to bilins of the biliverdin type (Figure 9.24). The larvae of non-biting midges (Diptera: Chironomidae) are sometimes known as blood worms, as they store both biliverdin and bilirubin in their fat body.

Cleavage of protoporphyrin IX between rings C and D gives pterobilin, first isolated from the larvae and pupae of the cabbage white butterfly *Pieris brassicae*. Irradiation of pterobilin causes rotation about the –CH= that joins rings A and B and then reaction of the vinyl group on ring B with ring A gives phorcabilin, first found in the green-banded swallowtail *Papilio phorcas* (Papilionidae) (Choussy *et al.*, 1975).

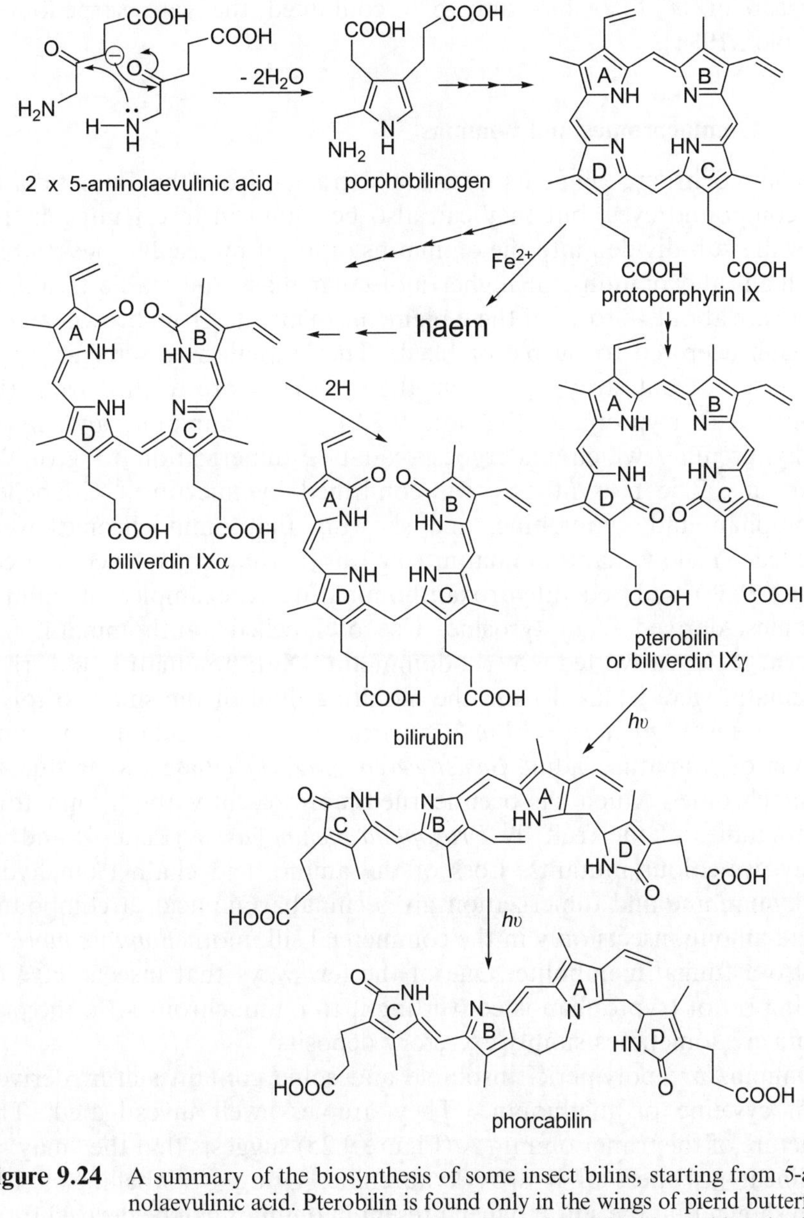

Figure 9.24 A summary of the biosynthesis of some insect bilins, starting from 5-aminolaevulinic acid. Pterobilin is found only in the wings of pierid butterflies, while phorcabilin is found in the wings of papilionid, arctid and microlepidoptera together with isophorcabilin and sarpedobilin (not illustrated).

Further reaction between the vinyl group of ring A and ring D also occurs to give sarpedobilin from *Papilio* (*Graphium*) *sarpedon* (Bois-Choussy and Barbier, 1983) (Plate 37). From a survey of 100 species of butterfly from ten families, pterobilin was found in 72% of them, phorcabilin in 15% but only 5% contained the rare sarpedobilin (Barbier, 1981).

9.6.6 Ommochromes and Ommins

The ommochrome pigments were first extracted from the ommatidia of the compound eyes, but they can also be found in insect integument. They are sub-divided into the ommatins of lower molecular mass, labile to alkali, and ommins, of higher molecular mass and stable to alkali. There are about 10 to 15 of these pigments in insects, from brown-yellow through deep-red to purple or black. They function as screening pigments to cut out stray light from the eye. They are formed from the amino acid tryptophan (Section 9.3.2) *via* kynurenine and hydroxykynurenine, which undergoes oxidative dimerization to give the ommatins. The biosynthesis was confirmed by injecting ^{14}C-labelled tryptophan and kynurenine, and showing that ommochromes were labelled. Yellow xanthommatin (the most frequently encountered, Figure 9.25) and red dihydroxanthommatin are examples of ommochromes, derived from tyrosine. Use of labelled xanthommatin has shown it is converted to rhodommatin. Xanthommatin and rhodommatin were isolated from the moulting fluid of the small tortoiseshell butterfly *Aglais* (= *Vanessa*) *urticae* (Nymphalidae). The pink colour of immature adult *Schistocerca gregaria* (Plate 38) is due to ommochromes. Much has been learned more recently about ommatins from studies of the fruit fly *Drosophila melanogaster* (Plate 4) and its many eye-colour mutants. Loss of the amino acid chain from hydroxykynurenine and dimerization gives cinnabarinic acid, a compound found among insects only in the commercial silk moth *Bombyx mori*. It is also a fungal metabolite. One of the few ways that insects have of getting rid of tryptophan is converting it to ommochromes, so the pigments are sometimes simply excretory deposits.

Ommins are polymeric, insoluble and some contain sulfur, derived from cysteine or methionine. They are less well investigated. The structure of the trimer ommin A (Figure 9.25) suggests that they may all be linear polymers of hydroxykynurenine. Using ^{35}S-labelling, it has been shown that the sulfur when present in ommins can be derived from cysteine or methionine, but not from sulfate, thiocyanate or sulfide.

9.6.7 Anthocyanins and Flavones

Anthocyanins and flavones are water-soluble flower pigments. Insects are probably the only animal group that sequester flavonoids from

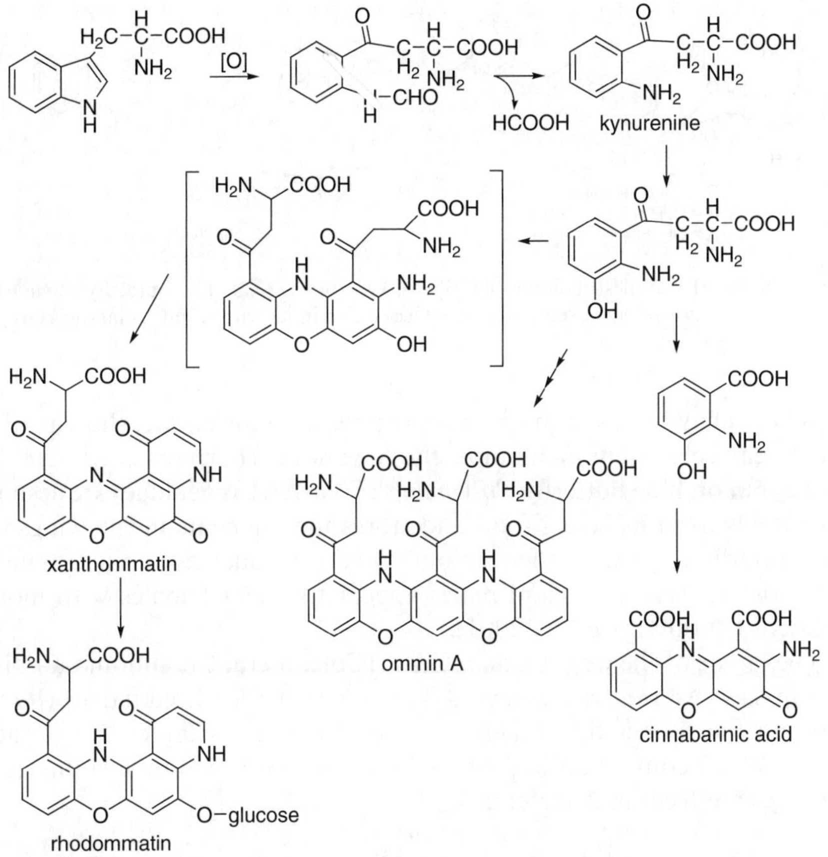

Figure 9.25 The formation of ommochromes and ommins from tryptophan *via* kynurenine.

plants. Little work has been done on their identification for some years, and their sequestration and metabolism do not follow any simple pattern. For example, the pale grass blue butterfly *Pseudozizeeria maha* (Lepidoptera: Lycaenidae) feeds on the leaves of *Oxalis corniculata*, which contains three closely related C-glycosylflavones (iso-orientin, isovitexin and swertisin, Figure 9.26). The larvae glucosylates isovixetin to saponarin, but at the pupal stage converts that back to isovexitin, and the butterfly selectively sequesters only isovitexin in its wings, much more in the female butterfly (Mizokami *et al.*, 2008; Mizokami and Yoshitama, 2009). The flavonoids are also detected in the eggs. *Pieris brassicae*, reared on a cabbage variety (*Brassica oleracea* var. *costata*), contains a range of 20 flavonoids, but the main component in the butterfly larvae was a minor component in the plant, and only two other

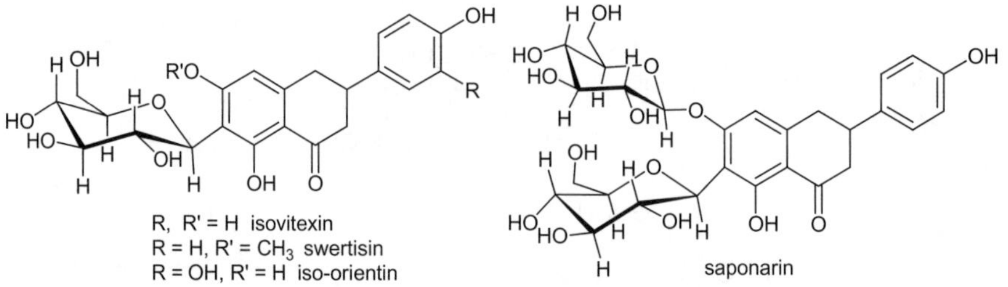

Figure 9.26 Examples of flavonoid glycosides from *Oxalis*. The butterfly *Pseudozizeeria maha* sequesters only isovitexin in its wings, but its larvae convert it to saponarin.

significant flavonoids in the plant were present in the larvae, showing the larvae can selectively metabolize the flavonoids (Ferreres *et al.*, 2007). The common blue butterfly *Polyommatus icarus* (Lycaenidae) sequesters flavonoids from its larval food and stores the pigments in the wings of the butterfly as part of their colour. Again, females accumulate much more of the flavonoids, and males appear to prefer females with more pigment (Burghardt *et al.*, 2001).

The larvae of the weevil *Cionus olens* (Coleoptera, Curculionidae), the turnip saw-fly *Athalia spinarum* (Hymenoptera, Tenthredinidae) (Plate 39) and the robber fly *Machimus* (*Asilus*) *chrysitis* (Diptera, Asilidae) (Plate 40) all contain anthocyanins. For most insects anthocyanins and flavones are feeding deterrents.

9.7 CHITIN, CUTICLE AND SCLEROTISATION

The subject of insect cuticle formation and sclerotisation or tanning is a large one (Chapman, 1998b), well beyond the scope of this book, but a brief examination of the small molecules involved is appropriate in the consideration of aromatic compounds and of pigment formation.

The cuticle of insects consists of a matrix of proteins and chitin. When the protein is removed by drastic alkaline extraction, a white residue of chitin remains. It is essentially a linear polymer of N-acetyl glucosamine (2-acetamido-2-deoxy-D-glucose) linked $(1 \rightarrow 4)$-β (Figure 9.27). All arthropod chitins have the same structure but not the same crystalline form. The molecular mass of the polymer is about 1×10^6 to 2×10^6, depending on the method of measurement, *i.e.* about 5000 to 10,000 acetylglucosamine units. The mass of chitin produced by living systems is second only to the mass of cellulose produced annually. Chitin can be regarded as a derivative of cellulose in which N-acetylglucosamine (with some glucosamine) replaces glucose, and chitin and cellulose have

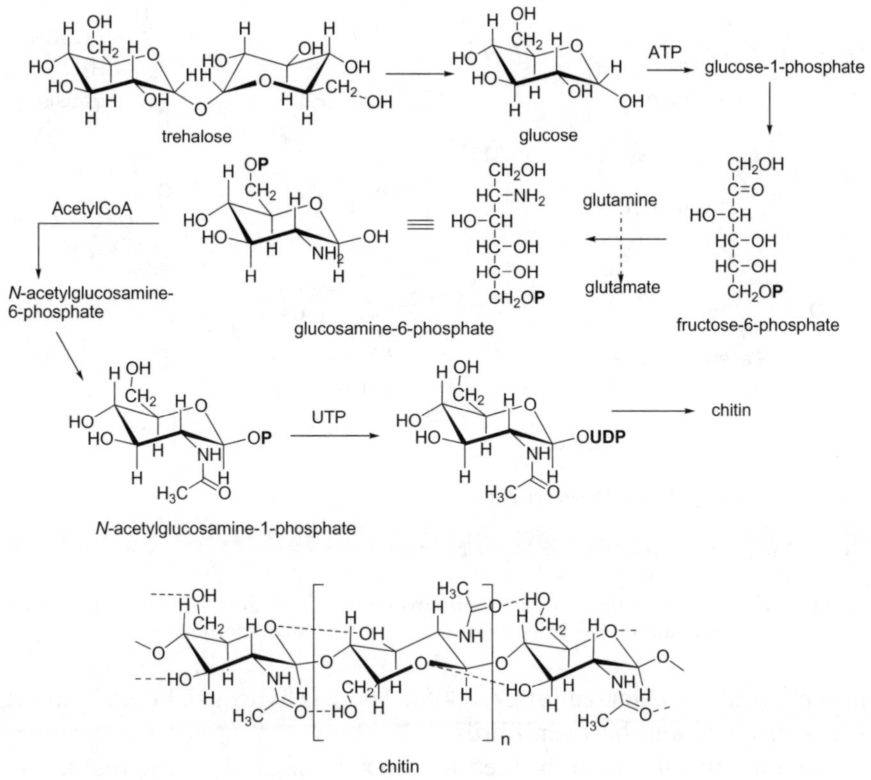

Figure 9.27 Biosynthesis of chitin from trehalose. The hydrogen bonds in chitin are shown as broken lines.

similar functions. Chitin can be regarded as a primary product of insect metabolism; its formation is important in insect development and several insecticides act by inhibiting chitin synthesis.

The biosynthesis of chitin may be considered to begin from trehalose, the principal sugar circulating in the haemolymph of most insects (Figure 9.27), although the full picture of the synthesis is still unclear. Hydrolysis of trehalose to glucose, phosphorylation and re-arrangement through fructose phosphate to glucose 6-phosphate, and amination by glutamine (which is converted to glutamic acid) gives glucosamine 1-phosphate, which is then acetylated. N-Acetylglucosamine exchanges the phosphate ester for a uridine diphosphate group, and this is the material which is polymerized to chitin, catalysed by chitin synthase, releasing uridine triphosphate. Chitin synthase is a membrane-integral enzyme. The polymer, as it forms, is moved from the cytoplasm across the membrane, and the single polymer strands spontaneously assemble

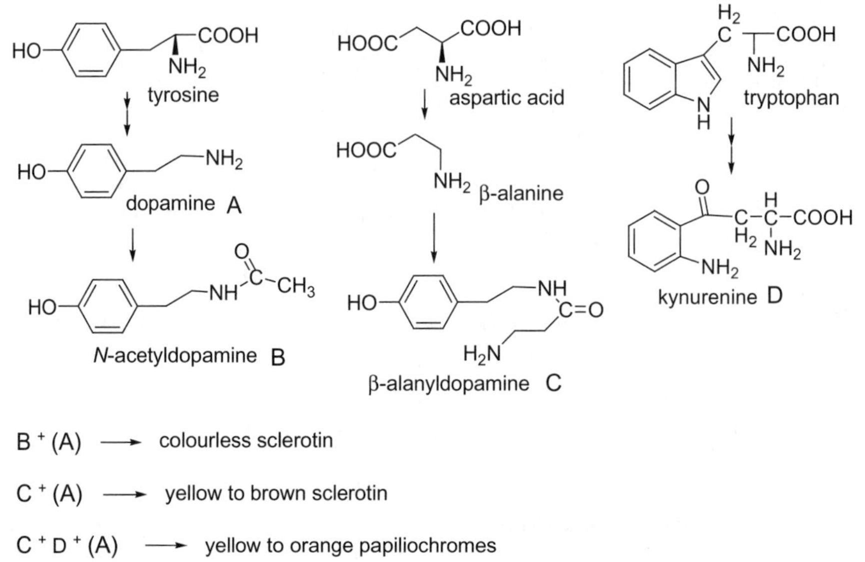

Figure 9.28 Intermediates in sclerotin formation. It is not clear to what extent dopamine (A) is incorporated with the other compounds.

B + (A) ⟶ colourless sclerotin

C + (A) ⟶ yellow to brown sclerotin

C + D + (A) ⟶ yellow to orange papiliochromes

into microfibrils (Merzendorfer, 2006). There is hydrogen bonding along the microfibril and between fibrils.

The microfibrils are embedded in a matrix of a very large number of different proteins, which may even vary over different parts of the body. The cuticle formed is rather like glass-fibre reinforced plastic, with the chitin microfibrils representing the glass fibres and the protein the plastic resin. Cuticle can vary from flexible, *e.g.* in inter-segmental membranes, to rigid in sclerites, mandibles, sting lances and other special structures.

Sclerotisation has parallels with melanin formation. Tyrosine, required for conversion to DOPA, dopamine, *N*-acetyldopamine and *N*-β-alanyldopamine, is stored in preparation for the event. Because of its low solubility tyrosine is often stored as a conjugate. In *Manduca sexta* it is stored as a glucoside. In the cuticle the DOPA and dopamine products are oxidized to quinones by phenoloxidase (Section 9.6.1). Melanin particles are formed, which strengthen the cuticle, but there is also cross-linking between the DOPA products and proteins and *via* oxygen atoms, to the chitin fibrils. Sclerotisation is not necessary to give hard cuticle, because the albino mutant of the desert locust *Schistocerca gregaria* has hard, white cuticle. It has been suggested that inclusion of *N*-acetyldopamine results in nearly colourless cuticle, while *N*-β-alanyldopamine gives dark-brown cuticle, with lighter-brown cuticle if both

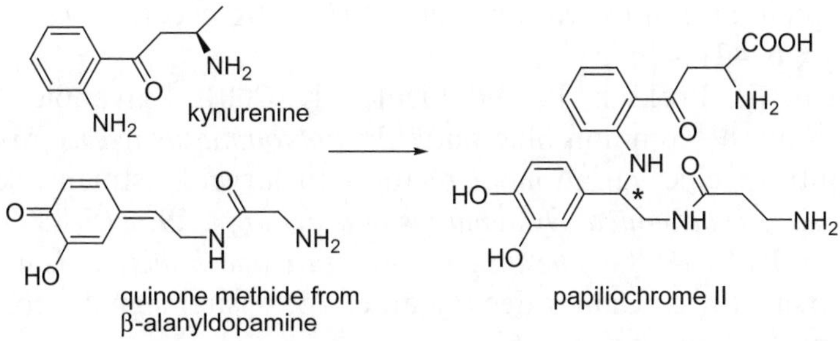

Aromatic Compounds

Figure 9.29 The formation of papiliochrome II from β-alanyldopamine, *via* the quinone methide and kynurenine. The asterisk indicates a chiral carbon atom: both enantiomers are present in the pigment.

are used. Mutants that cannot form *N*-β-alanyldopamine produce totally black melanin (Sugumaran, 2009). Papiliochromes are yellow wing pigments of butterflies that incorporate kynurenine with dopamine in sclerotisation. The structure of only one, papiliochrome II, is known (Figures 9.28 and 9.29).

REFERENCES

Aneshansley, D. J., Eisner, T., Widom, J. M. and Widom, B. 1969. Biochemistry at 100 °C: explosive secretory discharge of bombardier beetles (Brachinus). *Science*, **165**, 61–63.

Barbier, M. 1981. The status of blue-green bile pigments of butterflies, and their phototransformations. *Experientia*, **37**, 1060–1062.

Bel, Y., Porcar, M., Socha, R., Nemec, V. and Ferre, J. 1997. Analysis of pteridines in *Pyrrhocoris apterus* (L) (Heteroptera, Pyrrhocoridae) during development and in body-color mutants. *Archives of Insect Biochemistry and Physiology*, **34**, 83–98.

Blenau, W. and Baumann, A. 2001. Molecular and pharmacological properties of insect biogenic amine receptors: lessons from *Drosophila melanogaster* and *Apis mellifera*. *Archives of Insect Biochemistry and Physiology*, **48**, 13–38.

Bois-Choussy, M. and Barbier, M. 1983. Biosynthesis of the bile pigment sarpedobilin from [C-14]-labeled pterobilin by *Papilio sarpedon* (Lepidoptera). *Biochemical and Biophysical Research Communications*, **110**, 779–782.

Brown, K. S. 1975. Chemistry of aphids and scale insects. *Chemical Society Reviews*, **4**, 263–288.

Brown, K. S., Cameron, D. W. and Weiss, U. 1969. Chemical constituents of the bright orange aphid *aphis nerii* Folscombe. Part 1.

Neriaphin and 6-hydroxymusizin 8-*O*-β-D-glucoside. *Tetrahedron Letters*, **6**, 471–476.

Burghardt, F., Proksch, P. and Fiedler, K. 2001. Flavonoid sequestration by the common blue butterfly *Polyommatus icarus*: quantitative intraspecific variation in relation to larval hostplant, sex and body size. *Biochemical Systematics and Ecology*, **29**, 875–889.

Chapman, R. F. 1998a. *The Insects, Structure and Function*, Cambridge University Press, Cambridge, Chapter 25, Visual signals: color and light production, pp. 657–679.

Chapman, R. F. 1998b. *The Insects, Structure and Function*, Cambridge University Press, Cambridge, Chapter 16, Integument, pp. 415–440.

Choussy, M., Barbier, M., Vuillaume, M. and Vuillaume, M. 1975. Biosynthesis of phorcabilin, a blue bile pigment from *Actias selene* (Lepidoptera, Attacidae). *Biochimie*, **57**, 369–373.

Diehl, P. A., Guerin, P., Vlimant, M. and Stuellet, P. 1991. Biosynthesis, production site, and emission rates of aggregation-attachment pheromone in males of two *Amblyomma* ticks. *Journal of Chemical Ecology*, **17**, 833–847.

Dillon, R. J., Vennard, C. T. and Charnley, A. K. 2002. A Note: gut bacteria produce components of a locust cohesion pheromone. *Journal of Applied Microbiology*, **92**, 759–763.

Duffey, S. S, Underhill, E. W. and Towers, G. H. N. 1974. Intermediates in biosynthesis of HCN and benzaldehyde by a polydesmid millipede, *Harpaphe haydeniana* Wood. *Comparative Biochemistry and Physiology*, **47B**, 753–766.

Duffey, S. S., Aldrich, J. R. and Blum, M. S. 1977. Biosynthesis of phenol and guaiacol by hemipteran *Leptoglossus phyllopus*. *Comparative Biochemistry and Physiology*, **56**, 101–102.

Eisner, T. and Aneshansley, D. J. 1999. Spray aiming in the bombardier beetle: Photographic evidence. *Proceedings of the National Academy of Sciences, USA*, **96**, 9705–9709.

Eisner, T. and Meinwald, J. 1966. Defensive secretions of arthropods. *Science*, **153**, 1341–1350.

Eisner, T., Kluge, A. F., Carrel, J. E. and Meinwald, J. 1971. Defense of phalangid: liquid repellent administered by leg dabbing. *Science*, **173**, 650–652.

Eisner, T., Jones, T. H., Hicks, K., Silbergleid, R. E. and Meinwald, J. 1977. Defense-mechanisms of arthropods. 53. Quinones and phenols in defensive secretions of neotropical opilionids. *Journal of Chemical Ecology*, **3**, 321–329.

Eisner, T., Nowicki, S., Goetz, M. and Meinwald, J. 1980. Red cochineal dye (carminic acid): its role in nature. *Science*, **208**, 1039–1042.

Ferreres, F., Sousa, C., Valentão, P., Pereira, J. A., Seabra, R. M. and Andrade, P. B. 2007. Tronchuda cabbage flavonoids uptake by *Pieris brassicae*. *Phytochemistry*, **68**, 361–367.
Francke, W. and Dettner, K. 2005. Chemical signaling in beetles. *Topics in Current Chemistry*, **240**, 85–166.
Horikawa, M., Noguchi, T., Takaoka, S., Kawase, M., Sato, M. and Tsunoda, T. 2004. Furanaphin: a novel naphtho[2,3-*c*]furan-4(1*H*)-one derivative from the aphid *Aphis spiraecola* Patch. *Tetrahedron*, **60**, 1229–1234.
Horikawa, M., Hashimoto, T., Asakawa, Y., Takaoka, S., Tanaka, M., Kaku, H., Nishii, T., Yamaguchi, K., Masu, H., Kawase, M., Suzuki, S., Sato, M. and Tsunoda, T. 2006. Uroleuconaphins A_1 and B_1, two red pigments from the aphid *Uroleucon nigrotuberculatum* (Olive). *Tetrahedron*, **62**, 9072–9076.
Horikawa, M., Tanaka, M., Kaku, H., Nishii, T. and Tsunoda, T. 2008. Uroleuconaphins A_{2a}, A_{2b}, B_{2a}, and B_{2b}: four yellowish pigments from the aphid *Uroleucon nigrotuberculatum* (Olive). *Tetrahedron*, **64**, 5515–5518.
Itoh, M. T., Hattori, A., Nomura, T., Sumi, Y. and Suzuki, T. 1995. Melatonin and arylalkylamine N-acetyltransferase activity in the silkworm, *Bombyx mori*. *Molecular and Cellular Endocrinology*, **115**, 59–64.
Jones, T. H., Brunner, S. R., Edwards, A. A., Davidson, D. W. and Snelling, R. R. 2005. 6-Alkylsalicylic acids and 6-alkylresorcylic acids from ants in the genus *Crematogaster* from Brunei. *Journal of Chemical Ecology*, **31**, 407–417.
Kayser, H. and Palivan, C. G. 2006. Stable free radicals in insect cuticles: electron spin resonance spectroscopy reveals differences between melanization and sclerotization. *Archives of Biochemistry and Biophysics*, **453**, 179–187.
Kim, K. T., Hong, S. W., Lee, J. H., Park, K. B. and Cho, K. S. 1998. Mechanism of antibiotic action and biosynthesis of Centipedin purified from *Scolopendra subspinipes multilans* L. Koch (centipede). *Journal of Biochemistry and Molecular Biology*, **31**, 328–332.
Kuwahara, Y., Omura, H. and Tanabe, T. 2002. 2-Nitroethenylbenzenes as natural products in millipede defense secretions. *Naturwissenschaften*, **89**, 308–310.
Merzendorfer, H. 2006. Insect chitin synthases: a review. *Journal of Comparative Physiology B*, **176**, 1–15.
Messer, C., Dettner, K., Schulz, S. and Francke, W. 1999. Phenolic compounds in *Neanura muscorum* (Collembola, Neanuridae) and the role of 1,3-dimethoxybenzene as an alarm substance. *Pedobiologia*, **43**, 174–182.

Mizokami, H. and Yoshitama, K. 2009. Sequestration and metabolism of host-plant flavonoids by the Pale Grass Blue *Pseudozizeeria maha* (Lepidoptera: Lycaenidae). *Entomological Science*, **12**, 171–176.

Mizokami, H., Tomita, K., Tomita-Yokotani, K. and Yoshitama, K. 2008. Flavonoids in the leaves of *Oxalis corniculata* and sequestration of the flavonoids in the wing scales the pale grass blue butterfly *Pseudozizeeria maha*. *Journal of Plant Research*, **121**, 133–136.

Monro, A., Chadha, M., Meinwald, J. and Eisner, T. 1962. Defence mechanism of arthropods. *p*-Benzoquinone in the secretion of five species of millipedes. *Annals of the Entomological Society of America*, **55**, 261–262.

Mori, N., Kuwahara, T., Yoshida, T. and Nishida, R. 1995. Major defensive cyanogens from *Parafontaria laminata armigera* Verhoeff (Xystodesmidae: Polydesmida). *Applied Entomology and Zoology*, **30**, 197–202.

Nojima, S., Schal, C., Webster, F. X., Santangelo, R. G. and Roelofs, W. L. 2005. Identification of the sex pheromone of the German cockroach, *Blattella germanica*. *Science*, **307**, 1104–1106.

Oliver, J. E., Neal, J. W. and Lusby, W. R. 1987. Phenolic acetogenins secreted by rhododendron lace bug, *Stephanitis rhododendri* Horvath (Hemiptera, Tingidae). *Journal of Chemical Ecology*, **13**, 763–769.

Riley, P. A. 1997. Melanin. *International Journal of Biochemistry and Cell Biology*, **29**, 1235–1239.

Rilk-van Gessel, R. and Kaysera, H. 2007. Porphobilinogen synthase from the butterfly *Pieris brassicae*: purification and comparative characterization. *Journal of Insect Science*, **7**, Article No. 62.

Rutowski, R. L., Macedonia, J. M., Moorhouse, N. and Taylor-Taft, L. 2005. Pterin pigments amplify iridescent ultraviolet signal in males of the orange sulphur butterfly, *Colias eurytheme*. *Proceedings of the Royal Society B*, **272**, 2329–2335.

Schildknecht, H. 1971. Evolutionary peaks in the defensive chemistry of insects. *Endeavour*, **30**, 136–141.

Schildknecht, H. and Wenneis, W. F. 1967. Arthropod repellents. XXV Anthranilic acid as a precursor of the arthropod alkaloids glomerine and homoglomerine. *Tetrahedron Letters*, **19**, 1815–1818.

Sonenshine, D. E., Silverstein, R. M., Collins, L. A., Saunders, M., Flynt, C. and Homsher, P. J. 1977. Foveal glands, source of sex pheromone production in the ixodid tick *Dermacentor andersoni* Stiles. *Journal of Chemical Ecology*, **3**, 695–706.

Sugumaran, M. 2002. Comparative biochemistry of eumelanogenesis and the protective roles of phenoloxidase and melanin in insects. *Pigment Cell Research*, **15**, 2–9.

Sugumaran, M. 2009. Complexities of cuticular pigmentation in insects. *Pigment Cell and Melanoma Research*, **22**, 523–525.
Towers, G. H. N., Siegel, S. M. and Duffey, S. S. 1972. Defensive secretion: biosynthesis of hydrogen cyanide and benzaldehyde from phenylalanine by a millipede. *Canadian Journal of Zoology*, **50**, 1047–1050.
Tschinkel, W. R. 1969. Phenols and quinones from the defensive secretions of the tenebrionid beetle *Zophobas rugipes*. *Journal of Insect Physiology*, **15**, 191–200.
Tschinkel, W. R. 1975. Comparative study of chemical defensive system of tenebrionid beetles – chemistry of secretions. *Journal of Insect Physiology*, **21**, 753–783.
Walters, F. S., Mullin, C. A. and Gildow, F. E. 1994. Biosynthesis of sorbic acid in aphids. An investigation into symbiont involvment and potential relationship with aphid pigments. *Archives of Insect Biochemistry and Physiology*, **26**, 49–67.
Weatherstone, J. and Percy, J. E. 1976. The biosynthesis of phenethyl alcohol in the male bertha armyworm *Mamestra configurata*. *Insect Biochemistry*, **6**, 413–417.
Zagrobelny, M., Bak, S. and Møller, B. L. 2008. Cyanogenesis in plants and arthropods. *Phytochemistry*, **69**, 1457–1468.

CHAPTER 10
Alkaloids and Compounds of Mixed Biosynthetic Origin

10.1 ALKALOIDS

Alkaloids were formerly defined as basic, nitrogen-containing plant substances, but similar, and sometimes the same, compounds have been found also in fungi, marine organisms, amphibians, insects and even mammals. They frequently have toxic properties and physiological effects on animals. There are over 10,000 plant alkaloids known, all but a very few biosynthesized from amino acids, and the great majority of them are derived from aromatic amino acids. The structure and biosynthesis of alkaloids in plants has received a lot of study. Less is known about the biosynthesis and properties of the much smaller number of insect alkaloids, and the latter form a less homogeneous group. Some, derived from fatty acids and alkyl compounds, have already been discussed in Chapter 5, particularly the defensive compounds of Coleoptera (Section 5.4), and the piperidines and pyrrolidines of ant venoms.

All the available evidence suggests that plants make alkaloids to deter predators. Many of them, like the tobacco alkaloids, are strongly toxic to insects. Nicotine, anabasine and other related alkaloids are produced in the roots of the tobacco plant and translocated to the leaves. Formerly a crude preparation of nicotine was used as a commercial insecticide because of its high toxicity, but the larvae of the tobacco hornworm *Manduca sexta* (Lepidoptera: Sphingidae) have adapted to feed only on tobacco leaves; and others, like the cigarette beetle

Biosynthesis in Insects, Advanced Edition
By E. David Morgan
© E. David Morgan 2010
Published by the Royal Society of Chemistry, www.rsc.org

Lasioderma serricorne (Coleoptera: Anobiidae) live only on dried tobacco leaves.

10.2 INSECT ALKALOIDS

There is great variety in the chemical structures of alkaloids found in insects, many of them serving as defensive or offensive secretions, and some sex pheromones. These alkaloids are scattered across the types of aliphatic, alicyclic, aromatic and heterocyclic compounds, but biosynthetic studies have as yet touched very few of them. Ants of the subfamily Myrmicinae produce a wide variety of venom alkaloids, with a diversity of functions and are frequently of pharmacological interest.

A group of substituted pyrrolidines, different from the *Solenopsis* and *Monomorium* alkylpyrrolidines (Section 5.6.3.1), have been found in the venom glands of the slave-making ant *Harpegoxenus sublaevis* and its slaves *Leptothorax acervorum* and *L. muscorum* (Reder *et al.*, 1995; Koob *et al.*, 1997). They are possibly associated with sex attraction, and they all have 3*R* geometry. The amounts are very small, 10 ng down to 5 pg, but there is more in the slave-maker than in the slaves. The alkyl groups suggest amino acid origins (leucine, methionine, phenylalanine and isoleucine) and the carbon atoms of the pyrrole ring may come from isoprene (Figure 10.1).

Actinidine (Figure 10.2) is an iridoid alkaloid. It often accompanies iridoid monoterpenes in the defensive secretions of beetles (Section 7.4.1). It is the major constituent of the defensive secretion of *Philonthus* rove beetles (Staphylinidae) and the myrmicine ant *Pheidole biconstricta*; it is a defensive secretion in dolicoderine ants, the ponerine ant *Megaponera foetens* and the stick insect *Megacrania tsudai* (Phasmida:

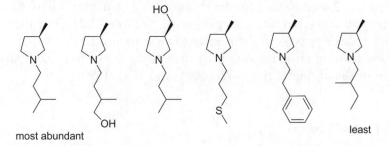

Figure 10.1 A group of substituted pyrrolidines from the venom of the slave-making ant *Harpegoxenus sublaevis* and its slaves. The compounds are arranged from the one in greatest quantity on the left to the least on the right.

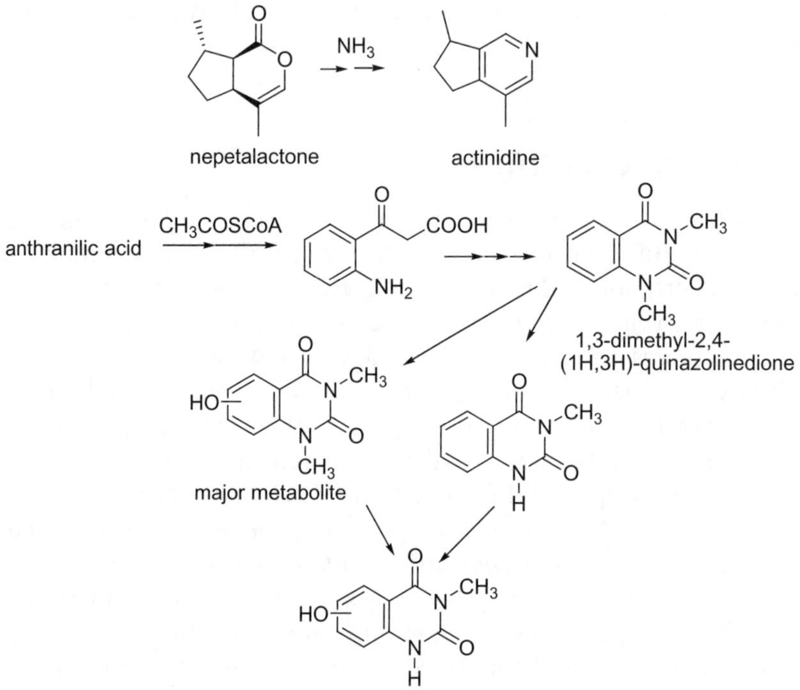

Figure 10.2 The structure of the iridoid alkaloid actinidine and its possible formation from nepetalactone. The dimethylquinazolinedione is the sex pheromone of *Phyllopertha diversa*, probably made from anthranilic acid, though that has not been studied, but its degradation on the male antennae has been shown to be by hydroxylation of the ring or loss of the 1-methyl group.

Phasmatidae), but is also found in plants. The condensation of chrysomelidial or nepetalactone with ammonia gives actinidine directly.

The pale-brown chafer beetle *Phyllopertha diversa* (Coleoptera: Scarabaeidae) uses an alkaloid as its female sex pheromone. The 1, 3-dimethyl-2,4-quinazolinedione (Figure 10.2) evidently is derived from anthranilic acid (compare with glomerine, Section 9.3.2). Only the males of this species possess in their antennae a cytochrome P450 specifically for deactivating this compound (by demethylation of N-1 and hydroxylation of the aromatic ring) (Wojtasek and Leal, 1999).

10.2.1 Ant Venoms

In addition to the ant venoms derived from fatty acids, already described in Section 5.6.3.1, there are other more complex heterocyclic amines in ant venoms. Some of these venoms have been studied

Alkaloids and Compounds of Mixed Biosynthetic Origin

biosynthetically, others have not been attempted. The tobacco alkaloids anabasine, anabaseine and bipyridyl are found in the venom glands of several species of *Messor* and *Aphenogaster* ants, but there is no evidence to suggest they use the same biosynthetic route as plants (from nicotinic acid and ornithine). In *Aphaenogaster rudis* anabasine, anabaseine and bipyridyl are all part of the trail pheromone. Alone, each compound shows no activity, but with N-isopentyl-2-phenylethylamine (Figure 10.3) they form the active pheromone. The anabasine in some *Messor* ants was shown to have the same $2'$-(S) configuration as it has in plants, but in some *Aphaenogaster* species anabasine was of mixed enantiomers (Leclercq *et al.*, 2001).

Three indolizidines from the venom gland of *Monomorium pharaonis* (Figure 10.4) and a pyrrolizidine from a *Solenopsis* ant are products

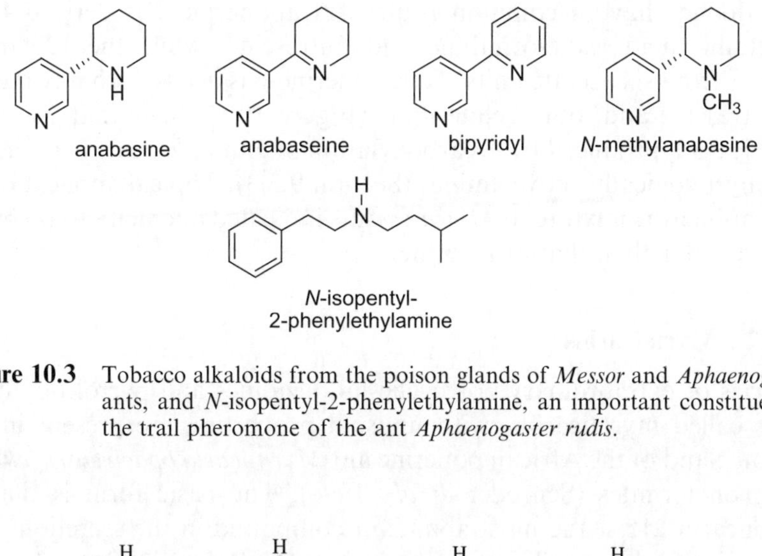

Figure 10.3 Tobacco alkaloids from the poison glands of *Messor* and *Aphenogaster* ants, and N-isopentyl-2-phenylethylamine, an important constituent of the trail pheromone of the ant *Aphaenogaster rudis*.

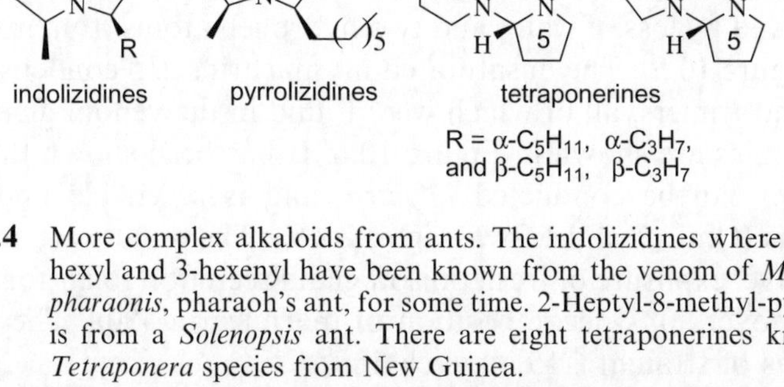

$R = \alpha\text{-}C_5H_{11}, \alpha\text{-}C_3H_7,$
and $\beta\text{-}C_5H_{11}, \beta\text{-}C_3H_7$

Figure 10.4 More complex alkaloids from ants. The indolizidines where R is butyl, hexyl and 3-hexenyl have been known from the venom of *Monomorium pharaonis*, pharaoh's ant, for some time. 2-Heptyl-8-methyl-pyrrolizidine is from a *Solenopsis* ant. There are eight tetraponerines known from *Tetraponera* species from New Guinea.

reminiscent of the coccinellid alkaloids. They would appear to be formed from acetogenins or fatty acids like the coccinellines, but have not been studied. More pyrrolizidines and indolizidines with two alkyl and alkenyl groups, but always with a straight carbon chain, have been found in *Monomorium* and *Solenopsis* species. *Tetraponerine* ants smear the venom of their poison glands onto prey. The venom consists of tricyclic alkaloids called tetraponerines. They are divided into two groups, designated 6,6,5 and 5,6,5 by the size of their three rings (Figure 10.4).

The biosynthesis of the tetraponerines has been studied using sodium [1-^{14}C]-acetate and [2-^{14}C]-acetate, [1,4-^{14}C]putrescine dihydrochloride and three compounds all uniformly labelled with ^{14}C (L-glutamic acid, γ-aminobutyric acid and L-ornithine) (Renson, *et al.*, 1994; Devijver *et al.*, 1997). The location of the radioactivity was followed by chemical degradation. Both types of tetraponerine were shown to have a mixed biosynthetic origin, from an amino acid and a chain of acetate units, but they do not have a common route. Tetraponerine-8 is derived from L-glutamic acid *via* L-ornithine and putrescine, while the 12-carbon chain is from six acetate units. Tetraponerine-6 is derived from two units of putrescine and four acetate units (Figure 10.5). Note that γ-aminobutyric acid, produced by decarboxylation of glutamic acid is another of the physiologically active amines (Section 9.3.3). Topical application of a tetraponerine mixture to *Myrmica* ants indicated a toxicity to them ten times greater than that of nicotine.

10.2.2 Myrmicarins

A series of extraordinary polycyclic indolizidines and pyrrolo-indolizidines called myrmicarins, with unusual properties, are present in the poison gland of the African ponerine ant *Myrmicaria opaciventris*, which preys on termites (Schröder *et al.*, 1996). The basic form is that of myrmicarin 215A, the most abundant compound in the secretion. It is accompanied by smaller amounts of myrmicarins 215B (with *trans* double bond) and saturated myrmicarin 217. The formation of 217 can be conceived by loss of water and two hydrogen atoms from myrmicarin 237A (Figure 10.5). The unsaturated myrmicarins 215 condense to form dimers and trimers, all of which were found in the venom glands of the ant; examples are shown in Figure 10.6. It has been shown that the dimerization can be conducted *in vitro*, and is reversible under acidic conditions (Ondrus and Movassaghi, 2009). The dimers are fragile and air-sensitive; exposure of fresh poison gland secretion to air for one hour results in over 90% decomposition of myrmicarin 430A They all have backbones of straight C15 alkyl chains.

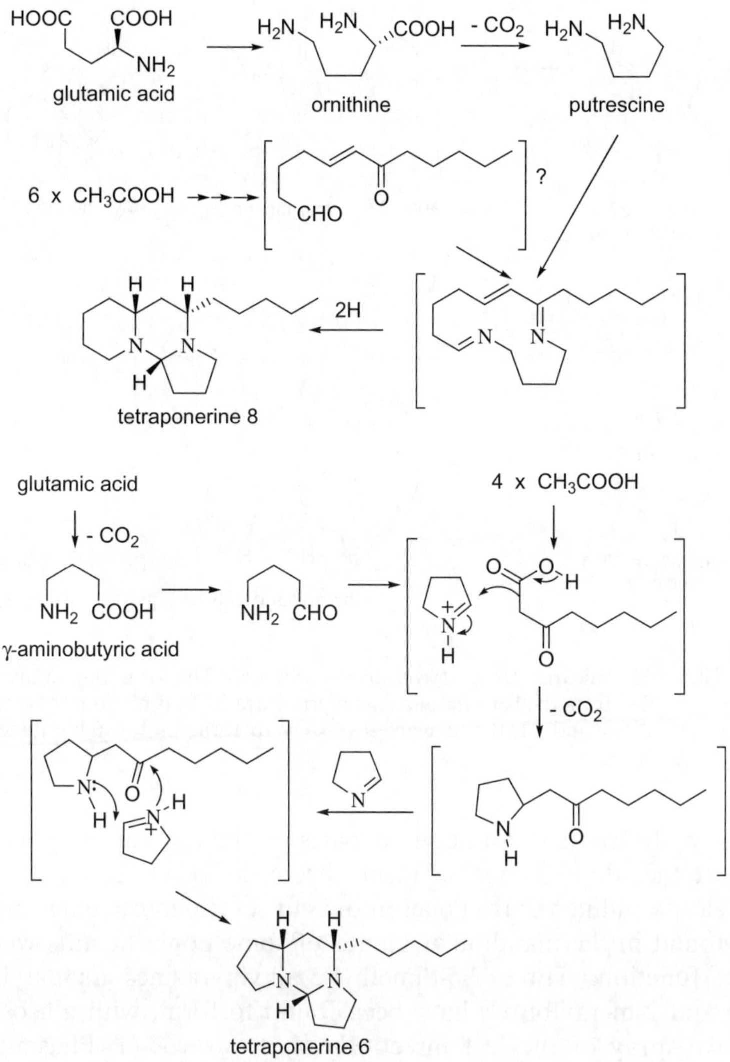

Figure 10.5 The formation of two types of tetraponerines from ants, requiring different synthetic routes. Glutamic acid, ornithine and γ-aminobutyric acid uniformly labelled with ^{14}C, sodium acetate labelled in C-1 and C-2 with ^{14}C and [1,4-^{14}C]putrescine were all used to show they require derivatives of glutamic acid and a straight carbon chain built from acetate units.

10.2.3 Other Examples

The alkaloids of Myrmicinae ants vary widely in their biological function and pharmacological interest.

Alkylpyrazines are strictly alkaloids, but they are better known among the flavours of food. They can be detected by both humans and

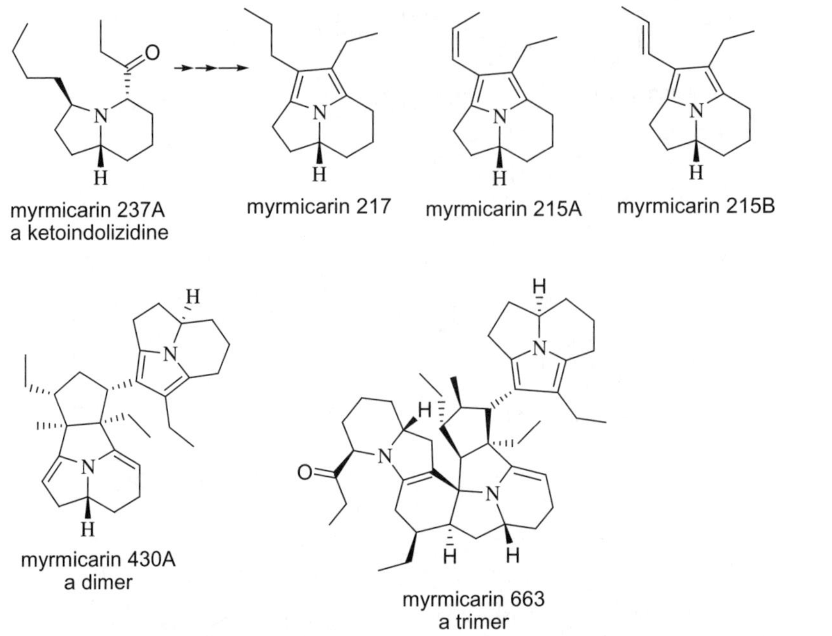

Figure 10.6 Myrmicarins from *Myrmicaria opaciventris*. The formation of mymicarin 217 from the keto-indolizidine myrmicarin 237A is obvious. Myrmicarins 215A and 215B can react reversibly to form the unstable dimers and trimers.

insects in the parts-per-million to parts-per-billion range. They contribute to the odours of cooked meat, cheese, coffee, chocolate and wine, as well as providing the trail pheromones of several myrmicine ants, and being found in the mandibular glands of some ponerine ants with unknown function. Three 2,5-dimethyl-3-alkylpyrazines (isobutyl, isopentyl and 2-methylbutyl) have been found to form, with glucose, the defensive spray of the leaf insect *Phyllium westwoodii* (Phasmitodea: Phyllidae) (Dossey *et al.*, 2009). Biosynthesis of these pyrazines has not been studied, but a reasonable prediction can be made from their known chemistry and the knowledge that the amino acid threonine decomposes by oxidation to a β-keto-acid and its decarboxylation to amino-acetone (Figure 10.7). Two molecules of amino-acetone condense to give 2,5-dihydro-3,6-dimethylpyrazine, which reacts with an aldehyde, in a way that has been demonstrated in the laboratory, to give a 2,5-dimethyl-3-alkylpyrazine. Most ant pyrazines have two methyl groups. The mandibular gland pyrazines of ponerine ants have trisubstituted pyrazines with C_3 to C_5 alkyl groups and tetrasubstituted pyrazines with an additional isopentyl group (hydroxylated or unsaturated).

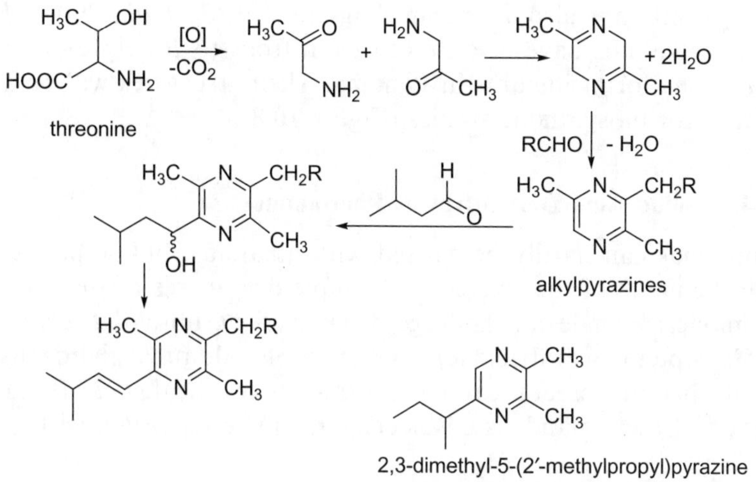

Alkaloids and Compounds of Mixed Biosynthetic Origin 301

Figure 10.7 The probable biosynthetic route to alkylpyrazines. The trisubstituted pyrazines have been converted to tetrasubstituted hydroxylated or unsaturated examples in the laboratory by addition of isopentanal, followed by $NaBH_4$ reduction and sometimes dehydration. All such compounds are found in nature. R is typically H, methyl, ethyl, propyl, isopropyl, 2-butyl or isobutyl. 2,3-Dimethyl-5-(2′-methylpropyl)pyrazine seems to have a different biosynthetic origin.

The trisubstituted pyrazines can be converted to the tetrasubstituted ones *in vitro* by free radical addition of isopentanal followed by reduction of the initially formed keto-group to an alcohol and possible dehydration to the unsaturated examples (Oldham and Morgan, 1993). This suggests the biosynthetic route. Some alkylpyrazines are of the 2,3-dimethyl type, with a less obvious biosynthetic origin. An example is 2,3-dimethyl-5-(2′-methylpropyl)pyrazine (Figure 10.7), the trail pheromone of the ant *Eutetramorium mocquerysi* from Madagascar.

Farnesylamine (Figure 10.8), long known as a potent synthetic pharmacologically active compound, has been found in nature for the first time in the ant *Monomorium fieldi* (Jones et al., 2003). Farnesylamine has been shown to inhibit arthropod moulting and reproduction, and have effects upon mammalian malignant cells.

The toxins of poisonous toads and frogs from South America, Madagascar and Australia have been recognized for some time, but only recently has it been discovered that some of these toads and frogs simply have an efficient system for accumulating the toxins from arthropod food. Frogs raised in captivity do not have the toxins. The piperidines, pyrrolidines, pyrrolizidines and indolizidines already mentioned from myrmicinae ants, plus quinolizidines and decahydroquinolines, also found in ants are also found in frogs (Saporito *et al.*, 2004). More complex alkaloids have been traced from frogs to *Brachymyrmex* and *Paratrechina* formicine ants in Panama. Their structures will be a new challenge for biosynthetic studies (Figure 10.8).

10.2.4 Amino Acid Derivatives as Pheromones

Amino acids can hardly be classed with alkaloids, but some mention should be made of the existence of simple derivatives of them used as pheromones. Females of the large black chafer grub *Holotrichia parallela* (Coleoptera: Scarabaeidae) have a pair of abdominal glands that are everted when the insect is calling to attract males (Leal *et al.*, 1992). The secretion of these glands is L-isoleucine methyl ester (Figure 10.9).

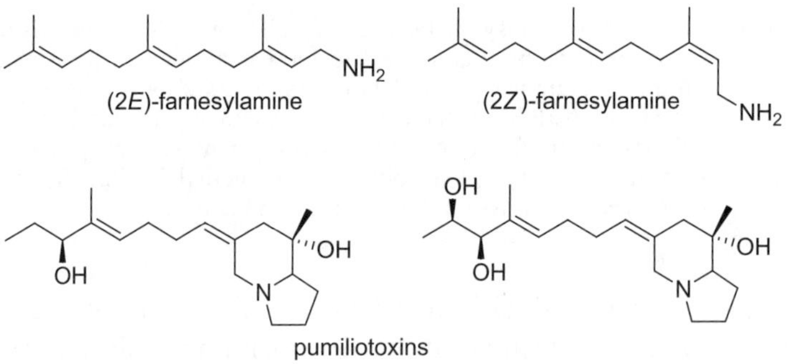

Figure 10.8 The two isomers of farnesylamine found in *Monomorium fieldi*, and two pumiliotoxins, non-linear alkaloids first known in frogs, but now recognized as produced by formicine ants.

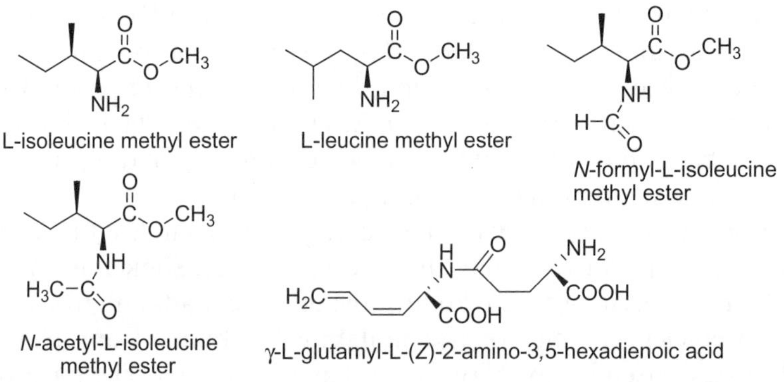

Figure 10.9 Some amino acid derivatives used by beetles as pheromones, and a dipeptide defensive compound from the Colorado potato beetle.

A curious situation exists with the genus *Phyllophaga* (Scarabaeidae). The cranberry white grub *Phyllophaga* (*Phytalis*) *anxia* consists of three races, females of one use L-valine methyl ester as their sexual pheromone, and females of the second use L-isoleucine methyl ester, while the third use a blend of both (Robbins *et al*., 2008) (Figure 10.9). *P. georgicum*, *P. gracilis* and *P. postrema* all use L-valine methyl ester. They appear to use some short-range method for recognition of species. *P. elenans* has a mixture of three compounds; L-isoleucine methyl ester, *N*-formyl L-isoleucine methyl ester and *N*-acetyl L-isoleucine methyl ester (Leal *et al*., 2003).

The Colorado potato beetle *Leptinotarsa decemlineata* (Chrysomelidae) does not sequester toxic potato alkaloids from its food, but produces its own toxic secretion: a dipeptide of L-glutamic acid and a non-protein amino acid, (Z)-2-amino-3,5-hexadienoic acid (Daloze *et al*., 1986), the dipeptide was shown to be toxic to ants.

10.3 COMPOUNDS OF MIXED BIOSYNTHETIC ORIGIN

Nature does not compartmentalize its biosynthetic methods and, as already found, compounds can be composed of fragments of different chemical types. Amino acids can be linked to fatty acids (epilachnine); blatellostanosides have a sugar bonded to sterols; ecdysone palmitate is a sterol hormone with a fatty acid attached. Some compounds of mixed biosynthetic origin do not fit readily into any simple type.

Some pharmacologically important venoms are of mixed type. The European beewolf *Philanthus triangulum* (Hymenoptera: Philanthinae) paralyses honeybees to provide food for its larvae. Their venom contains a number of philanthotoxins that do not fit the usual form of alkaloids (Piek *et al*., 1985). They have fatty acid, amino acid and diamine parts (Figure 10.10). The compounds of the group differ in the number of carbon atoms between the amine nitrogens. The diamine putrescine is formed by decarboxylating ornithine with pyridoxal phosphate. Reaction of putrescine with what is called decarboxylated *S*-adenosyl methionine (Figure 10.10) adds on the propylamine residue to give spermidine, and addition of another propylamine to spermidine gives spermine. Both spermidine and spermine are important compounds in normal cell function. Polyamines like spermine and spermidine are found in the venom of the funnel-web spider *Atrax robustus* (Araneae: Hexathelidae) (Plate 41), and have been identified in the venom of tarantula spiders (Araneae: Theraphosidae) some time ago. Compounds similar to the philanthotoxins have been identified in many spider venoms. The compound designated CNS 2103, based on tryptophan has

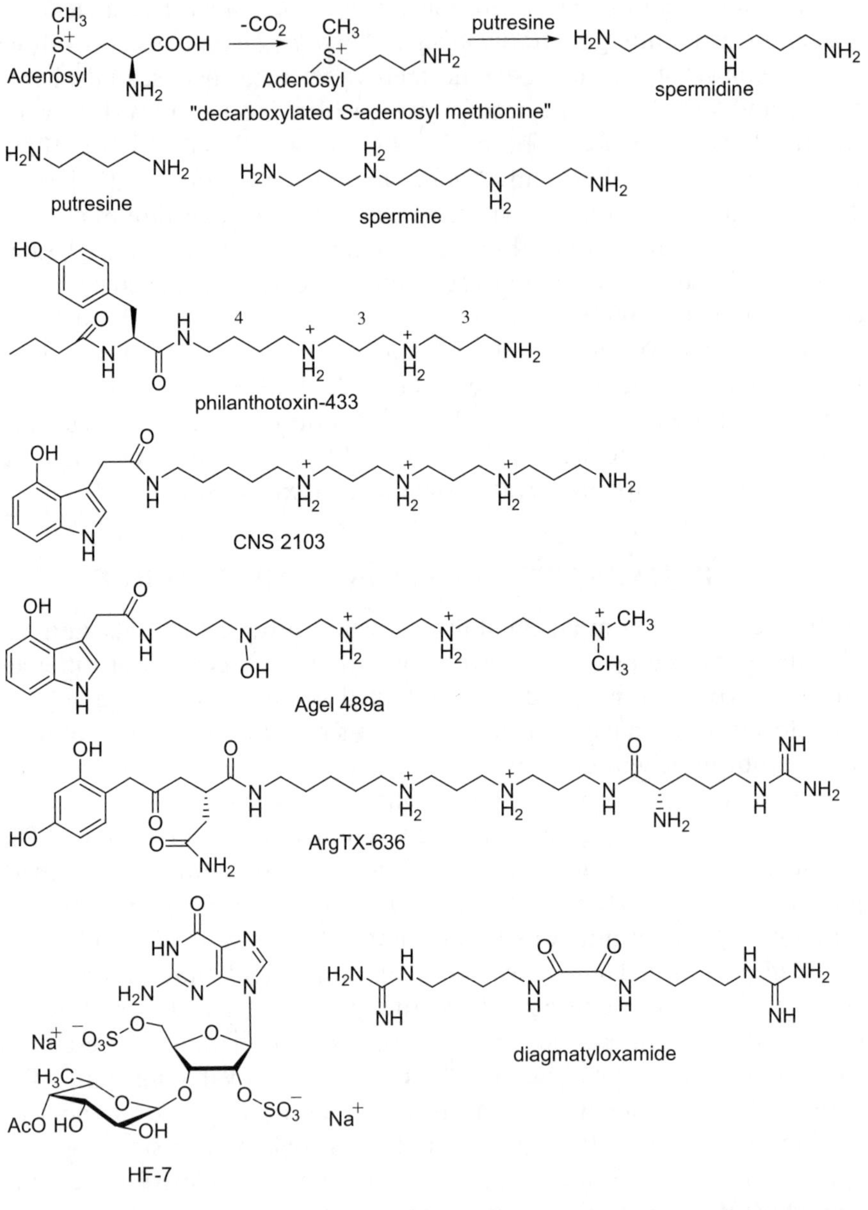

Figure 10.10 The formation of spermidine from putrescine, and structures of complex alkaloids found in wasp and spider venoms. Philanthotoxin is from wasp venom, CNS 2103, Agel 489a and ArgTX-636 are from spider venoms. The guanosine derivative designated HF-7 is from venom of *Hololena curta*, a funnel-web spider, and the amide of oxalic acid with agmatine (decarboxylated arginine) is from a hunting spider *Plectreurys tristis*.

been found in the venom of a fishing spider *Dolomedes okefinokensis* (Araneae: Pisauridae) (McCormick *et al.*, 1993). It consists of hydroxylated indoleacetic acid amide linked to cadaverine that has been extended with three propylamine units (Figure 10.10). About 70 such compounds are known in spider venoms, some have N-OH or N-CH$_3$ groups. Agel 489a from *Agelenopsis aperta* (Araneae: Agelenidae) is another example (Jasys *et al.*, 1992). Such compounds are described as polyaza-alkanes coupled to an aromatic head group. An equally unusual structure has been found in the venom of a funnel-web spider *Hololena curta* (Araneae: Agelenidae) (McCormick *et al.*, 1999). The funnel-web spiders are noted for their particularly deadly venom. The compound, designated HF-7, is a disulfate of guanosine, linked to an acetylated fucose sugar (Figure 10.10). It acts as many neurotoxins do by blocking calcium channels in cell membranes. This compound and the oxalic acid amide of agmatine, or decarboxylated arginine (Figure 10.10) from the primitive hunting spider *Plectreurys tristis* (Araneae: Plectreuidae), are further examples (see Section 6.7) of spiders adapting derivatives of primary metabolites (Quistad *et al.*, 1993).

Isoxazolinones were first known in leguminous plants, but two isoxazolinone glucosides have been discovered in chrysomelid beetles of the subtribes Chrysomelina and Phyllodectina (Figure 10.11). Moreover, some of these substances are further elaborated by addition of 3-nitropropionic esters. Both parts of the beetle compounds have been shown to be biosynthesized from L-aspartic acid (Randoux *et al.*, 1991). Leaves of aspen *Populus tremula* were coated with L-[U-^{14}C] aspartic acid and fed to adults of the poplar leaf beetle *Chrysomela tremulae* (Plate 42). The beetles were "milked" for the secretion with bits of filter paper, and the two compounds isolated each showed about 0.015% incorporation of radioactivity. No radioactivity was found in the glucose; all of the activity was associated with the isoxazolinone in the glucoside and 30% of it was associated with the nitropropionic acid in the ester. The proposed biosynthetic route (Figure 10.11) is the same as is found in plants.

A remarkable example of a pheromone is the oviposition deterrent of the cherry fruit fly *Rhagoletis cerasi* (Diptera: Tephritidae) (Plate 43). This substance is placed on the cherry fruit by the female fly after she has laid her egg in it (Ernst and Wagner, 1989). Its purpose is to deter other females of that species from laying competing eggs in the same fruit. The pheromone must be stable and non-volatile and remain intact in sun and rain. It consists of sugar, fatty-acid and amino acid portions, which together ensure a non-volatile, UV-transparent and insoluble substance (Figure 10.12). Both enantiomers at C-15 in the fatty-acid portion are

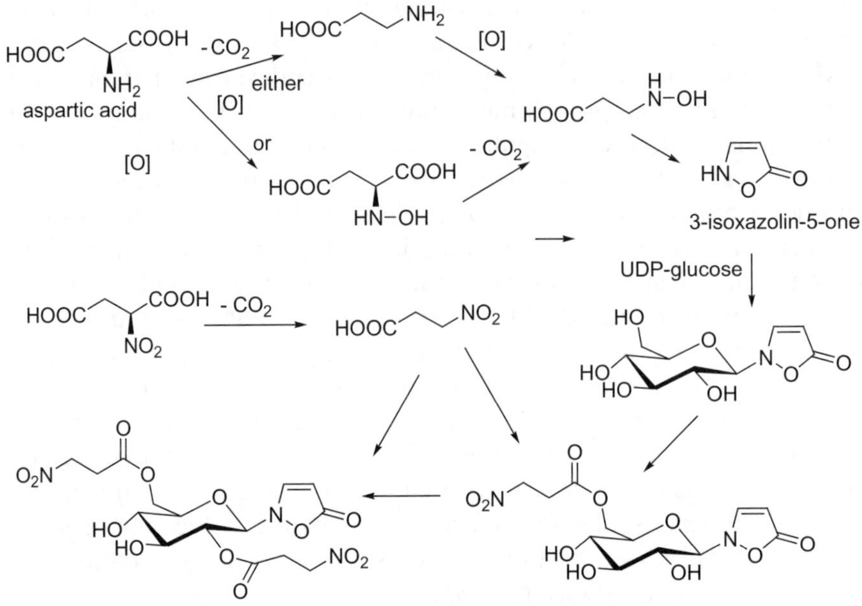

Figure 10.11 The formation of the defensive secretion of *Chrysomela tremulae* from labelled aspartic acid. Both the isoxazolinone part and the nitropropionic acid part are derived from aspartic acid. The 6′-nitropropionic ester as well as the 2′,6′-bis-(nitropropionic) ester shown are present in the secretion.

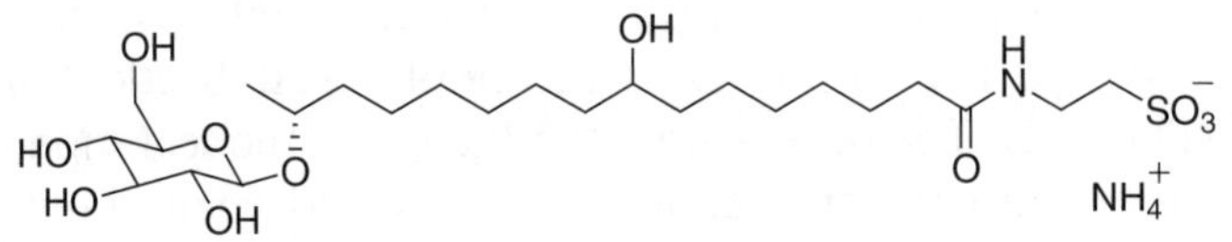

Figure 10.12 The oviposition deterrent pheromone of the cherry fruit fly. It consists of two enantiomers of a dihydroxypalmitic acid with a glucose molecule attached as an ether and an amide of the amino acid sarcosine.

present in the pheromone. The compound is sensed by tarsal contact chemoreceptors.

10.4 LUCIFERIN

Bioluminescence, the ability of living organisms to emit light, is widely distributed throughout more primitive orders. It occurs in insects (fireflies and glow-worms) among the Collembola, Hemiptera, Diptera, Lepidoptera and Coleoptera. It is found particularly in over 2000 species of three Coleoptera families: Lampydidae (fireflies) (Plate 44), Elateridae (click beetles) and Phengodidae (glow-worms). There is also a genus of bioluminescent millipedes. Eight species of *Motyxia* (Diplopoda: Xystodesmidae) have whole-body luminescence (Plate 45) (Shelley, 1997). They glow continuously. It has been suggested this is as a warning, which is very probable, since they also contain cyanogenic glycosides (Section 11.4).

All the bioluminescent insects examined share the identical luciferin (sometimes called firefly luciferin to distinguish it from light-emitting compounds from other organisms with quite different structures), and closely homologous luciferase enzymes to produce light. In the Lampyridae it is chiefly used for mate attraction. The light-emitting cells are sandwiched between thin transparent external cells and backed by reflector cells containing crystals of uric acid, the chief nitrogen excretory product of insects.

The bioluminescence of insects and other organisms has been the subject of a great mass of research. Luciferase has been isolated from a number of beetles, its amino acid sequence and its three-dimensional structure determined for a number of species, and it has been cloned and sequenced. It is located in the peroxisomes of the light-emitting organ cells. Astonishingly, the biosynthetic pathway for the formation of luciferin is unknown. Indeed, it is not even clear to which family of metabolites it is related (Day *et al.*, 2004). Luciferin is said to be synthesized by the fusion of benzoquinone (of unknown origin, Section 9.5), cysteine and a nitrogen source to give 2-cyano-6-hydroxybenzthiazole, but this has not been proved. The benzthiazole is said to condense with a second molecule of cysteine before oxidation and re-arrangement in an unexplained way to give luciferin (Day *et al.*, 2004) (Figure 10.13). Certainly uniformly labelled ^{14}C-cysteine is incorporated into luciferin

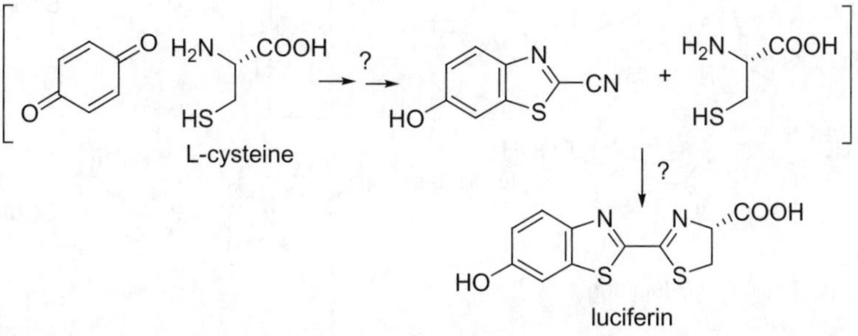

Figure 10.13 The proposed biosynthesis of luciferin. The involvement of cysteine has been demonstrated by radio-labelling.

by larvae of *Pyrearinus termitilluminans* (Elateridae) (Colepicolo *et al.*, 1988). L-Cysteine gives L-luciferin but only D-luciferin contributes to bioluminescence. Niwa *et al.* (2006) have shown that L-luciferin is isomerized to D-luciferin *via* luciferyl CoA and its enol form (Figure 10.14) by luciferase in the Japanese firefly *Luciola lateralis*, and confirmed it in another Japanese firefly *L. cruciata* and the North American firefly *Photinus pyralis* (all Lampyridae).

Luciferin reacts with oxygen very slowly, but many times faster in the presence of luciferase to oxyluciferin. In the catalysed reaction, luciferin reacts with adenosine triphosphate to give luciferyl-adenosine monophosphate (luciferyl-AMP), which reacts with molecular oxygen to give a peroxy ion which loses AMP and forms a dioxetane ring (Marques and de Silva, 2009). The function of the AMP is not to activate the molecule, rather to increase the acidity of the C-4 proton of the thiazole ring. The dioxetane in turn loses CO_2 to give oxyluciferin in the first excited state (Figure 10.14). These reactions have been studied with $^{18}O_2$ and with luciferin labelled with ^{18}O in the carboxyl group. Recent work on the crystal structure of oxyluciferin has shown that it exists with the thiazole ring in *trans* relationship and in the enol form (Naumov *et al.*, 2009).

The quantum yield is a maximum of 41%, the most efficient bioluminescence known. The wavelength of the emitted light varies in an unexplained way. The colour depends upon species. Lampyridae produce yellow-green light. Elateridae produce green to orange light and

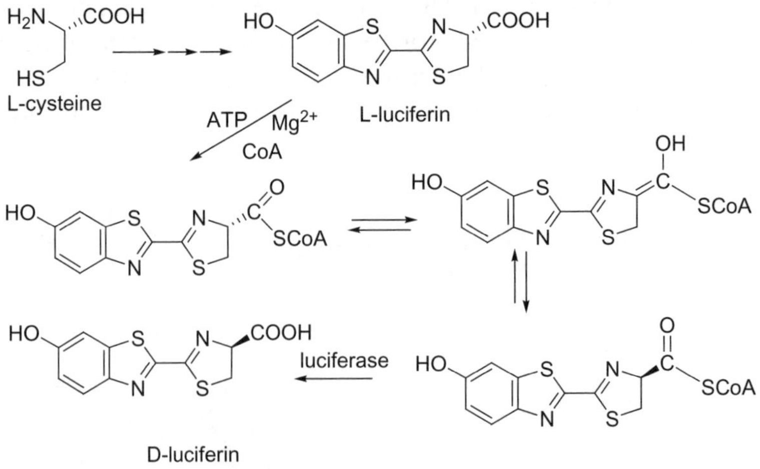

Figure 10.14 The isomerization of L-luciferin to D-luciferin through the CoA ester and the catalytic action of luciferase. Note the thiazole rings are now shown *trans* to each other.

Phengodidae emit green to red light. The current idea is that the wavelength emitted is dependant upon the micro-environment in the pocket of the enzyme (Ugarova and Brovko, 2002), but Naumov *et al.* (2009) show that oxyluciferin can exist in six different enolates and enolate ions, and the difference of colour may also be due to small pH differences. The phenolate ion of the enol form shows properties closest to the yellow-green emission of fireflies (Naumov *et al.*, 2009).

The oxyluciferin is converted to waste thioglycolic acid and 2-cyano-6-hydroxybenzthiazole, which with D-cysteine gives more luciferin (Figure 10.15). This is no proof, however, that the cyanohydroxybenzthiazole is the biosynthetic origin of luciferin. The luciferin regenerating enzyme has been purified from three species, cloned and expressed in *E. coli* (Gomi and Kajiyama, 2001).

Insects emitting light make themselves an easy target for prey. It is not surprising that lampyrid beetles also employ chemicals, lucibufagins (Section 8.7), to make themselves unpalatable.

10.5 PLANT VOLATILE ELICITORS

Plants are able to detect and respond to a broad range of substances produced by insects as they feed on the plants. These herbivore-produced compounds induce the plant on which they are feeding to emit volatile organic compounds that attract predators and parasitoids of the plant-eating insects (Wu and Baldwin, 2009). The first of these discovered was a glucosidase; since then, four general types of elicitors have been discovered, three of them based on fatty acids from the plant, altered by the metabolism of the insect (Felton and Tumlinson, 2008). Some lepidopteran larvae induce plants to release volatiles by the saliva or regurgitated juices from the mouths of the larvae. One of these, caused by the beet armyworm *Spodoptera exigua* (Lepidoptera: Noctuidae) feeding on corn seedlings, is called volicitin. Volicitin has been identified as *N*-(17*S*)-hydroxylinolenyl)-L-glutamine (Figure 10.16). The linolenic acid, taken from the plant, is hydroxylated by the insect and conjugated with insect glutamine. The saliva also contains 17-hydroxylinolenic acid, 17-hydroxylinoleic acid, linolenyl-glutamine and linoleyl-glutamine, but none of these show activity comparable to volicitin. There does not seem to be any obvious benefit to the insect nor is it clear how different insect species affect the bouquet of plant volatiles differently. Another type are the bruchins, based upon diol esters of 3-hydroxypropionic acid, and the more recent caeliferins containing sulfuric acid esters (Felton and Tumlinson, 2008).

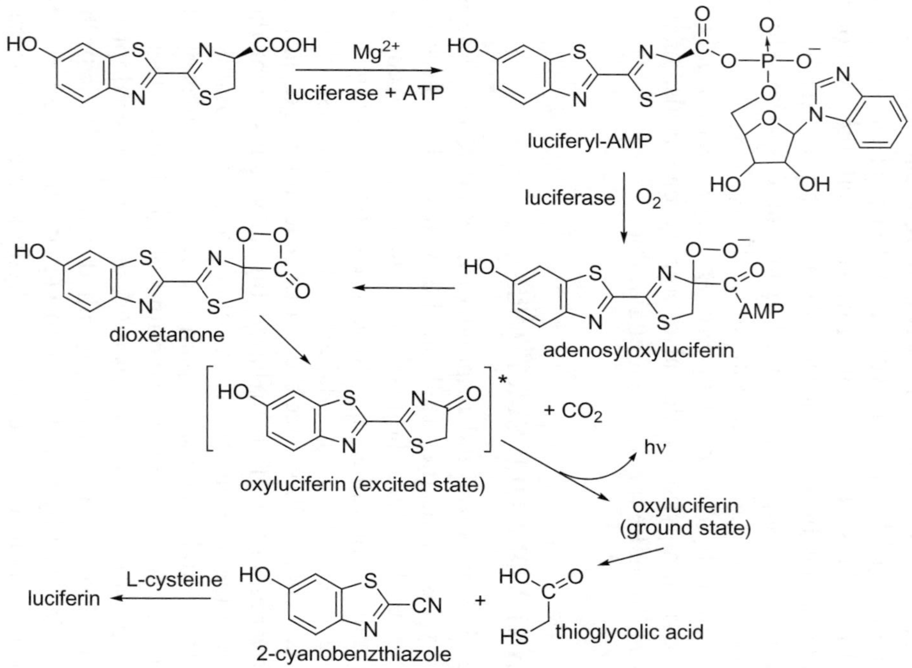

Figure 10.15 The formation of oxyluciferin in the excited state. When $^{18}O_2$ gas was used, labelling was found equally in CO_2 and oxyluciferin. The asterisks indicate the excited electronic state. Light is emitted on return to the ground state.

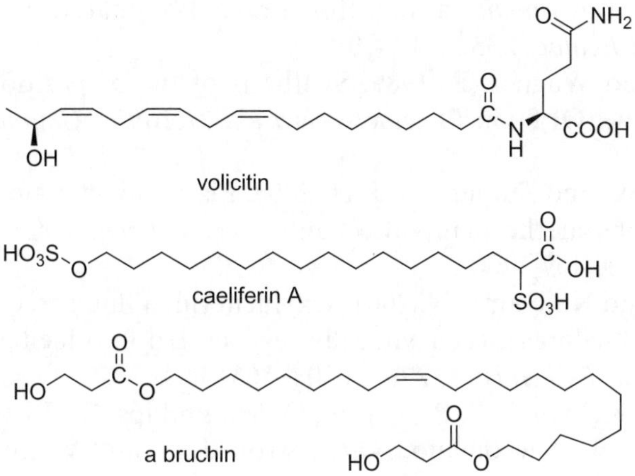

Alkaloids and Compounds of Mixed Biosynthetic Origin 311

Figure 10.16 Volicitin, caeliferin A and a bruchin, examples of insect-released substances, which stimulate plants to produce volatile compounds that attract predators and parasitoids of the insect.

BACKGROUND READING

Piek, T. (ed.) 1986. *Venoms of the Hymenoptera: Biochemical, Pharmacological and Behavioural Aspects*, Academic Press, London, 570 pp.

Schmidt, J. O. 1986. Ant venoms: chemistry, pharmacology and chemical ecology of ant venoms. In: *Venoms of the Hymenoptera: Biochemical, Pharmacological and Behavioural Aspects*, Piek, T. (ed.), Chapter 5, Academic Press, London, pp. 425–508.

REFERENCES

Colepicolo, P., Pagni, D. and Bechara, E. J. H. 1988. Luciferin biosynthesis in larval *Pyrearinus termitilluminans* (Coleoptera: Elateridae). *Comparative Biochemistry and Physiology*, **91B**, 143–147.

Daloze, D., Braekman, J. C. and Pasteels, J. M. 1986. A toxic dipeptide from the defense glands of the Colorado beetle. *Science*, **233**, 221–223.

Day, J. C., Tisi, L. C. and Bailey, M. J. 2004. Evolution of beetle bioluminescence: the origin of beetle luciferin. *Luminescence*, **19**, 8–20.

Devijver, C., Braekman, J. C., Daloze, D. and Pasteels, J. M. 1997. Biosynthesis of tetraponerine-6. Evidence that two different pathways are operating in the biosynthesis of the two tetraponerine skeletons. *Chemical Communications*, 661–662.

Dossey, A. T., Gottardo, M., Whitaker, J. M., Roush, W. R. and Edison, A. S. 2009. Alkyldimethylpyrazines in the defensive spray of

Phyllium westwoodii: a first for order Phasmatodea. *Journal of Chemical Ecology*, **35**, 861–870.

Ernst, B. and Wagner, B. 1989. Synthesis of the oviposition-deterring pheromone (OPD) in *Rhagoletis cerasi* L. *Helvetica Chimica Acta*, **72**, 165–171.

Felton, G. W. and Tumlinson, J. H. 2008. Plant–insect dialogs: complex interactions at the plant–insect interface. *Current Opinion in Plant Biology*, **11**, 457–463.

Gomi, K. and Kajiyama, N. 2001. Oxyluciferin, a luminescence product of firefly luciferase, is enzymically regenerated into luciferin. *Journal of Biological Chemistry*, **276**, 36508–36513.

Jasys, V. J., Kelbaugh, P. R., Nason, D. M., Phillips, D., Rosnack, K. J., Forman, J.T., Saccomano, N. A., Stroh, J. G. and Volkmann, R. A. 1992. Novel quaternary ammonium salt-containing polyamines from the *Agelenopsis aperta* funnel-web spider. *Journal of Organic Chemistry*, **57**, 1814–1820.

Jones, T. H., Clark, D. A., Heterick, B. E. and Snelling, R. R. 2003. Farnesylamine from the ant *Monomorium fieldi* Forel. *Journal of Natural Products*, **66**, 325–326.

Koob, R., Rudolph, C. and Veith, H. J. 1997. The absolute configuration of 3-methylpyrrolidine alkaloids from poison glands of ants Leptothoracini (Myrmicinae). *Helvetica Chimica Acta*, **80**, 267–272.

Leal, W. S., Matsuyama, S., Kuwahara, Y, Wakamura, S., and Hasegawa, M. 1992. An amino-acid derivative as the sex pheromone of a scarab beetle. *Naturwissenschaften*, **79**, 184–185.

Leal, W. S., Oehlschlager, A. C., Zarbin, P. H. G., Hidalgo, E., Shannon, P. J., Murata, Y,, Gonzalez, L., Andrade, R. and Ono, M. 2003. Sex pheromone of the scarab beetle *Phyllophaga elenans* and some intriguing minor components. *Journal of Chemical Ecology*, **29**, 15–25.

Leclercq, S., Charles, S., Daloze, D., Braekman, J. C., Aron, S. and Pasteels, J. M. 2001. Absolute configuration of anabasine from *Messor* and *Aphaenogaster* ants. *Journal of Chemical Ecology*, **27**, 945–952.

Marques, S. M. and de Silva, J. C. E. 2009. Firefly bioluminenscence: a mechanistic approach of luciferase catalysed reactions. *IUBMB Life*, **61**, 6–17.

McCormick, K. D., Kobayashi, K., Goldin, S. M., Reddy, N. L. and Meinwald, J. 1993. Characterization and synthesis of a new calcium antagonist from the venom of a fishing spider. *Tetrahedron*, **49**, 11155–11168.

McCormick, J., Li, Y. B., McCormick, K. D., Duynstee, H. I., van Engen, A. K., van der Marel, G. A., Ganem, B., van Boom, J. H. and

Meinwald, J. 1999. Structure and total synthesis of HF-7, a neuroactive glyconucleoside disulphate from the funnel-web spider *Hololena curta*. *Journal of the American Chemical Society*, **121**, 5661–5665.

Naumov, P., Ozawa, Y., Ohkubo, K. and Fukuzumi, S. 2009. Structure and spectroscopy of oxyluciferin, the light emitter of the firefly bioluminescence. *Journal of the American Chemical Society*, **131**, 11590–11605.

Niwa, K., Nakamura, M. and Ohmiya, Y. 2006. Stereoisomeric bio-inversion key to biosynthesis of firefly D-luciferin. *FEBS letters*, **580**, 5283–5287.

Oldham, N. J. and Morgan, E. D. 1993. Structures of the pyrazines from the mandibular gland secretion of the ponerine ant *Dinoponera australis*. *Journal of the Chemical Society, Perkin Transactions I*, **1993**, 2713–2716.

Ondrus, A. E. and Movassaghi M. 2009. Total synthesis and study of myrmicarin alkaloids. *Chemical Communications*, **2009**, 4151–4165.

Piek, T., Dunbar, S. J., Kits, K. S., van Marle, J. and van Wilgenburg, H. 1985. Philanthotoxins, a review of the diversity of actions on synaptic transmission. *Pesticide Science*, **16**, 488–494.

Quistad, G. B., Lam, W. W. and Casida, J. E. 1993. Identification of bis(agmatine)oxalate in venom from the primitive hunting spider *Plectreurys tristis* (Simon). *Toxicon*, **31**, 920–924.

Randoux, T., Braekman, J. C., Daloze, D. and Pasteels, J. M. 1991. De novo biosynthesis of Δ3-isoxazolin-5-one and 3-nitropropanoic acid derivatives in *Chrysomela tremulae*. *Naturwissenschaften*, **78**, 313–314.

Reder, E., Veith, H. J. and Buschinger, A. 1995. Novel alkaloids from the poison glands of ants Leptothoracini. *Helvetica Chimica Acta*, **78**, 73–79.

Renson, B., Merlin, P., Daloze, D., Braekman, J. C., Roisin, Y. and Pasteels, J. M. 1994. Biosynthesis of tetraponerine-8, a defense alkaloid of the ant *Tetraponera* sp. *Canadian Journal of Chemistry*, **72**, 105–109.

Robbins, P. S., Cash, D. B., Linn, C. E., and Roelofs, W. L. 2008. Experimental evidence for three pheromone races of the scarab beetle *Phyllophaga anxia* (LeConte). *Journal of Chemical Ecology*, **34**, 205–214.

Saporito, R. A., Garraffo, H. M., Donnelly, M. A., Edwards, A. L., Longino, J. T. and Daly, J. W. 2004. Formicine ants: an arthropod source for the pumiliotoxin alkaloids of dendrobatid poison frogs. *Proceedings of the National Academy of Sciences, USA*, **101**, 8045–8050.

Schröder, F., Franke, S., Francke, W., Baumann, H., Kaib, M., Pasteels, J. M. and Daloze, D. 1996. A new family of tricyclic alkaloids from *Myrmicaria* ants. *Tetrahedron*, **52**, 13539–13546.

Shelley, R. M. 1997. A re-examination of the millipede genus *Motyxia* Chamberlain, with a re-diagnosis of the tribe Xystcheirini and remarks on the bioluminescence. *Insecta Mundi*, **11**, 331–351.

Ugarova, N. N. and Brovko, L. Y. 2002. Protein structure and bioluminescent spectra for firefly bioluminescence. *Luminescence*, **17**, 321–330.

Wojtasek, H. and Leal, W. S. 1999. Degradation of an alkaloid pheromone from the pale brown chafer, *Phyllopertha diversa* (Coleoptera: Scarabaeidae), by an insect olfactory cytochrome P450. *FEBS Lettres*, **458**, 333–336.

Wu, J. and Baldwin, I. T. 2009. Herbivory-induced signalling in plants: perception and action. *Plant Cell and Environment*, **32**, 1161–1174.

CHAPTER 11
Plant Substances Altered and Sequestered by Insects

11.1 INTRODUCTION

The concept of insects using substances in their food as semi-raw materials for necessary metabolites is already familiar. Examples are the modification of plant or animal sterols for hormones or defensive compounds, and metabolism of aromatic amino acids for a range of purposes. Plant toxins have been evolved to deter herbivorous insects while some insects have in turn evolved ways of metabolizing the toxins, or turning them to their own use, by storing the toxins in their bodies, or converting them to other toxins and, in certain cases, to pheromones. Butterflies and moths, followed by beetles, are prominent insect sequesterers of plant chemicals. Day-flying, brightly coloured butterflies and moths without protective toxins would otherwise be particularly attractive prey. The plant substances may be unchanged, or changed so much that their connection with the plant may not be immediately apparent. The subject of sequestration of plant metabolites by insects has been comprehensively reviewed by Opitz and Müller (2009). More than 250 species of insect are known to sequester toxic plant compounds from at least 40 families of plants. Some insects seek out plants that are not their normal food source, specifically to obtain such toxic compounds. This practice is called pharmacophagy, and the insects that seek the substances, pharmacophagous species (Boppré, 1984). The phenomenon has already been encountered in cantharidiniphilous insects that eat dead beetles that contain cantharidin (Section 7.5.2.2). The defensive secretion may contain one or a few compounds or may consist of a mixture of several types of compound from different biosynthetic routes. How insects avoid the toxic consequences of such ingestion, generally, we do not know, but studies on this have begun, and the difference in the genes of sequestering species and the amino acid sequences of important enzymes discovered (Labeyrie and Dobler, 2004).

In some cases insects partially metabolize food toxins to create their own versions for protection. They use a variety of enzymes to degrade or modify the plant products, including oxidases, reductases, hydrolases, esterases and transferases. This metabolizing of toxins constitutes further biosynthesis. Metabolism and its result vary from species to species, and are not easy to summarize. Examples only can be given.

The extent of sequestering can be along a range of possibilities. The plant substances can be partially metabolized or remain unchanged in the gut, and pass through; therefore analysis of the whole insect for the plant substance does not necessarily show the substance has been sequestered. Substances may be absorbed into the haemolymph. Still others may be passed to special organs or vacuoles, and some, especially in Coleoptera, may be stored in defensive glands of the pronotum and elytra. Generally, less polar substances tend to pass across the gut lining more easily.

11.1.1 Nuptial Gifts

The passing of defensive material from the male to female and then to the eggs has already been described for the passage of cantharidin from male beetles to females at copulation and the consequent investment of eggs with cantharidin (Section 7.5.2.2). Males of many species provide females with nutritional contributions during courtship and mating. The passage of materials with the sperm in this way is called by the rather romantic term of nuptial gifts. These nuptial gifts can include captured prey, nutritional substances manufactured by male accessory glands and defensive compound in the spermatophore (Vahed, 1998).

11.2 CARDIAC GLYCOSIDES

The first example of the collection of toxic plant compounds by insects for protection was demonstrated by Miriam Rothschild and collaborators. They showed that the North African grasshopper *Poekilocerus bufonius* (Orthoptera: Pyrgomorphidae), feeding on the giant milkweed *Calotropis procera*, contained cardiac glycosides, similar to those of the plant (von Euw *et al.*, 1967). Cardiac glycosides are steroidal compounds with sugar residues attached. They consist of two classes, cardenolides, with the side-chain converted into a five-membered ring lactone, and bufadienolides with the side-chain as a six-membered ring lactone. The steroidal parts also occur in plants without sugar residues, in which case the name ends in -genin. Of six cardiac glycosides in the giant milkweed, only two cardenolides, calotropin and calactin (Figure 11.1), were retained in its poison glands by the grasshopper. Calotropin is highly toxic to vertebrates but evidently has no ill effects on the insect. The cardiac glycosides are largely acquired during feeding by nymphs, but they persist into adults, and are found in the haemolymph, fat body, integument and muscles, and are passed into the eggs. Cardiac glycosides can also be biosynthesized by *Chrysolina* beetles from sterols (Section 8.7).

Rothschild and her collaborators next showed that the larvae of the monarch butterflies *Danaus plexippus* and *D. chrysippus* (Nymphalidae), feeding on another milkweed *Asclepias curassavica*, absorb various cardenolides (mostly calotropin, calactin and calotropagenin), which are retained by the larvae and through the pupal stage, making the adult butterflies unpalatable to predators like birds (Reichstein *et al.*, 1968).

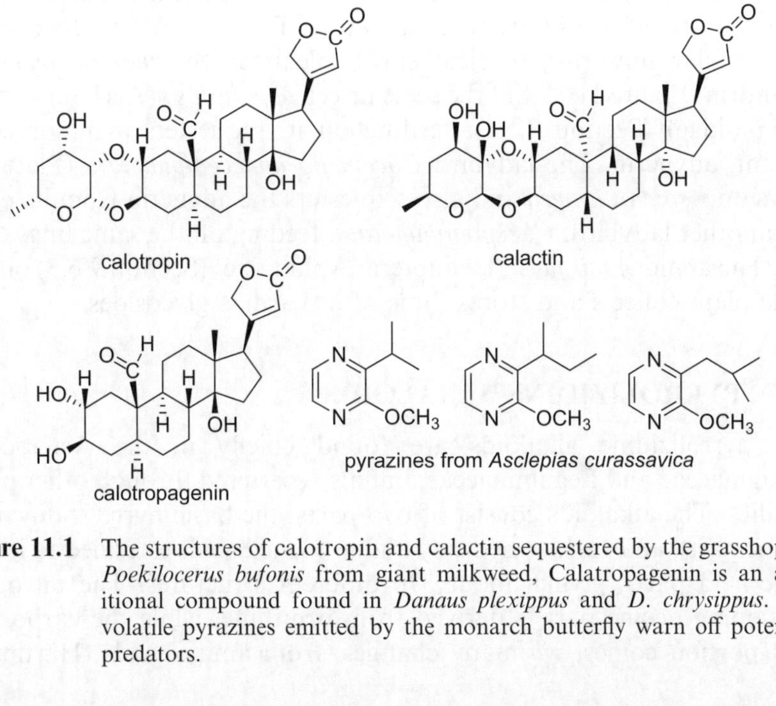

Figure 11.1 The structures of calotropin and calactin sequestered by the grasshopper *Poekilocerus bufonis* from giant milkweed, Calotropagenin is an additional compound found in *Danaus plexippus* and *D. chrysippus*. The volatile pyrazines emitted by the monarch butterfly warn off potential predators.

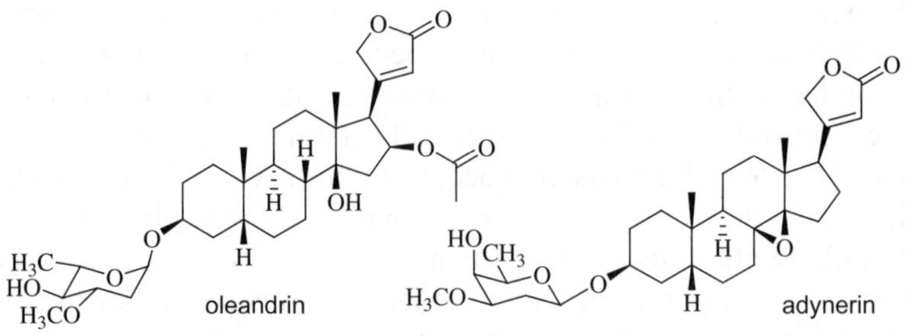

Figure 11.2 Two cardiac glycosides from oleander used by insects. Oleandrin is not sequestered by the aphid *C. undecempunctata*, but only the minor product adynerin is accumulated. The sugar in oleandrin is oleandrose, in adynerin it is digitalose.

In all stages of these insects the toxins are distributed through the body. In addition to calotropin, the butterfly also stores three volatile alkyl-methoxypyrazines (Figure 11.1) from the plant. The warning odour of these volatile compounds, associated with calotropin, is enough to deter a bird on close approach from trying to eat a butterfly charged with calotropin. Pyrazines appear to function as a general warning signal for many insect species (Nishida, 2002).

It does not follow that because an insect feeds on a toxic plant it will sequester the toxins, nor does the possession of the cardenolides by an insect ensure protection from all predators. The principal cardenolide in the popular flowering tropical shrub oleander (*Nerium oleander*) is oleandrin (Figure 11.2), but the scale insect *Aspidiotus nerii* (Homoptera: Diaspididae) (Section 7.5.1), feeding on it, sequesters a minor component, adynerin. The ladybird *Coccinella undecempunctata* (Plate 49) preys on *Aspidiotus nerii* and also sequesters the adynerin from its prey, but another ladybird, *C. septempunctata*, feeding on the same bugs does not. The aphid *Aphis nerii* (Hemiptera: Aphididae) (Section 9.6.3) on the same plant collects and stores three of its cardiac glycosides.

11.3 PYRROLIZIDINE ALKALOIDS

The pyrrolizidine alkaloids are found chiefly in the Asteraceae, Boraginaceae and Leguminaceae, but also scattered through other plant families. The alkaloids consist of two parts, the basic pyrrolizidine (the necine part), and various hydroxy and branched acids esterified to it (the necic acid part), giving about 370 different structures. The nitrogen-containing necine part is derived from ornithine, while the carboxylic acid portion comes, *via* many changes, from amino acids (Hartmann and Ober, 2000). Taxonomically unrelated insect groups (but chiefly in the Coleoptera and Lepidoptera) feed on unrelated plants, which contain these alkaloids and store the compounds in their bodies. A feature of pyrrolizidine alkaloids is that they exist in two interchangeable forms, the non-toxic N-oxides, and the pro-toxic free bases. In plants they are usually synthesized, transported and stored as N-oxides. They only become toxic when the free bases are metabolized to highly reactive pyrroles by cytochrome P450 oxidases. Many insects reduce them in the gut to pro-toxic free bases but then they are rapidly re-oxidized by a specific mono-oxygenase back to the N-oxides for storage. If insects can avoid attacking the free base forms with their cytochrome oxidases, it is possible for the alkaloids to be stored in their bodies (Narberhaus *et al.*, 2005). The alkaloids that are retained may then be metabolized to simpler compounds and stored, in the whole body in Lepidoptera, or in defensive glands in Coleoptera. But there are still many exceptions to these general statements.

11.3.1 Pyrrolizidines in Lepidoptera

The caterpillars of *Utetheisa* moths (Lepidoptera: Arctiidae) feed on *Crotolaria* plants (Fabaceae) and absorb the pyrrolizidine alkaloids, which protect them from predators. The alkaloids are retained in the adult and found in the scales of the wings (Rossini *et al.*, 2003). The ornate moth *Utetheisa ornatrix* (Plate 46) degrades pyrrolizidine alkaloids like monocrotaline N-oxide, producing (R)-(–)-hydroxydanaidal (Figure 11.3), and uses the hydroxydanaidal (which is non-toxic) as a sex attractant for females (Eisner and Meinwald, 1995). The female favours males that have much of the hydroxydanaidal in the brush-like structures (called coremata) on the abdominal tip, which males expose to the female. The amount of hydroxydanaidal is a measure of the quantity of alkaloid in the male that the female can acquire at copulation (del Campo *et al.*, 2007). The alkaloids passed by the male in the spermatophore, and supplemented by more from the female, are endowed in the eggs, and protect them from predation (Bezzerides and Eisner, 2002).

While monocrotaline (with ($7R$)-configuration) is metabolized to (R)-hydroxydanaidal, heliotrine with the opposite configuration at C-7 also gives (R)-hydroxydanaidal. Deuteration studies showed the heliotrine was oxidized to the C-7 ketone and reduced to (R)-hydroxydanaidal (Schulz *et al.*, 1993) (Figure 11.3).

Larvae of danaid butterflies feed on non-toxic plants, but the adult males collect pyrrolizidine alkaloids like seneciphylline from ragwort, *Senecio jacobaea*, by licking the surface of leaves; part of the alkaloid

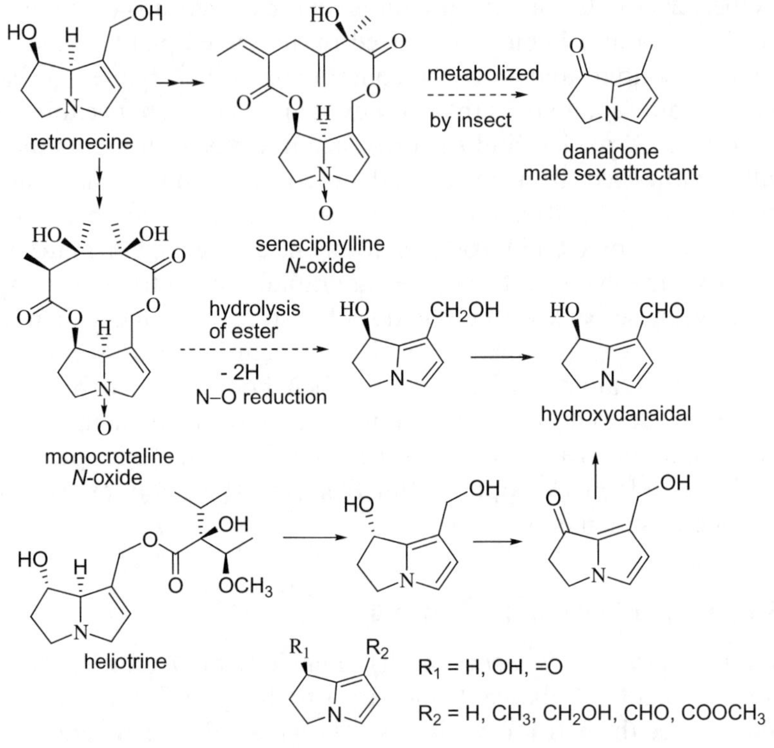

Figure 11.3 Some examples of pyrrolizidine alkaloids and the sex attractants made from them by danaid butterflies and ornate moths. Generally the pyrrolizidine alkaloids exist in plants as *N*-oxides, as shown for monocrotaline and seneciphylline. At the bottom are summarized the structures of compounds used as male pheromones.

they store, and part degrades to danaidone (Figure 11.3) to use as a sex attractant. The danaidone is presented on corematal organs called hairpencils. Female butterflies are able to detect which males have more toxin by the amount of danaidone they secrete, and choose males with most alkaloid. The males pass on the alkaloids with their sperm, and the females invest their eggs with the alkaloids in turn.

11.3.2 Pyrrolizidines in Coleoptera

Two groups of chrysomelid beetles have evolved different ways of dealing with the pyrrolizidine alkaloids. The genus *Oreina* feed on *Senecio* and *Adenostyles* plants (both Asteraceae). *Oreina* beetles also make their own cardiac glycosides (Section 8.7), and accumulate pyrrolizidines as N-oxides, in their haemolymph and body generally. One species has only pyrrolizidines and no cardenolides. Generally the insects do not alter the alkaloids, but two examples, one of hydrolysis and one of epoxidation by beetles, are known.

The genus *Platyphora* has a different mechanism for handling pyrrolizidines of the lycopsamine type. These open-chain compounds are partially metabolized and the processed alkaloids concentrated in defensive secretion. *Platyphora* store the alkaloids as the tertiary amines, not the N-oxides, and only in the haemolymph of larvae and defensive glands of adults (Figure 11.4). Studies with [^{14}C]rinderine showed that the larvae and adults of *Platyphora boucardi* (Plate 47) can equally absorb amine and *N*-oxide forms, but accumulate only the tertiary amine form (Pasteels *et al.*, 2003). Rinderine is rapidly converted to intermedine and found in all tissues and on the surface of larvae. *P. boucardi* was also shown to be able to take ingested retronecine and esterify it with propionic, lactic and α-hydroxyisovaleric acids to make new alkaloids. The beetles also make saponins *de novo* (Section 8.7.1).

There are isolated examples of pyrrolizidine alkaloid sequestration among other insect orders; the grasshopper *Zonocerus variegates* (Orthoptera: Pyrgomorphidae) (Plate 48) and the aphid *Aphis jacobaeae*

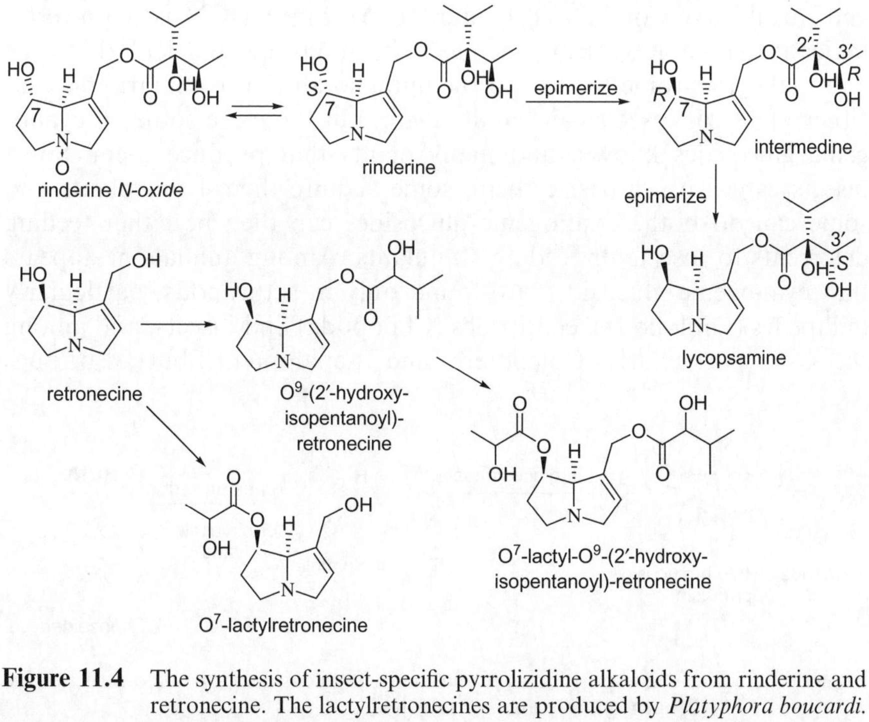

Figure 11.4 The synthesis of insect-specific pyrrolizidine alkaloids from rinderine and retronecine. The lactylretronecines are produced by *Platyphora boucardi*.

(Homoptera: Aphididae) sequester pyrrolizidines while the ladybird *Coccinella septempunctata* (Coleoptera: Coccinellidae), which feeds on the aphids, in turn sequesters their alkaloids. The first report of a scale insect *Ceroplastes albolineatus* (Hemiptera: Coccidae) feeding on *Pittocaulon* (*Senecio*) *praecox* has shown that all life stages, females, eggs, crawling larvae and wax covers are endowed with pyrrolizidines, all as free bases (Loaiza *et al.*, 2007).

It has been suggested that the toxic effects of pyrrolizidine alkaloids and cyanogenic glycosides (Section 11.4) do not really protect insects from bird or spider predators, because the toxicity only comes into effect when the compound is metabolized, which is well separated in time from the eating or tasting of the insect, rather it is only the bitterness of the taste of the compounds that offers protection to the insect (Kassarov, 2002).

11.4 CYANOGENIC GLUCOSIDES

Many plants store cyanogenic glucosides for their protection. When the plant is damaged, a glucosidase is released which cleaves the glucoside, giving a hydroxynitrile, which either uncatalysed above pH 6, or catalysed by an α-hydroxynitrile lyase below pH 6, gives hydrogen cyanide and an aldehyde or ketone (Figure 11.5). Since HCN is a powerful toxicant to complexed iron, such as in haem groups, particularly to the terminal cytochrome oxidase in the mitochondria, it is remarkable that either plants or insects can handle them, but there are some 60 cyanogenic glucosides known, and many plants that produce them. Some insects can also synthesize them, some acquire them from plants and some can do both. Cyanogenic glucosides can then be either feeding deterrents to insects or feeding stimulants. Among animals, it appears that cyanogenic glucosides are found only in arthropods, particularly millipedes (Diplopoda), centipedes (Chilopoda) and insects, and among the last, particularly Coleoptera and Lepidoptera, but also bugs

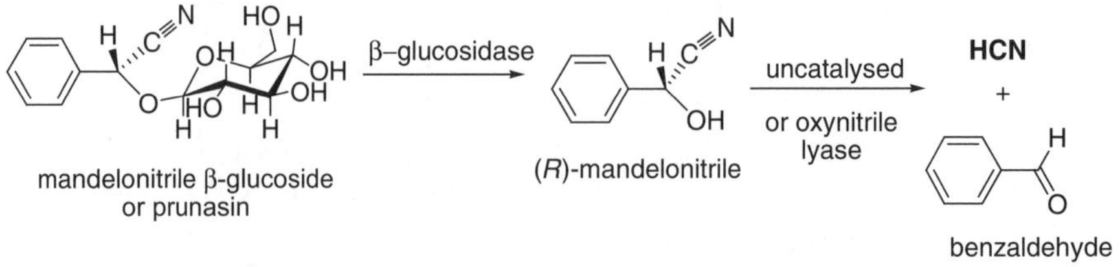

Figure 11.5 The decomposition of prunasin, the β-glucoside of mandelonitrile to give hydrogen cyanide and benzaldehyde.

(Heteroptera). Lepidoptera make aliphatic cyanogenic glucosides while the other groups make aromatic derivatives. Like many of the other defensive compounds, the cyanogenic glucosides are extremely bitter-tasting.

The millipede *Oxidus gracilis* (Diplopida: Paradoxosomatidae) converts racemic [2-^{14}C]phenylalanine to HCN, but does not make HCN from [2-^{14}C]tyrosine (Towers *et al.*, 1972). Because the glucosides are so easily decomposed, and a hydroxynitrile like mandelonitrile decomposes to aldehyde and HCN on gas chromatography, the exact storage form of the products that release hydrogen cyanide in millipedes and centipedes is not known. The secretion in *O. gracilis*, as analysed, is essentially mandelonitrile (Taira *et al.*, 2003). The steps from phenylalanine to mandelonitrile were studied in the millipede *Harpaphe haydeniana* by feeding a wide range of potential precursors, from which the biosynthetic scheme shown in Figure 9.3 (Section 9.3) was constructed (Duffey *et al.*, 1974).

Cyanogenic glucosides are synthesized in plants from the amino acids valine, leucine, isoleucine, phenylalanine and tyrosine (Zagrobelny *et al.*, 2008). A cytochrome P450 catalyses two sequential *N*-hydroxylations followed by a dehydration, a decarboxylation and isomerization to the corresponding *Z*-aldoxime. An NADPH-dependent dehydration and a second P450 hydroxylation to introduce the C-OH is followed by glycosylation, catalysed by uridine diphosphate glucose-dependant glycosyl transferase (Figure 11.6). The synthesis in arthropods probably follows the same or similar pathway (Zagrobelny *et al.*, 2004).

Hydrogen cyanide is detoxified by reaction with cysteine, catalysed by β-cyanoalanine synthase in mitochondria. The β-cyanoalanine produced can be further converted to asparagine. Cyanide ion can alternatively react with thiosulfate ion in the presence of the enzyme rhodanese

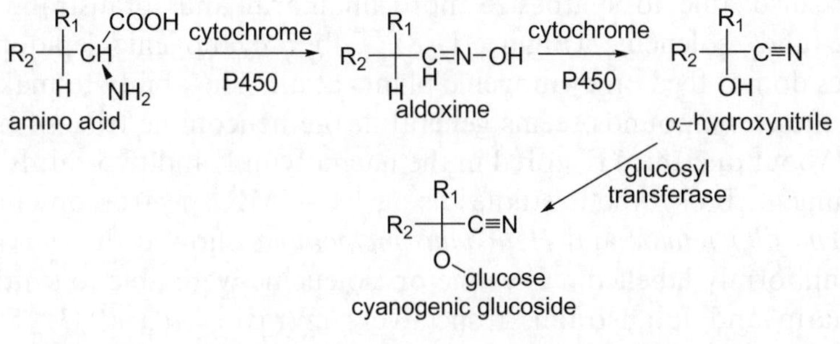

Figure 11.6 A summary of the accepted route from amino acids to cyanogenic glycosides in plants.

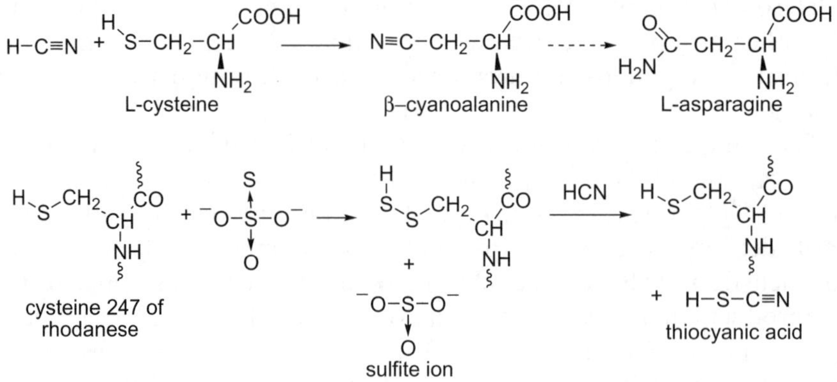

Figure 11.7 Two alternative routes for the destruction of hydrogen cyanide in plants and insects. In the second reaction, cysteine 247 of the enzyme rhodanese reacts first with thiosulfate to give a disulfide which then reacts with cyanide.

(thiosulfate sulfurtransferase) to thiocyanate (Figure 11.7). The reaction requires participation of cysteine no. 247 in the rhodanese molecule.

Only three species of Coleoptera have been identified as producers of cyanogenic glycosides; *Paropsis atomaria*, *Chrysophtharta variicollis* and *C. amoena* (all Chrysomelidae). The Australian chrysomelid beetle *Paropsis atomaria* contains prunasin and mandelonitrile (Figure 11.5) in all life stages. The beetles feed on *Eucalyptus* trees that do not contain cyanogens (Nahrstedt and Davis, 1986). The larvae emit HCN from glands on the seventh and eighth tergites.

Cyanogenic species are much more common among Lepidoptera, with at least 200 species identified as producers, and many larvae are able to synthesize the cyanogens. The bright red-and-black burnet moth *Zygaena trifolii* (Lepidoptera: Zygaenidae) (Plate 50) obtains the cyanogenic glycosides linamarin and lotaustralin from *Lotus corniculatus*, and is also able to synthesize more linamarin and lotaustralin from valine and isoleucine (Figure 11.8). Other cyanogenic lepidopteran species do not feed on cyanogenic plants at all. The ability to make and store these compounds seems general in the heliconiine tribe of butterflies. About one-third is stored in the haemolymph and two-thirds in the integument. Biosynthetic studies using ^{13}C-NMR spectroscopy in both *Zygaena filipendulae* and *Heliconius melpomone* showed that larvae fed with uniformly labelled ^{13}C-valine or isoleucine were able to synthesize linamarin and lotaustralin, respectively (Wray *et al.*, 1983). Studies showed that the carbon skeletons of the amino acids were retained intact except for the carboxyl groups. The biosynthetic enzymes in these

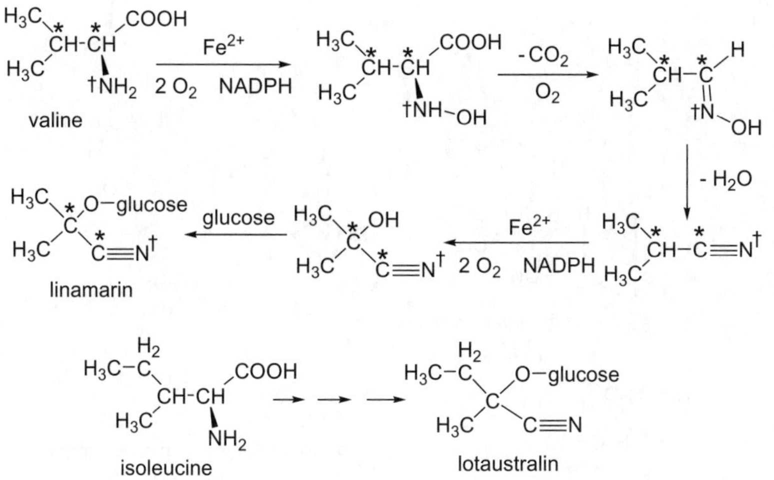

Figure 11.8 The biosynthesis of linamarin from valine and lotaustralin from isolucin in the burnet moth. Asterisks on carbon and the dagger on nitrogen indicate isotopically labelled atoms that were retained in place in the products.

families have a higher affinity for isoleucine and so more linamarin than lotaustralin is produced. The glucosides are retained in all life stages. Adult *Zygaena* moths when attacked emit secretion from around the mouthparts containing 2-methoxy-3-alkylpyrazines (Section 11.2) and haemolymph smelling of hydrogen cyanide from the legs.

In *Zygaena filipendulae* the biosynthetic step from 2-methylbutyronitrile to α-hydroxynitrile is known to be catalysed by a membrane-bound cytochrome P450. The changes in content of glycosides through the life cycle of *Z. filipendulae* have been recorded (Zagrobelny et al., 2008). The glycosides are produced in the larval integument and transferred to other tissues. Since nitrogen is required for chitin synthesis, and cyanogenic glucoside levels fall at pupation, it is possible some of the cyanide nitrogen is used for that purpose.

11.5 VERATRUM ALKALOIDS

The veratrum alkaloids are a group of particularly toxic steroidal alkaloids. The specialist sawfly *Rhadinoceraea nodicornis* (Hymenoptera, Tenthredinidae) stores in its haemolymph ceveratrum alkaloids from its host plant, false, or white, hellibore *Veratrum album*. The alkaloids in sawflies can be either directly sequestered, partly metabolized and sequestered, excreted intact or destroyed (Schaffner et al., 1994). The

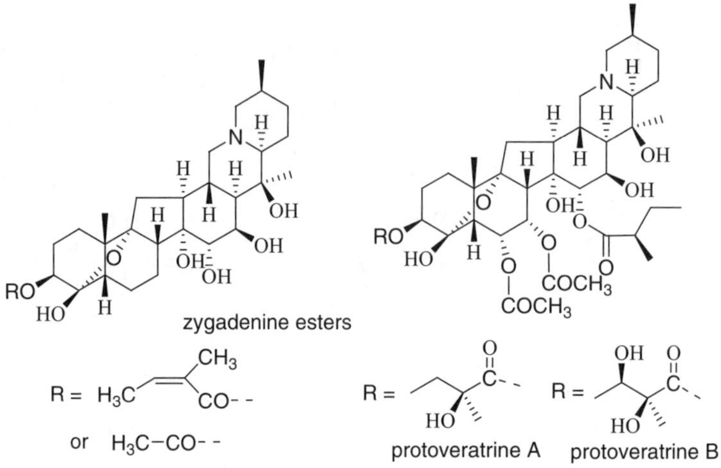

Figure 11.9 Alkaloids from *Veratrum* plants. The angeloyl ester of zygadenine is present in the plant, the acetyl ester in the insect. Protoveratrines are, at the same time, metabolized completely by the sawfly *R. nodicornis*.

principal alkaloid stored in the haemolymph is 3-acetylzygadenine (Figure 11.9). 3-Angeloylzygadenine of the plant is probably hydrolysed in the gut to zygadenine and then acetylated. At the same time protoveratrines A and B (Figure 11.9) are degraded. Another generalist sawfly, *Aglaostigma* sp. (Tenthredinidae), fed on false hellibore leaves had no alkaloids in its haemolymph and did not excrete any (Schaffner *et al.*, 1994).

11.6 OTHER EXAMPLES

There are some thousands of plant toxins available to arthropods for ingestion and sequestration for defence, but only a small fraction of them are known to be used by insects. Nishida (2002) lists 17 types of plant substances sequestered by Lepidoptera alone. Many are absorbed unchanged, like the glucosinolates (mustard oil glycosides) of Cruciferae, the phenanthrene derivatives aristolochic acids from Aristolochiaceae and cycasin from cycad plants (Figure 11.10). Other compounds are metabolized a little. For example, the chrysanthemum beetle *Diabrotica speciosa* (Coleoptera, Chrysomelidae) feeding on cucumbers (Cucurbitaceae), which contain bitter triterpenoid cucurbitacins, converts cucurbitacin B to cucurbitacin D by removing an acetate, then reducing a side-chain double bond to dihydrocucurbitacin D, and carry out glucosylation and desaturation reactions (Andersen *et al.*,

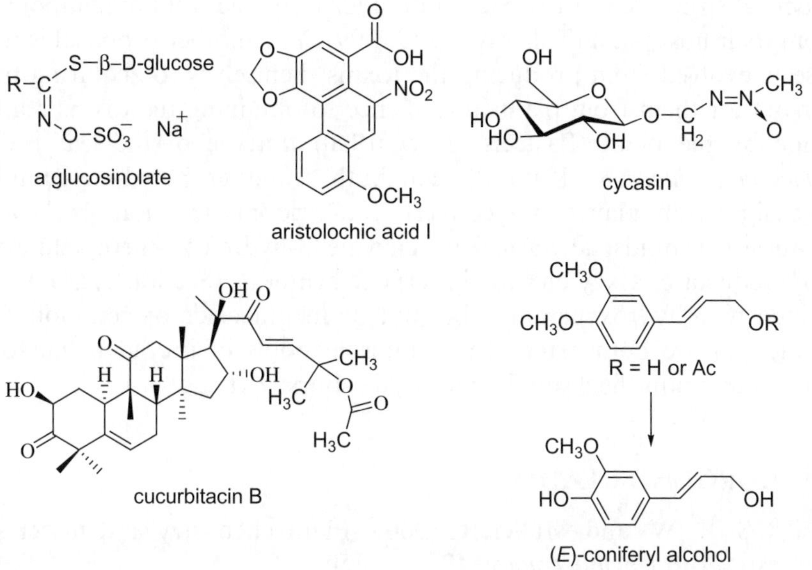

Figure 11.10 Three examples of sequestered compounds from plants not altered by insects, and two, cucurbitacin B and dimethoxycinnamyl alcohol and its acetate, which are minimally changed.

1988). Males of the oriental fruit fly *Bactrocera dorsalis* (Diptera: Tephritidae) feed on the fragrant flowers of *Fagraea berteriana*, which contain (*E*)-3,4-dimethoxycinnamyl alcohol and smaller amounts of (*E*)-3,4-dimethoxycinnamyl acetate. The male flies convert these to (*E*)-coniferyl alcohol and store that in rectal glands. The coniferyl alcohol is very attractive to females (Nishida et al., 1997). The larvae of *Chrysomela tremulae* feed on willow (*Salix*) trees, ingest salicin and, with the aid of a glucosidase and an oxidase, convert it to salicylaldehyde, which they release from abdominal and thoracic glands (Brückmann et al., 2002). Further examples of relatively small biosynthetic changes are known (Nishida, 2002; Opitz and Müller, 2009).

11.7 ORIGINS OF SEQUESTRATION

It has been proposed that some insects seek out toxic plants because ancestrally these insects fed on the plant, but have evolved away to feed on other available plants, but still seek the toxins. It is called the ancestral host hypothesis. An alternative proposal suggests that lack of specificity in taste receptors has provided opportunities for new compounds, including plant toxins, to elicit a feeding response in an insect, when the chemical structure of the toxic compounds resembles that of compounds from their host plant (Tallamy *et al.*, 1999). Yet another proposal is that insects evolved from producing the toxins themselves to acquiring precursors for them from plants and thence to obtaining the toxins ready-made by the plant (Pasteels *et al.*, 1990). Larvae of the leaf beetle *Phaedon cochleariae* (Plate 18) can both sequester 8-hydroxygeraniol glucoside from plants and convert it to iridoids (Section 7.4.1) and synthesize iridoids *de novo*. The enzyme 3-hydroxy-3-methylglutaryl-CoA reductase, a key enzyme in terpene synthesis (Section 7.2.1), is inhibited by 8-hydroxygeraniol, but not by its glucoside or geraniol. The larvae can use both external and internal pools of iridoid precursors, depending upon their supply and needs (Burse *et al.*, 2008).

BACKGROUND READING

Opitz, S. E. W. and Müller, C. 2009. Plant chemistry and insect sequestration. *Chemoecology*, **19**, 117–154.

Schulz, S. 1998. Insect-plant interactions – metabolism of plant compounds to pheromones and allomones by Lepidoptera and leaf beetles. *European Journal of Organic Chemistry*, **1998**, 13–20.

REFERENCES

Andersen, J. F., Plattner. R. D. and Weisleder, D. 1988. Metabolic transformations of cucurbitacins by *Diabrotica virgifera virgifera* Leconte and *D. undecimpunctata howardi* Barber. *Journal of Chemical Ecology*, **18**, 71–77.

Bezzerides, A. and Eisner, T. 2002. Apportionment of nuptial alkaloidal gifts by a multiply-mated female moth (*Utetheisa ornatrix*): eggs individually receive alkaloid from more than one male source. *Chemoecology*, **12**, 213–218.

Boppré, M. 1984. Redefining pharmacophagy. *Journal of Chemical Ecology*, **10**, 1151–1154.

Brückmann, M., Termonia, A., Pasteels, J. M. and Hartmann, T. 2002. Characterisation of an extracellular salicyl alcohol oxidase from larval defensive secretions of *Chrysomela populi* and *Phratora vitellinae* (Chrysomelina). *Insect Biochemistry and Molecular Biology*, **32**, 1517–1523.

Burse, A., Frick, S., Schmidt, A., Büchler, R., Kunert, M., Gershenzon, J., Brandt, W. and Boland, W. 2008. Implication of HMGR in homeostasis of sequestered and *de novo* produced precursors of the

iridoid biosynthesis in leaf beetle larvae. *Insect Biochemistry and Molecular Biology*, **38**, 76–88.
del Campo, M. L., Possner, S. T. and Eisner, T. 2007. Corematal function in *Utetheisa ornatrix* (Lepidoptera: Arctiidae): hydroxydanaidal is devoid of intrinsic defensive potency. *Chemoecology*, **17**, 19–22.
Duffey, S. S., Underhill, E. W. and Towers, G. H. N. 1974. Intermediates in the biosynthesis of HCN and benzaldehyde by a polydesmid millipede, *Harpaphe haydeniana* (Wood). *Comparative Biochemistry and Physiology*, **47B**, 753–766.
Eisner, T. and Meinwald, J. 1995. Defence mechanisms of arthropods 129. The chemistry of sexual selection. *Proceedings of the National Academy of Sciences, USA*, **92**, 50–55.
Hartmann, T. and Ober, D. 2000. Biosynthesis and metabolism of pyrrolizidine alkaloids in plants and specialized insect herbivores. *Topics in Current Chemistry*, **209**, 207–243.
Kassarov, L. 2001. Do cyanogenic glycosides and pyrrolizidine alkaloids provide some butterflies with a chemical defence against their bird predators? A different point of view. *Behaviour*, **138**, 45–67.
Labeyrie, E. and Dobler, S. 2004. Molecular adaptation of *Chrysochus* leaf beetles to toxic compounds in their food plants. *Molecular Biology and Evolution*, **21**, 218–221.
Loaiza, J. C. M., Cespedes, C. L., Beuerle, T., Theuring, C. and Hartmann, T. 2007. *Ceroplastes albolineatus*, the first scale insect shown to sequester pyrrolizidine alkaloids from its host plant *Pittocaulon praecox*. *Chemoecology*, **17**, 109–115.
Nahrstedt, A. and Davis, R. H. 1986. (*R*)-Mandelonitrile and prunasin, the sources of hydrogen-cyanide in all stages of *Paropsis atomaria* (Coleoptera, Chysomelidae). *Zeitschrift fur Naturforschung C*, **41**, 928–934.
Narberhaus, I., Zintgraf, V. and Dobler, S. 2005. Pyrrolizidine alkaloids on three trophic levels – evidence for toxic and deterrent effects on phytophages and predators. *Chemoecology*, **15**, 121–125.
Nishida, R. 2002. Sequestration of defensive substances from plants by Lepidoptera. *Annual Review of Entomology*, **47**, 57–92.
Nishida, R., Shelley, T. E. and Kaneshiro, K. Y. 1997. Acquisition of female-attracting fragrance by males of Oriental fruit fly from a Hawaiian lei flower, *Fagraea berteriana*. *Journal of Chemical Ecology*, **23**, 2275–2285.
Opitz, S. E. W. and Müller, C. 2009. Plant chemistry and insect sequestration. *Chemoecology*, **19**, 117–154.
Pasteels, J. M., Duffey, S. and Rowell-Rahier, M. 1990. Toxins in chrysomelid beetles – possible evolutionary sequence from *de novo*

synthesis to derivation from food-plant chemicals. *Journal of Chemical Ecology*, **16**, 211–222.

Pasteels, J. M., Theuring, C., Witte, L. and Hartmann, T. 2003. Sequestration and metabolism of protoxic pyrrolizidine alkaloids by larvae of the leaf beetle *Platyphora boucardi* and their transfer via pupae into defensive secretions of adults. *Journal of Chemical Ecology*, **29**, 337–355.

Reichstein, T., von Euw, J., Parsons, J. A. and Rothschild, M. 1968. Heart poisons in monarch butterfly, some aposematic butterflies obtain protection from cardenolides present in their food plants. *Science*, **161**, 861–866.

Rossini, C., Bezzerides, A., Gonzalez, A., Eisner, M. and Eisner, T. 2003. Defense mechanisms of arthropods. Paper 186. Chemical defense: incorporation of diet-derived pyrrolizidine alkaloid into the integumental scales of a moth (*Utetheisa ornatrix*). *Chemoecology*, **13**, 199–205.

Schaffner. U., Boeve, J. L., Gfeller, H. and Schlunegger, U. P. 1994. Sequestration of veratrum alkaloids by specialist *Rhadinoceraea nodicornis* Konow (Hymenoptera, Tenthredinidae) and its ecoethological implications. *Journal of Chemical Ecology*, **20**, 3233–3250.

Schulz, S., Francke, W., Boppré, M., Eisner, T. and Meinwald, J. 1993. Defense-mechanisms of arthropods. 117. Insect pheromone biosynthesis – stereochemical pathway of hydroxydanaidal production from alkaloidal precursors in *Creatonotos transiens* (Lepidoptera, Arctiidae). *Proceedings of the National Academy of Sciences, USA*, **90**, 6834–6838.

Tallamy, D. W., Mullin, C. A. and Frazier, J. L. 1999. An alternate route to insect pharmacophagy: The loose receptor hypothesis. *Journal of Chemical Ecology*, **25**, 1987–1997.

Taira, J., Nakamura, K. and Higa, Y. 2003. Identification of secretory compounds from the millipede, *Oxidus gracilis* C. L. Koch (Polydesmida: Paradoxosomatidae) and their variation in different habitats. *Applied Entomology and Zoology*, **38**, 401–404.

Towers, G. H. N., Siegel, S. M. and Duffey, S. S. 1972. Defensive secretion: biosynthesis of hydrogen cyanide and benzaldehyde from phenyalanine by a millipede. *Canadian Journal of Zoology*, **50**, 1047–1050.

Vahed, K. 1998. The function of nuptial feeding in insects: review of empirical studies. *Biological Reviews*, **73**, 43–78.

von Euw, J., Fishelson, L., Parsons, J. A., Reichstein, T. and Rothschild, M. 1967. Cardenolides (heart poisons) in a grasshopper feeding on milkweeds. *Nature*, **214**, 35–39.

Wray, V., Davis, R. H. and Nahrstedt, A. 1983. Biosynthesis of cyanogenic glycosides in butterflies and moths: incorporation of valine and isoleucine into linamarin and lotaustralin by *Zygaena* and *Heliconius* species (Lepidoptera). *Zeitschrift für Naturforschung*, **38C**, 583–588.

Zagrobelny, M., Bak, S., Rasmussen, A. V., Jørgensen, B., Naumann, C. M. and Møller, B. L. 2004. Cyanogenic glucosides and plant-insect interactions. *Phytochemistry*, **65**, 293–306.

Zagrobelny, M., Bak, S. and Møller, B. L. 2008.Cyanogenesis in plants and arthropods. *Phytochemistry*, **69**, 1457–1468.

Looking Ahead

TRENDS IN RESEARCH

There are trends in research in insect biochemistry and insect chemical ecology, but prediction is a dangerous business, and has little more accuracy than astrology. Improvements in instrumental techniques steadily make advances in what can be accomplished. Changes in personnel, with new entrants coming into the field, and many established figures retiring or approaching retirement, will have unpredictable effects, as will funding policies by governments. Molecular biological approaches are capable of transforming the way biosynthesis research is conducted. The cost of equipment and resources make this route only available to the most generously funded, but costs should come down, and as more young researchers with training in molecular biology become established, more research will be conducted through that route.

Methods and equipment in chromatography and mass spectrometry continue to make advances that make it easier to work with the very small quantities insects often provide. Nuclear magnetic resonance spectroscopy particularly provides new methods to overcome difficulties in detecting heavy isotopes, resolving spectra and determining structure. The hyphenating, or coupling, of several spectroscopic and chromatographic techniques, pioneered by Wilson and Nicholson (Wilson *et al.*, 2000), produces quantities of data (Phalaraksh *et al.*, 1999; Wilson *et al.*, 1999; Phalaraksh *et al.*, 2008) from very small samples, but has not generally been followed up in other laboratories, partly due to the cost of equipment. Although chiral chromatography is very useful in determining chirality of many pheromone substances, a new method of assigning chirality would be a great help for larger molecules.

Looking Ahead

Agendum

Are differences in linear alkane chains detectable by insect receptors? Or, are only alkenes and methylalkanes important for insect recognition? The clarification of this question will have importance for the deeper understanding of cuticle lipid biosynthesis and the answer should not be far off. In methyl-branched alkanes, the methyl groups are inserted at exact positions. It would be a great discovery to learn how this is specified, but may require detailed knowledge of the structure of the enzymes. Curiosity about the very-long-chain internal lipids of some insects should excite attention to their production and purpose.

The question of whether some insects possess polyketide synthases should soon be settled, but a positive answer will leave another of how the necessary genes were acquired. The Phalangida or opilionids are a large group ripe for further biosynthetic studies of their defensive compounds, both aliphatic and aromatic. Study of biosynthesis of simple aromatic compounds is not the most exciting or far-reaching of objectives but will provide training opportunities for many aspiring researchers, and re-enliven a neglected sector. Compounds such as 2-nitrostyrene (Section 9.3.1) deserve study. If the biosynthetic routes to some aromatic compounds, such as *p*-hydroxybenzoic acid, were better known, it might explain why one route is used in some circumstances, and another in others. One expects there must be some order in the sources of beetle quinones.

Biosynthesis of terpene compounds has received little attention, with the exception of bark beetle pheromones, and the clear exception of juvenile hormone, which was fully solved almost 40 years ago. The synthesis of diterpenes particularly has lagged. The identification of terpenoid sex pheromones from scale insects (Section 7.5.1), important and difficult pest species, will stimulate activity in that field, while the wax scale sesterpenes should provide an easy study, but with difficult experimental material.

Melanin is a very active subject of research but other insect pigments are relatively neglected in recent years, having been actively pursued some decades ago. Aphid pigments should provide great challenges to organic chemists. Luciferin should quickly reveal its secrets to a well-planned investigation.

The determination of the biosynthetic route to alkylpyrazines and the way in which the cherry fruit fly *Rhagoletis cerasi* oviposition deterrent is assembled would give great personal satisfaction.

The studies on biosynthesis of cyanogenic glucosides in arthropods were mostly conducted some decades ago, and would benefit from a renewed investigation.

Of all the aspects of insect biosynthesis, sequestration of plant toxins seems to be one with many unanswered or half-answered questions that decades of research have not settled. How did toxin sequestration begin? How do insects metabolize some and retain others? How are the substances transported to glands? How is their toxic effect avoided? It has already been indicated in discussing experimental methods (Section 3.5) that cloning methods are already making a large difference to what can be studied or achieved. Comparison of the genome of related species that can and cannot metabolize plant toxins or sequester them should make much more understandable their differences, leading from there to understanding the mechanism of sequestration. There is some indication of how the tobacco hornworm *Manduca sexta* avoids the toxic effects of nicotine (Eastham *et al.*, 1998), but the answer is far from complete.

There is always the unexpected and unpredictable, like the discovery of an aggression pheromone, (Z)-11-vaccenyl acetate, in males of *Drosophila*.

A fertile ground awaits any young scientist with an enquiring mind and the skills to apply them here.

REFERENCES

Eastham, H. M., Lind, R. J., Eastlake, J. L., Clarke, B. S., Towner, P., Reynolds, S. E., Wolstenholme, A. J. and Wonnacott, S. 1998. Characterization of a nicotinic acetylcholine receptor from the insect *Manduca sexta*. *European Journal of Neuroscience*, **10**, 879–889.

Phalaraksh, C., Lenz, E. M., Lindon, J. C., Nicholson, J. K., Farrant, R. D., Reynolds, S. E., Wilson, I. D., Osborn, D. and Weeks, J. M. 1999. NMR spectroscopic studies on the haemolymph of the tobacco hornworm, *Manduca sexta*: assignment of H-1 and C-13 NMR spectra. *Insect Biochemistry and Molecular Biology*, **29**, 795–805.

Phalaraksh, C., Reynolds, S. E., Wilson, I. D., Lenz, E. M., Nicholson, J. K. and Lindon, J. C. 2008. A metabonomic analysis of insect development: H-1-NMR spectroscopic characterization of changes in the composition of the haemolymph of larvae and pupae of the tobacco hornworm, *Manduca sexta*. *Science Asia*, **34**, 279–286. .

Wilson, I. D., Morgan, E. D., Lafont, R., Shockcor, J. P., Lindon, J. C., Nicholson, J. K. and Wright, B. 1999. High performance liquid chromatography coupled to nuclear magnetic resonance spectroscopy and mass spectrometry applied to plant products: identification of ecdysteroids from *Silene otites*. *Chromatographia*, **49**, 374–378.

Wilson, I. D., Lindon, J. C. and Nicholson, J. K. 2000. Advancing hyphenated chromatographic systems-although expensive, HPLC/NMR/MS may be the best way to unequivocally characterize complex mixtures. *Analytical Chemistry*, **72**, 534A–542A.

APPENDIX 1
Common Abbreviations

Ac	acetyl
AcCoA	acetyl co-enzyme A
ADP	adenosine diphosphate
ATP	adenosine triphosphate
B:	a basic group on an enzyme
cDNA	complementary DNA, sometimes coding or copy DNA (all the same material)
Ci	Curie unit of radioactivity
CoA-SH	co-enzyme A (free form)
D	deuterium (^{2}H)
DAHP	3-deoxy-D-arabinoheptulosonic acid 7-phosphate
DMAPP	dimethallyl diphosphate
DOPA	3,4-dihydroxyphenylalanine
DXP	1-deoxy-D-xylulose 5-phosphate
Enz	enzyme
FAD	flavin adenine dinucleotide
FADH$_2$	reduced flavin adenine dinucleotide
GC	gas chromatography
GC-MS	linked gas chromatography and mass spectrometry
GTP	guanosine triphosphate
HMG-CoA	β-hydroxy-β-methylglutaryl co-enzyme A
HPLC	high performance liquid chromatography
IPP	isopentenyl diphosphate
JH	juvenile hormone
Kbp	kilobase pairs, 1000 base pairs of nucleotides

Biosynthesis in Insects, Advanced Edition
By E. David Morgan
© E. David Morgan 2010
Published by the Royal Society of Chemistry, www.rsc.org

Appendix 1

LC-MS	linked liquid chromatography (HPLC) and mass spectrometry
MALDI	matrix assisted laser desorption-ionization (mass spectrometric technique)
MEP	methylerythritol phosphate
MH	moulting hormone
MVA	mevalonic acid
NAD^+, $NADP^+$	nicotinamide adenine dinucleotide (phosphate)
NADH, NADPH	nicotinamide adenine dinucleotide (phosphate), reduced form
NMR	nuclear magnetic resonance
PBAN	pheromone biosynthesis activating neuropeptide
PCR	polymerase chain reaction
PG	prostaglandin
PLP	pyridoxal phosphate
pp_i	inorganic diphosphoric acid
Pr	propyl
PTTH	prothoracotropic hormone
3'RACE	rapid amplification of cDNA 3'-ends
T	tritium (^3H)
TOF	time of flight (mass analysis technique)
TPP	thiamine diphosphate
U	uniformly labelled (in reference to isotopic labelling)

APPENDIX 2
Glossary of Terms

(MB indicates special terms of molecular biology)

Allelochemicals, semiochemicals that carry signals between different species (Greek *allelon* = of each other). Allelochemicals are divided into allomones, *q.v.*, and kairomones, *q.v.*

Allomones, semiochemicals or allelochemicals where the signal is to the advantage of the emitter, *e.g.* a warning odour to deter potential predators.

Amplification, (MB) producing additional copies of part of a chromosome or DNA.

Aposematic colour, prominent or gaudy coloration of some species to warn potential predators of a poisonous or distasteful defence.

Asymmetric induction, the creation of a chiral product from an achiral compound, frequently found in enzyme-catalysed reactions.

Axenic, growing an organism in a sterile environment, especially, free of any internal micro-organisms.

Base pair, (MB) two nucleic acid bases, hydrogen-bonded together in a DNA double helix or a DNA-RNA hybrid.

Biotransformation, the conversion by an organism of a substance which is not produced by it.

Chelicerata, a subphylum of arthropods including spiders, ticks, mites, scorpions and king crabs.

Clone, (MB) a group of identical cells, containing identical recombinant DNA molecules.

Complementary DNA (cDNA), (MB) DNA copied from mRNA using the enzyme reverse transcriptase and the PCR reaction.

Corpus allatum, (plural, corpora allata) endocrine gland behind the brain of insects that secretes juvenile hormone.

Biosynthesis in Insects, Advanced Edition
By E. David Morgan
© E. David Morgan 2010
Published by the Royal Society of Chemistry, www.rsc.org

Cosmid, (MB) very approximately, a large plasmid (*q.v.*) which can accept larger portions of DNA.
Cytosol, the soluble portion of the cytoplasm in cells.
Cytoplasm, the interior of the cell, within the plasma or cell membrane, including the cytosol and organelles but excluding the nucleus.
Diapause, a period of delayed growth or development in unfavourable conditions like winter or drought.
Eclosion, emergence of the adult from the pupal case.
Elytra, the hardened fore-wings of beetles that serve as wing covers.
Enantioselective, enantiospecific, a reaction that is selective or specific for one enantiomer over the other or from a mixture of the two.
Endoplasmic reticulum, a system of membranes in the interior of cells, usually involved in the synthesis and transport of compounds, particularly proteins and steroids.
Endopterygota, a synonym for Holometabola, *q.v.*
Eukaryote, organism with a distinct nucleus and multiple chromosomes (all living organisms except bacteria and algae), *see* prokaryote.
Exon, (MB) a segment of a gene which is translated into RNA.
Exopterygota, a synonym for Hemimetabola, *q.v.*
Expressed sequence tag (MB), a short, unique section of DNA in a gene, useful for identifying the gene that is being expressed in a cell at a particular time.
Expression, (MB) the process of transcription and translation from a gene to a protein.
Fat body, mass of fatty tissue filling the abdomen of insects with functions of storage and metabolism.
Fibroin, a structural protein of the β-keratin type made by insects; it is the main component of the silk of pupal cases (cocoons) of silkworms and the thread of spiders.
Functional genomics, (MB) the study of the function of genes, the proteins for which they code and what function those proteins serve.
GenBank, (MB) one of a number of computer databases of nucleotide sequences in genes and gene fragments.
Gene library, (MB) a number of DNA fragments made from the genome of an organism.
Genome, (MB) the complete set of genes of an organism.
Haemocoel, the blood-filled area around many of the organs of insects and other arthropods.
Haemocyte, cells circulating in the blood of insects.

Haemolymph, the blood-like fluid circulation in the haemocoel.

Hemimetabola, insects that change gradually in form with each moult, from nymph to adult without going through a pupal stage.

Holometabola, insects that undergo metamorphosis, starting with larval stages, then a pupal stage, to emerge in quite different form as adults. Includes Coleoptera, Diptera, Hymenoptera, Lepidoptera and other orders.

Homologous gene, (MB) a gene in another animal with similar function and similar nucleotide sequence.

Hybridize, (MB) pairing of complementary DNA and RNA strands to give a DNA-RNA hybrid.

Imago, the sexually mature stage of an insect, the final instar.

Instar, the form of an insect between moults, a development stage on the way to adult or imago.

Intron, (MB) a segment of a gene not translated into RNA.

Kairomone, a semiochemical or allelochemical where the signal is to the advantage of the receiver, *e.g.* the odour of a host detected by a potential parasite.

Malpighian tubules, a part of the excretory and water-regulating system of insects and some other arthropods. The tubules are branches off the alimentary canal, closed at the far end, that absorb wastes and excess water from the haemolymph. Aphids and the primitive springtails (Collembola) lack the tubules.

Metabolon, a multi-enzyme complex.

Microarray, (MB) a system of thousands of microscopic spots of DNA oligonucleotides. For more, see Wikipedia.

Microsomes, fragments of endoplasmic reticulum of cells (*q.v.*) produced by homogenizing cells.

Nymph, an immature stage of a hemimetabolous insect, between egg and adult.

Oenocytes, specialized cells in the insect fat body, important in metabolism.

Ommatidium, *pl.* ommatidia, a single light-sensing unit of the compound eye of an insect. It contains several photoreceptor cells, surrounded by support cells and pigment cells, but only one nerve axon to the brain.

Ootheca, a pouch containing a mass of insect eggs.

Opiliones, another name for Phalangida, *q.v.*

Patrilines, a generation produced by one male. In species where the female mates with more than one male, her offspring belong to several patrilines.

Peroxisomes, organelles in almost all eucaryotic cells containing a number of oxidative enzymes, using molecular oxygen, and including those catalysing β-oxidation of fatty acids and peroxidases.

Phalangida, a group of arthropods that include daddy longlegs and harvestmen. They are close to spiders (order Arachnida) but have long legs and no silk glands. Also called Opiliones.

Pharmacophagy, pharmacophagous insects are those that seek for, and ingest, specific compounds in plants or animals, not their normal food, and use them for a purpose such as defensive secretions.

Pheromones, semiochemicals which carry signals within a species. They are sub-divided into releaser pheromones which change behaviour, and primer pheromones which change physiology.

Plasmid, (MB) a ring-shaped piece of DNA, usually from a bacterium, into which some foreign DNA can be inserted.

Polymerase chain reaction, (MB) a process by which many copies of a portion of DNA can be created using a DNA polymerase enzyme.

Polyphagous, feeding on different kinds of food.

Primer, (MB) a short sequence of nucleotides used to prime, that is, to start building a second strand of DNA guided by a polymerase enzyme.

Prokaryote, single-cell organism with a single chromosome and no discrete nucleus, essentially bacteria.

Prothoracic gland, an endocrine gland in the prothorax of insects that secretes moulting hormone.

Pseudogene, (MB) inactive part of the genome derived by mutation from a former active gene.

Pygidium, in insects, the rear part of the abdomen.

Quorum sensing, the use of chemicals (pheromones) for communication and decision-making in single-cell animals. It is also used of decision-making on nest choice in some social insects.

Recombinant, (MB) an organism with genes from another organism incorporated into its genome.

Regiospecific, a reaction where a specific part of a molecule is the target of a reaction. Many enzyme-catalysed reactions are regioselective or regiospecific.

Restriction enzyme, (MB) an enzyme that can cut DNA molecules into smaller fragments.

Reverse transcription, (MB) synthesis of DNA from an RNA template.

Royal jelly, secretion of the hypopharyngeal glands of honeybee workers. It is the only food for the first four days for larvae destined to become queens. Smaller amounts are given to larvae that become worker bees.

Sclerite, a hardened plate of cuticle covering a body segment.

Sclerotin, a darkly pigmented, rigid, polymer of the cuticle of insects.

Sclerotization, the hardening and colouring of newly formed cuticle.

Semiochemicals, substances that carry a message or signal. In chemical ecology they consist of allelochemicals and pheromones *q.v.*

Sequencing, (MB) process of determining the order of the nucleotides in DNA or RNA.

Specificity (of enzymes), the ability of an enzyme to recognize the molecules of a substance and catalyse their transformation into only one other substance.

Spermatophore, a packet containing sperm, and possibly nutrients and, for some species, defensive compounds transferred from male to female at mating.

Spiracle, the opening of a trachea (*q.v.*) on the cuticle of an insect. There are normally two spiracles on each body segment, one on each side.

Stereoselective, when an enzyme is able to select one enantiomer over other isomers as substrate, then it is behaving stereoselectively.

Stereospecific, when an enzyme is able to catalyse the formation of a single enantiomer, the reaction is stereospecific.

Symbiosis, the close association of two dissimilar species. Usually the smaller, dependent partner is called the symbiont and the larger is called the host. Frequently the symbiont is a micro-organism living inside the gut or inside the cells of an insect.

Synthase, a general trivial name for enzymes, chiefly transferases (Class 2 enzymes; EC 2) that transfer groups *e.g.* methyl, acyl, sugar or phosphate by the general reaction $AB + C \rightarrow AC + B$.

Synthetase, a specific trivial name used only for ligases (Class 6; EC 6). Ligases catalyse bond formation coupled to breakdown of ATP: *e.g.* $A + B + ATP \rightarrow AB + ADP + P_i$.

Tergite, the dorsal part of a body segment.

Trachea, plural, tracheae, the aeration system of insects; branched tubule through which air enters the body of insects. They connect with the surface through openings called spiracles.

Trail pheromone, chemical laid down by an insect returning from a food source or a new nest for other insects to follow to find the food or nest.

Transcriptome, (MB) the set of all RNA molecules in a cell or organism. Because it includes all mRNA, it reflects the genes that are being actively expressed in that cell or organism.

Transgenic animal, (MB) an animal into which foreign genes have been transferred.

Appendix 2

Vector (cloning vector), (MB) a plasmid or other piece of DNA into which foreign DNA can be spliced to alter genes, and inset them into an organism.

Xenobiotic, any compound foreign to an organism, not a product of normal metabolism. Pesticides and environmental pollutants are examples.

Subject Index

Bold numbers indicate figure numbers.
Some longer chemical names have been simplified in the index by omitting chirality or position of substituents.

Abbreviations, 336
Acanthoscelides obtectus, 112, **5.13**
Acetic acid,
  doubly labelled, 53, 206
  phosphorylation, 20, **2.8**
  acetyl coenzyme A, 20, **2.8, 2.14**
  acetylenase, 75
Acetogenins, 150–151, 266–267, 268, 298
  aceto-propiogenins, 151, 192
  low substrate specificity, 152, 159
  and pheromone activity, 159
  pheromones, 150–151
  use of butyrate, 155–159
Acetyl CoA, 20, 68, 146, 180–181, **2.8, 2.14, 4.4, 6.3, 7.2, 10.2**
  synthase, 20, **2.8**
N-Acetylglucosamine, 286, **9.27**
3-Acetylzygadenine, **11.9**
Achaeta domestica, 76, 236
Acromyrmex octospinosus, 236
Acryloyl CoA, **4.17**
Actinidine, 295–296, **7.18, 10.2**
Acyl carrier protein, 20, 67, 68, 71, **2.8, 4.4, 4.6**

Acyrthosiphon pisum, 60
Adalia bipunctata, 117, Plate 13
Adaline, 115, **5.18**
Adalinine, 115, **5.18**
Adenosine triphosphate, 19–20, **2.6**
S-Adenosyl homocysteine, 26, 28, **2.18**
S-Adenosyl methionine, 26–28, 47, **2.18**
  'decarboxylated', 303, **10.10**
Adrenalin, 263
Adynerin, 318, **11.2**
Aedes aegypti, 60, 21
Aenictus ants, 261
Agabus affinis, 244
Agabus guttatus, 244
Agel 489a, **10.10**
Agelonopsis aperta, 148, 305
Aglais urticae, 284
Aglaostigma sawflies, 326
Agriotes beetles, 220
Albocerol, **8.4**
Albolic acid, **8.5**
Albolineol, **8.5**
Alcohol dehydrogenase, 22, **2.11**

Alcohols, very long chain, 89
Aldol reaction, 181, **7.3**
Aldolase, 30
Aleochara curtula, 111
Alkaloids 114–122, 295–302
Alkenes, long chain, 90–91, **4.22**
  methyl branched, 90
  terminal, 91, **4.22**
Allelochemicals, 8
Allohormone, 237
Allomones, 8
Alydus eurinus, 81, 150
Amblyomma ticks
  A. americanum, 79, 268
  A. maculatum, 268
  A. variegatum, 268
Amines, aromatic, 263–265
Amino-acid derivatives, as pheromones, 302, **10.9**
γ Aminobutyric acid, **10.5**
5-Aminolaevulinic acid, 281, **9.24**
Amitermes, 137, 205, **5.41**
Amitinol, 188, **7.8**
Ammonia, 25
Amyrin, 242–243, **8.22**
Anabaseine, 297, **10.3**
Anabasine, 297, **10.3**
Anabolism versus catabolism, 22, 71, **4.4**
Ancistrodial, 204, **7.27**
Ancistrofuran, 204, **7.27**
Ancistrotermes cavithorax, 204
Anisomorpha buprestoides, 198
Anisomorphal, 198, **7.18**
Anisops bugs, 282
Anomala cuprea, 113, **5.14**
Anopheles gambiae, 60, 237
Anthocharis cardamines, 281, Plate 35
Anthocyanins, 284–285
Anthonomis eugenii, 195

Anthonomis grandis, 89, 194, 207, Plate 9
Anthranilic acid, 261, **9.5**, **9.6**, **10.2**
Anthraquinones, 275–276, **9.18**
Anthrones, **6.22**
Anthroquinones, **6.22**
Antitrogus parvulus, 87–88, **4.19**
Ant venoms, 296–298
Aonidiella citrina, 204, **7.26**
Aphaenogaster ants,
  A. albisetosus, 153
  A. cockerelli, 260
  A. rudis, 297
Apheloria corrugata, 259
Aphins, 276–277, **9.19**
Aphis aphids,
  A. jacobaeae, 321
  A. neri, 277, 318
  A. spiraecola, 277
Apis mellifera, 58, 60, 126–128, 150, 228, 235
  comb wax, 128
  Nasonov gland, 195, **7.17**
  queen retinue pheromone (queen substance), 126, **5.28**, **5.29**
  royal jelly, 126, 127
Apterygota, 7
Arachidonic acid, 76–79, **4.11**, **4.12**
Arctia caja, 110, 264
ArgTX-636, **10.10**
Arginine phosphate, 19, **2.7**
Argoporus beetles,
  A. alutacea, 270
  A. bicolour, 270
  A. rufipes, 270
Argyrotaenia citrana, 107
Argyrotaenia velutinana, 103, **5.6**
Aristolochic acid 326, **11.10**
Arthropods, 7

Ascotis selenaria cetacea, 99–100
Asparagine, 32, 323, **2.21**
Aspartic acid, 305, **9.28, 10.11**
Aspidiotus nerii, 203, 318, **7.26,** Plate 21
Astaxanthin, 247, **8.26**
Astigmatid mites, 230
Asymmetric induction, 32–34, **2.23, 2.24**
Athalia spinarum, 286, Plate 39
Atrax robustus, 303, Plate 41
Atta ants, 153
   A. cephalotes, 153
   A. sexdens, 153
   A. texana, 153
Attagenus megatoma, 112
Autoradiography, whole body, 40

Bactrocera fruit flies
   B. cacuminata, 168, **6.26**
   B. cucumis, 168, **6.27**
   B. dorsalis, 327, **11.10**
   B oleae, 167–169, **6.25, 6.26**
   B. tryoni, 169, **6.28**
Bark beetles, 186–194
Becquerel unit, 40
Bees, 126–129
   colletine, 128, **5.30**
   halictine, 128
Benzaldehyde, 258, 259, 260, **9.2, 9.3, 11.5**
Benzoic acid, 260
Benzoquinone, 269, 270, 307, **10.13**
Biacetyl, 277
Bifloratriene, 223, **8.3**
Bilins, 281–284, **9.24**
Bilirubin, 282, **9.24**
Biliverdin, 282, **9.24**
Bioluminescence, 19, 306–309
   quantum yield, 308
Biotin, 68–69, 280, **4.5, 9.22**

Bipyridyl, 297, **10.3**
Bishomogeranyl diphosphate, 210, **7.32**
Bishomomanicone, **6.8**
Blattella germanica, 76, 81, 83, 86, 87, 137, 239, 270, **5.42**
Blattellostanosides, 239, **8.19**
Blattodea, 137–138
Bombardier beetles, 270–271, **9.14**
   Brachinus explodens, Plate 30
Bombykol, 96, **5.4, 5.8**
Bombyx mori, 58, 60, 106, 109, 228, 229, 230, 236, 237, 265, 284. **5.8,** Plate 12
   genetic defect, 247
Borneol, **7.5**
Brachymyrmex ants, 302
endo-Brevicomin, 190, 193, **7.14**
exo-Brevicomin, 190, 192, **7.14**
Brucins, 309, **10.16**
Bucculatrix thurberiella, 108–109
Buenoa bugs, 282
Buibuilactone, 112, **5.14**
Butyric acid derivatives, 148–150, 206
   butyl butyrate, 150, **6.4**
   crotyl butyrate, 150, **6.4**

^{13}C Double labelling, 53, 206
^{13}C-^{13}C coupling, 51–53, 164, **3.10, 6.20**
^{13}C-^{2}H coupling, 49, **3.8**
^{36}Cl labelling, 268
CNS 2103, 303, **10.10**
Caeliferins, 309, **10.16**
Cahn-Ingold-Prelog rules, 32, 33, 34, **2.22, 2.24, 2.25**
Calactin, 317, **11.1**
Calliphora eythrocephala, 227
Callococcithrips fuscipennis, 201
Calotropagenin, 317, **11.1**

Calotropin, 317, **11.1**
Calvia quattuordecimguttata, 115
Calvine, 115, **5.17**
Campestrol, 226, 235
Camplyomma verbasci, 150
Camponotus ants, 164
  C. atriceps, 161
  C. clarithorax, 260
  C. herculaneus, 161
  C. ligniperda, 161
  C. nearticus, 261
  C. pennsylvanicus, 29
  C. socius, 160
  C. vagus, 160
Cane grubs, 111–112
Cantao parentum, 171, **6.30**, Plate 18
Cantharenone, 221, **8.2**
Cantharidin, 206–208
  biosynthesis, 207, **7.30**
Cantharidimide, **7.30**
Cantharidiniphiles, 208, 315
Cantharis livida, 221
Caparrapi oxide, 205, **7.27**
Carboxypeptidase 15, **2.3**
Carcinus maenas, 236
Cardenolides, 240–242, 317, **8.20**
Cardiac glycosides, 240–242, **8.20**
  sequestered, 316–318
Carminic acid, 276, **9.18**
Carotenes, 245–249, **8.26**
Carpophilus beetles, 156–160
  C. davidsoni, **6.13**
  C. freemani, **6.13**
  C. hemipterus, **6.18**
  C. mutilatus, **6.13**
Carvone, 32, **2.21**, **7.1**
Caryophyllene oxide, 206, **7.29**
Catalase, 271
Cedrene, 209, **7.31**

Cembrene, 221, **8.2**, **8.3**
Centipedin, **6.21**, **9.11**
Ceratitis capitata, 31
Ceriferol, **8.4**
Ceroplastes scale insects, 224
  C. albolineatus, 224, 322
  C. ceriferus, 224
  C. destructor, Plate 25
Ceroplastols, **8.5**
Ceutorhyncus assimilis, 81
Chalcogran, 166, **6.24**
Chelicerata, 7
Chemical communication, 8
Chemical parsimony, 136
Chemochromes, 272
Chilasa butterflies, 206
Chilocorus cacti, 117–118
Chilocorus renipustulatus, 117
Chilocorines, 117–118, **5.19**
Chiloglottone A, 130, **5.32**
Chirality, 31–32, 53–54, 98, **2.21**, **2.22**, **2.23**
  and gas chromatography, 54
  and NMR spectroscopy, 54
  effect on odour and taste, 32, **2.21**
  chiral induction, *see* asymmetric induction
  in methyl-branched hydrocarbons, 87, **4.19**
Chironomus midges, 282
Chitin, 286–288, **9.27**
5β–Cholestan-3-one, 239, **8.19**
Cholesterol, 226, 232, **8.7**
Cholesteryl oleate, 239, **8.19**
Chorismic acid, 256, 261, 262, **9.1**
  derivatives in insects, 261–262
Choristoneura fumiferana, **5.4**
Chortoglyphis arcutus, 160
Chortolure, 160, **6.17**
Chromones, 267, **9.10**
Chrysarobin, 165, **6.22**

Chrysolina carnifex, 240
Chrysolina coerulans, 240
Chrysomela lapponica, 80
Chrysomela tremulae, 305, 327, **10.11**, Plate 42
Chrysomeliadial, 199, 201, **7.18, 7.19**
Chrysophanol, 165, **6.22**
Chrysophtharta amoena, 324
Chrysophtharta variicollis, 324
Cinnabarinic acid, 284, **9.25**
Cinnamic acid, 257, **9.2**
Cionus olens, 286
Citric acid cycle, 23, 25, **1.2**
Citronellol, **7.1, 7.5**
Claisen condensation, 67, 130, 181, 281, **5.32, 7.2**
Coccinella septempunctata, 318, 322, Plate 49
Coccinella undecempunctata, 318
Coccinellid defensive compounds, 114–122
Coccinellines, 115, **5.16**
Cochineal bugs, *see Dactylopius*
Cochliomyia hominivorax, 89
Cochylis hospes, 89
Co-enzymes, 12, 18–28
  co-enzyme A, 20–21, **2.8**
Coleoptera, 110–122
  defence, 114–122
  hydrocarbons, 111
Colias eurytheme, 272, Plate 31
Collie, J. N., 66–67
Collembola, 268
Coniferyl alcohol, 327, **11.10**
Coniine, 134, **5.39**
Conophthorin, 167, **6.24**
Conophthorus beetles, 190
Convergine, **5.16**
Coptotermes formosanus, 31, Plate 2

Coremata, 319
Corpus allatum, 212, 232
Cortexone, **8.24**
Corticosterones, 244
Crematofuran, 221, **8.2**
Crematogaster ants, 131, **5.33**
  C. difformis, 268
  C. inflata, 268
m-Cresol, 163
Crinosterol, 230, **8.10**
Crustacea, 7
Cryptocerus punctulatus, 31
Cryptolaemus montrouzieri, 118, **5.19**, Plate 14
Cryptolestes beetles, 113,
  C. ferrugineus, 203, **7.26**
Cryptoxanthin, 250, **8.29**
Cubitermes umbratus, 223
Cucurbitacins, 326, **11.10**
Cucujolides, 113, 203, **5.15, 7.26**
Curculio caryae, 195
Curie unit, 40
β–Cyanoalanine, 323, **11.7**
Cyanogenic glucosides, 258, 322–325, **11.5, 11.6**
  bitter taste, 323
2-Cyano-6-hydroxybenzthiazole, 307, 309, **10.13**
Cybister limbatus, 244
Cybisterone, **8.23**
Cycasin, 326, **11.10**
Cylindrocopturus adspersus, 89
Cysteine, 180, 181, 274, 307, 323, **9.16, 10.13**, (Chap 9,10,11)
5-Cysteinyldopa, 274, **9.16**
Cytochromes, 17–18, 50, 84, 168, 187, 214, 233, 235, 242, 296, 319, 323, **2.5**

DOPA, 263, **9.7**
DOXP pathway, 186

Dactylopius coccus, 275, Plate 32
Danaidone, 320, **11.3**
Danaus chrysippus, 317, **11.1**
Danaus plexippus, 317, **11.1**
Dehydratase, 68–70, **4.6**
3-Dehydroecdysone, **8.14**
Dendroctonus beetles 190–191
   D. frontalis, 190, 192–193
   D. jeffreyi, 56, 191
   D. ponderosae, 191, 192
   D. rufipennis, 191
   D. simplex, 192
   D. terebrans, 31
   D. vitei, 150
Dendrolasin, 205, **7.28**
Dermacentor variabilis, 239
Dermestes vulpinus, 227, 228
Desaturase, 56, 57, 72–76, 102–106, **5.5**
   of bombykol, 104, **5.8**
   trans unsaturtion, 103–104
   triple bonds,
Deuterium labelling, 5, 22, 24, 47, 98, 112, 120, 156, 184, 198, 206, 229, **3.6 4.8, 4.9, 4.15, 5.13. 5.21, 6.9, 6.21**
1,17-Diamino-9-octadecene, 121, **5.24**
Diabrotica speciosa, 326
Diagmatyloxamide, **10.10**
Diatraea grandiosella, 89
2,6-Dichlorophenol, 268, **9.10**
2,4-Dihydroxyacetophenone, 164, **6.20**
Dimethallyl diphosphate, 182–183, **7.4, 7.7**
1,3-Dimethoxybenzene, 269
4,8-Dimethyldecanal, 154, **6.9**
3,5-Dimethylhexanolide, 161, **6.19**
Dimethyl disulphide, 43, **3.3**
Dimethylnonacosan-2-one, 87, 137, **5.42**

Dimethyloctatetracontanol, 89, **4.21**
1,3-Dimethyl-2,4-quinazolinedione, 296, **10.2**
Dimethyl trisulphide, 43, **3.3**
Dinutes beetles, 205
Diploptera punctata, 214
Disparlure, 100, **5.3**
Diterpenes, 179, 220–223
3-Dodecen-1-ol, **5.40**
3,6-Dodecadien-1-ol, **5.40**
Dolichodial, 201, **7.18**
Dolichols, 224, **8.6**
Dolomedes okefinokensis, 305
Dopamine, 263, 264, **9.7**
Dopaquinone, 273
Dorcus rectus, 31
Doru taeniatum, 270
Double bonds,
   location, 53
   terminal 91, **4.22**
Drosophila buzzatii, 41
Drosophila melanogaster, 60, 83, 123–124, 182, 235, 249, **5.26** Plate 4
   feminised, 124
   mutants, 284
Drosophila pseudoobsura, 90
Dysdercus fasciatus, 236
Dytiscidae, 243–244
   Dytiscus marginalis, 266, **9.10**

Ecdysone, 231, 235, **8.12, 8.14**
   20-mono-oxygenase, 231, **8.12**
Ecdysonoic acid, 238, **8.17**
Ecdysteroids, 231–239,
   biosynthesis, 232–235, **8.13, 8.14**
   conjugates, 236, **8.16**
   inactivation and excretion, 238–239 **8.18**
   in adult insects, 236–237

Ecdysteroids (*continued*)
in other phyla, 236
in plants, 232
Ectatomma ruidum, 220
Eicosanoids, 76–79, **4.11, 4.12**
20-Eicosanolide, 128, **5.30**
Eicosatrienoic acid, 99, 101, **4.11**
Eleodes beamerii, 111
Eleodes longicollis, 111, 270
Emodin, **9.18**
Epiblema scudderiana, 31
Ericerus pela, 224
Eriococcus scale insect, **9.18**
Erythrose 4-phosphate, 255–256, **1.2, 9.1**
Enoyl reductase, 68–70
Enzymes, 12–18
  activation and inactivation, 13
  active site, 12, **2.3, 2.11**
  allosteric enzymes, 13
  catalysis, 4–5, 12
  inhibitors, 39
  substrate, 12
Epicauta pestifera, 207, Plate 24
Epilachna beetles,
  E. borealis, 121, **5.23**
  E. signatipennis, 121, **5.24**
  E. varivestis, 119–120, 134, 230, **5.21**
Epilachnine, **5.21**
Epilachnadiene, **5.21**
Epiphyas postvittana, 104, **5.7**
Epoxide hydrolase, 129
7,8-Epoxy-2-methyloctadecane, 100, **5.3**
9,10-Epoxytricosane, **5.25**
Erannis bajaria, 98, **5.2**
Ergosterol, 226, 230
Erythropterin, 281, **9.22**
Erythrose 4-phosphate, 255, **1.2**
Estigmene acrea, 100
Ethacrylic acid, 46

Ethanolamine, 115, 121, **5.22, 5.23, 5.24**
7-Ethyl-3,5-dimethyl-2,4,6,8-undecatetraene, 156, **6.11**
6-Ethyl-4-methyl-3,5,7-decatriene, 156, **6.11**
5-Ethyl-3-methyl-2,4,6-nonatriene, 156, **6.11**
p-Ethylphenol, 266
Eucerceris wasps, 149, **6.4**
Eucondylodesmus elegans, 261
Eumelanin, 274
Eupoecilia ambiguella, 110
Eurycotis floridana, 138–139, 228
Eurygaster integriceps, 211
Euschistus heros, 159, **6.15**
Eutetramorium mocquerysi, 301
Exochomine, 117, **5.19**
Exochomus quadripustulatus, 117
Expressed sequence tags, 58, 60

Faranal, **7.35**
Farnesene, 202, **7.24**
Farnesoic acid, 214
Farnesol, 196, 202, 207, **7.1, 7.24**
Farnesylamine, 301, **10.8**
Farnesyl diphosphate, 202, **7.24**
  synthase, 202
Fatty acids, **4.1**
  anteiso-acids, 79.81
  biosynthesis, 66–72, **4.44, 4.6**
  branched, 79–82, 85, **4.13, 4.14**
  chain lengthening, 71–83, **4.15**
  decarboxylation, 83, **4.15**
  desaturase, 72–76, **4.7, 4.8, 4.9, 4.10**, Plate 5
  epoxidase, 74
  hydroxylase, 74, **4.10**
  iso-acids, 79, 81
  synthase, 68, **4.3**
  unsaturated, 72
Fire ants, *see Solenopsis*

Subject Index

Fireflies, *see Photinus*
Flavin adenine dinucleotide, 23, **2.12**
Flavones, 284–285, **9.26**
Flocerene, **8.4**
Flocerol, **8.4**
Floceric acid, **8.4**
Floridenol, **8.4**
Floridenone, **8.4**
6-Fluoromevalolactone, 184–185, **7.6**
Formic acid biosynthesis, 29, **2.19**
Formica ants
  F. exsecta, 83
  F. rufa, 164
  F. truncorum, 85
Frontalin, 190–191, 192, **7.12, 7.13**
Fucosterol, 226
Functional genomics, 55, 58–60, **3.13**, Plate 51

Galeruca tanaceti, 165, **6.22**
Galleria mellonella, 249
Gas chromatography, 41, 44, 46, 48, 53, 323
  chiral, 54
Gaterophilis intestinalis, 282, **9.23**
Gastrophysa viridula, 198, 199
Genome, 31
  gene library, 56
Gentisylquinone isovalerate, 270, **9.13**
Geraniol, 183, 196, **7.1, 7.5, 7.17**
Geranylcitronellol, 220, **8.1**
Geranyl diphosphate, 56, 59, 202, **3.11, 7.24**
  synthase, 182, 187, **7.4, 7.8**
  bifunctional enzyme, 58–59, 187
Geranylfarnesol, 224

Geranylfarnesyl diphosphate, **8.4, 8.5**
Geranylgeraniol, **8.1**
Geranylgeranyl diphosphate, 220, 224, 246, **8.1, 8.2**
  synthase, 223
Geranyllinalool, 201
Germacrene, 206, **7.29**
Glomerine, 261, **9.6**
Glomeris marginata, 261–262
Glossina flies
  G. austeni, 125, **5.27**
  G. morsitans, 125, **5.27**
  G. pallidipes, 125, **5.27**
Glucose metabolism, 29–30, **2.20**
Glucosinolates, 326, **11.10**
Glutamine, 115, 287
Glutamic acid, 168, 298, **10.5**
γ-Glutamyl-2-amino-3,5-hexadienoic acid, 303, **10.9**
Glyceraldehyde 3-phosphate, 30, **1.2, 2.20**
Glycerol 3-phosphate, 33, 185, **2.23, 7.7**
Glycine, 25, 26, 29, 174, 281
Gnamptogenys pleurodon, 163
Gnamptogenys striatula, 210
Gnathotrichus beetles, 190
Gonepteryx rhamni, 278, Plate 34
Grandisol, 194–195, **7.16**
Grandlure, 194
Grapholitha molesta, 164, 260
Green leaf volatiles, 138–141, **5.43**
Gryllus bimaculatus, 238
Gryllus pennsylvanicus, 88
Guaiacol, 261, **9.4**
Guanosine 5'-phosphate, 280
Guanosine triphosphate cyclohydrolase, 278, **9.22**
Gyrinidal, 205, **7.27**
Gyrinidione, 205, **7.27**

Gyrinidone, 205, **7.27**
Gyrinus beetles, 205

HF-7, 305, **10.10**
^{2}H labelling, *see* Deuterium labelling,
HMG-R, 56, 57, 181, **7.3**
Habrobracon hebetor, 221
Haem, 15–16, 281, **2.4, 2.5**
Haemoglobin, 281–282
Haemolymph, 79, 83, 85, 97, 114, 117, 172, 199, 207, 213, 272, 287, 316
Haloween genes, 235
Halyzia sedecimguttata, 115
Harpagoxenus sublaevis, 295
Harpaphe haydeniana, 259
Heliconius melpomone, 324
Helicoverpa zea, 78, 89, **5.4**
Heliothis virescens, 82, 89
Heliotrine, 319, **11.3**
Hemiterpenes, 189
7,11-Heptacosadiene, **5.26**
10-Heptadecanone, 41, **3.1**
2-Heptanol, **6.5**
2-Heptanone, **6.5**
2-Heptyl-8-methylpyrrolizidine, **10.4**
10,12-Hexadecadien-1-ol (bombykol) **5.4**
7,11-Hexadecadienyl acetate, **5.4**
9-Hexadecenal, **5.4**
11-Hexadecenal, **5.4**
13-Hexadecen-11-ynyl acetate, 105, **5.8**
Hexadecanal, **5.4**
Hexamethyldocosane, 88, **4.19**
Hexapoda, 7
Hippodamine, 117, **5.16,**
Histidine, 26,29, 74
  histidine box, 56
Hololena curta, 305, **10.10**

Holomelina aurantiaca, 97
Holomelina lamae, 97
Holotrichia parallela, 302
Homoeosoma electellum, 89
Homofarnesol, **7.32**
Homogentisic acid, 266, 269, **9.10, 9.12**
Homogeranyl diphosphate, 210, **7.32**
Homoglomerine, 261–262, **9.6**
Homomanicone, 153, **6.8**
Homomevalonic acid, 209, 221, 224
  biosynthesis, 209–210, **7.32**
Homo-ocimene, **7.33**
Homosesquiterpenes, 211, **7.34**
Homospringene, 221, **8.2**
Homoterpenes, 209–214
Homotrinervitane, 223, **8.3**
Hormones, 8
  in pest control, 9
Hospitalitermes umbrinus, 223
Hyalophora cecropia, 75
Hycleus lunata, 208
Hydrocarbons, allenic, 111–112, **5.13**
  cuticular, 82–92
  biosynthesis, 83–91, **4.14**
  branched, 85–90
  chain length, 85
  in internal lipids, 89
  melting points, 91
  odd and even number of carbon atoms, 85, 88
  physical state, 91–92
Hydrogen cyanide, 322
  detoxified, 323
  reaction with thiosulfate, 323–324, **11.7**
p-Hydroxybenzoic acid, 266
3-Hydroxybutyric acid, **6.4**
  esters, 149

3-(3'-Hydroxybutyroyl)butyric acid, **6.4**
Hydroxydanaidal, 319, **11.3**
5-Hydroxydecanolide, 129, **5.31**
10-Hydroxy-2-decenoic acid, 126, **5.28, 5.29**
20-Hydroxyecdysone, 123, 231, **8.12**
  22-acetate, 240
8-Hydroxygeraniol, 181, **7.21**
  glucoside, 200, **7.21**
8-Hydroxyisocoumarin, 164, **6.21**
3-Hydroxy-3-methyl-2-butanone, 189, **7.10**
2-Hydroxy-6-methylacetophenone, 42, 163, **3.2**
6-Hydroxymusazin, 277, **9.21**
Hydroxynitriles, 322
Hydroxylase, 74
3-Hydroxyretinal, 250
Hygrobiidae beetles, 243–244
Hylurgopinus rufipes, 186
Hyperaspine, 118, **5.20**
Hyperaspis campestris, 118
Hypoclinia ants, 163

Ilybius fenestratus, 244, Plate 26
Indole, 261, **9.5**
Indole-5,6-quinone, **9.15**
Indolizidines, 297–298
Insects,
  antiquity, 1, **1.1**
  cold hardy, 31
  colour vision, 249, 272
  cuticle, 286–289
  immunity, 79
  mammalian hormones in, 243
  number of species, 1
  pigments, 271–286
  vision and pigments, 249–250, 272, **8.28, 8.29**
  wound healing, 79, 272
Intermedine, 321, **11.4**

Invictolide, 161, **6.19**
Ips beetles,
  I. amitinus, 188
  I. confusus, 187, 188
  I. duplicatus, 188
  I. paraconfusus, 56, 191, 194
  I. pini, 56, 58, 59, 188, 191, 194
  I. tridens, 191
  I. typographus, 189, 211, Plate 21
Ipsdienol, 59, 187, 188, **3.12, 7.8**
Ipsdienone, 59
Ipsenol, 187, 188, **7.8**
Iridoids, 196–201, 328
  glycosides, 197
  noriridoids, 199
  sequestered, 200
  thioglucosides, 199
Iridodial, **7.18**
Iridomyrmecin, 197, **7.18**
Isobutyric acid, 46, 80, 81, 206, **3.5, 4.13**
Isocoumarin, 164–165
  from symbionts? 164–165
Isocrematofuran, 221, **8.2**
Isogyrinidal, 205, **7.27**
Isoleucine, 47, 80, 137, 209, 324, 325, **3.6, 4.13, 11.8**
  N-acetyl methyl ester, **10.9**
  N-formyl methyl ester, **10.9**
  methyl ester, **10.9**
Isomero-oxygenase, 249
Isopentenyl diphosphate, 181–183, 186, 202, 221, **7.3, 7.4, 7.7, 7.24, 8.1**
  isomerase, 182, **7.4**
N-Isopentyl-2-phenylethylamine, 297, **10.3**
Isoprene, **7.1**
Isopsylloborine A, **5.17**
Isoptera, 135–137

Isotopes, table, 38
Isotope effects, 45–47
　kinetic, 45,46, 47
　on NMR spectra, 45, 46, 49,
　　50–51, **3.8, 3.10**
　^{13}C-^{13}C coupling, 51–53
Isotopic labelling, 38,
　double labelling, 43, **3.3**
　enrichment, 43–45, 49
　examples, 46–51
　heavy isotopes, 43–45
　isotope ratio mass spectrometry
　　44
　radio-isotopes, 39–43, **3.2, 3.3**
　specific labelling, 39
　uniform labelling, 39
Isovaleric acid, 80, 88, **4.13, 4.20**
Isovixetin, 285, **9.26**
Isoxazolinone glucosides, 305,
　10.11

Jalaric acid, 209, **7.31**
Japonilure, 112, **5.14**
Juvenile hormone, 32, 138,
　212–214
　biosynthesis, 214
　epoxidase, 214
　inactivation, 214
　JH 0, JH I, JH II, 213, **7.36**,
　　JH III, 154, 194, 195, 213,
　　7.36
　JH B$_3$, 213, **7.36**
　4-methyl-JH I, 213, **7.36**

Kairomones, 8
Keiferia lycopersicella, 104
Kermesic acid, 275, **9.18**
Kermes ilicis, 275
Ketoreductase, 68–71
β–Ketosynthase, 68, 69, 148, 174,
　4.3, 4.4, 6.3, 6.32
Kynurenine, 284,

Labidus praeditor, 211
Laccaic acids, 209, 276, **9.18**
Laccijalaric acid, 209, **7.31**
Lac insects, 208–209, 275–276
　lac, 209, **7.31**
Lambdina athasaria, **5.1**
Lambdina fiscellaria lugubrosa, **5.1**
Lanierone, 190, **7.11**
Lanosterol, 226, **8.7**
Lardoglyphus konoi, 160
Lasioderma sericorne, 160, 228, 295
Lasius fuliginosus, 164, 205,
　Plate 17
Lepidopteran defence, 110
Lepidopteran sex pheromones,
　96–110
　blends, 109–110
　chain shortening, 103, 107–108
　characteristics, 98
　chirality, **5.1**
　desaturases, 100–108, **5.5**
　epoxide, 99–100
　epoxidase regiospecificity, 100
　even number of carbon atoms,
　　98
　hydrocarbons, 97–98
　oxidation, 109, **5.11**
　reductases, 108–109
　type I , type II, 100, **5.4**
Leptinotarsa decemlineata, 82, 247,
　303
Leptogenys diminuta, 153
Leptoglossus bugs, 261
　L. phyllopus, 261, 265, Plate 27
　L. zonatus, 138, **5.44**
Leptothorax acervorum, 295
Leptothorax muscorum, 295
Leucine, 88, 79, 97, 191, **4.13, 4.20**
　methyl ester, **10.9**
Leucopterin, 278, **9.22**
Limnephilus lunatus, 153, Plate 15
Limonene, **7.1, 7.5, 7.23**

Lineatin, 188, **7.9**
Linepithema humile, 131, 196, **5.34**
Linalool, 183, **7.1, 7.5**
Linalyl diphosphate, 183, **7.4**
Linamarin, 324, **11.8**
Linoleic acid, 5, 75, 83, 90, **4.1**
Linolenic acid, 5, 75, 98, 100, 110, **4.1, 4.11**
Linyphia spiders, 149
Lipids, 66, 81, 82, 83, 85,
  internal, 89, **4.21**
Lipoamide, 25, **2.14**
Lipophorin, 66, 83, 85, 88, 98
Lithobius forficatus, 237
Locusta migratoria, 247, 272
Lotaustralin, 324, **11.8**
Lucibufagins, 241, 309, **8.21**
Lucidida atra, 241
Luciferase, 307
Luciferin, 306–310, **10.13, 10.14, 10.15**
Luciferyl AMP 308, **10.15**
Luciferyl CoA, 308, **10.14**
Lucilia cuprina, 213
Luciola cruciata, 308
Luciola lateralis, 308
Luehdorfia puzibi, 206
Lutein, 247
Lutzomyia longipalpis, 212
Lycopene, 247, **8.26**
Lycopsamine, 321
Lymantria dispar, 88, **5.3**, Plate 11
Lysozyme, 14–15, 79, **2.1, 2.2**

Machimus chrysitis, 286, Plate 40
Macrosiphon liriodendri, 248, **8.27**
Macrosiphon rosae, 277, Plate 33
Makisterone A, 235, **8.15**
24-*epi*-Makisterone A, 236
Makisterone C, 236, **8.15**
Malacosoma moths, 239
Malonyl-acetyl transferase, 68–69

Malonyl CoA, 68–69, **4.5**
Mamestra configurata, 260
Mammalian hormones, 243
Mandelonitrile, 258, 259, **9.3, 11.5**
Manduca sexta, 77, 78, 79, 89, 106,
  109, 224, 235, 237, 238, 282, 288,
  5.10, Plate 6
  cholesterol metabolites, 245,
  8.25
Manica rubida, 153, 211
Manicone, 153, **6.8**
Mannich reaction, 117
Marginalin, 266, **9.10**
Mass spectrometry, 44, 47, 85,
  198, **3.7**
Mastotermes darwiniensis, **7.25**
Mayolenes, 110, **5.12**
Megacrania tsudai, 295
Megalomyrmex ants, 134
Megaponera foetens, 295
Megarhyssa nortoni nortoni, 168,
  169–171, **6.29**
Melanin, 272–275, **9.15**
Melanotus beetles, 112, 220
Melatonin, 265, **9.8**
Mellein, 164, 268, **6.20, 9.11**
Meloe proscarabaeus, 207
Menippe mercenaria, 236
Menthol, **7.1**
Messor ants, 134
Metabolism,
  dynamic nature, 5
  secondary, 1, **1.2**
Metalloenzymes, 15–17, **2.3**
Metamasius hemipterus sericeus,
  191
Methacrylic acid, 46, **3.5**
Methionine, 26, 43, 86, 137, 161,
  164, 23
Methylbenzoquinone
  (toluquinone), 269, **9.12**
3-Methylbutanol, 189, **7.10**

2-Methyl-3-buten-2-ol, 189, **7.10**
3-Methyl-3-buten-1-ol, 189, **7.10**
2-Methylbutenyl butyrate, 81, 50, **4.14**
2-Methylbutyl butyrate, 81, 150, **4.14**
2-Methylbutyric acid, 46, 80 **4.13**
Methyl 2,4-decadienoate, 112
3-Methylenepentyl diphosphate, 210, **7.32**
Methylerythritol phosphate pathway, 185–186, 246, **7.7**
Methyl farnesoate, 214
  as hormone, 214
9-Methylgermacrene B, **7.35**
2-Methyheptadecane, 97
4-Methyl-3-heptanone, 153, **6.7**
6-Methyl-5-hepten-2-one, 190
14-Methyl-8-hexadecenol, **4.14**
3-Methyl-α-himachalene, **7.35**
Methylmalonate (methylmalonic acid), 86, 87, 138, 146, 152–153, **4.17, 5.42, 6.3, 6.6**
3-Methyl-7-methylene-1,3,8-nonatriene, 211, **7.33**
Methyl 14-methyl-8-hexadecenoate, **4.14**
2-Methyl-7-octadecene, 88
6-Methyl-3-octanone, **6.6**
Methyl oxidase, 75
6-Methylpelletierine, **5.20**
13-(1'-Methylpropyl)tridecanolide, 131, **5.34**
6-Methylsalicylic acid, 146, **6.2**
  methyl ester, 47, 161–163, **3.7, 3.8**
Methyl 2,4,5-tetradecatrienoate, **5.13**
Methyl transferase, 174

Methyl 2,6,10-trimethyltridecanoate, 159, **6.15**
Mevalonic acid, 39, 180, 207, **1.2, 7.6**
  pathway, 180–185, **7.2, 7.3, 7.4**
Mevalonolactone, 184, 191, 198, 223, **7.6, 7.19**
Michael addition, 130, 174, **5.32**
Microlinyphia spiders, 149
Mixed function oxidases, 16–18
Model insects, 55, 60, 62
Molecular biology, 55–62
Monocrotaline, 319, **11.3**
Monoterpenes, 179–183
  biosynthesis, 180–186
  degraded, 189–190, **7.11**
  pheromones, 186–196
Monomorium ants
  M. fieldi, 301
  M. pharaonis, 212, 221, 297, **10.4**
  M. rothsteini, 134
Morpho butterflies,
  M. didius, 272
  M. peleides, 230
  M. rhetenor, 272
Motyxia millipedes, 307, Plate 45
Moulting hormone, 231–239
  see also ecdysteroids
Multistriatin, 190, 192, 193
Musca autumnalis, 123
Musca domestica, 81, 83, 89, 227
  sex pheromone, 122–123, **5.25**
Musizin, 277, **9.21**
Myrcene, 56, 184, **3.11, 3.12, 7.8, 7.23**
Myriapoda, 7
Myrmica ants, 152–153, 211, **6.6**
Myrmicaria opaciventris, 298
Myrmicarins, 298, **10.6**
Myrrhine, **5.16**

NMR spectroscopy, 39, 43, 44, 45, 46, 49, 50–51, 85, 324, **3.8**, **3.10**
^{13}C-^{13}C coupling, 51–53, 164, **3.10**, **6.20**
chiral, 54
Nannotrigona testaceicornis, 220
Naphthoquinones, 270, 275–276, **9.13**, **9.18**
Narceus gordanus, 270
Nasonia vitripennis, 129, **5.31**
Nasutitermes corniger, 222
Nasutitermes octopilis, 223
Nasutitermes takasagoensis, 223
Nasutitermitinae, 222
 frontal glands, 222, **8.3**
Neanura muscorum, 269, Plate 29
Nemobius fasciatus, 82, 88
Neoclytus accuminatus accuminatus, 111
Neozeleboria cryptoides, 130
Neozeleboria monticola, 130
Nepetalactone, 197, **7.18**, **10.2**
Neriaphin, 277, **9.21**
Neriene spiders, 149
Nerol, **7.1**
Nerolidol, **7.24**
Nicotinamide adenine dinucleotide, 21–22, 30, **2.10**
Nitrogen excretion, 25
 fixation, 31
o-Nitrophenol, 268
3-Nitropropionic acid, 305, **10.11**
2-Nitrostyrene, 261, **9.4**
Nitulid beetles, 155–159
Non-mevalonate pathway, 186
Norgeraniol, 198
Normanicone, 13, **6.8**
Nothomyrmecia macrops, 90, Plate 10
Nuptial gifts, 316, 319

^{18}O labelling, 44–45, 50, 113, 168, 169, 192, 203, 308, **3.4**, **5.15**, **6.26**, **7.26**, **10.15**
Ochtodenol, 194, **7.16**
Ocimene, **7.1**, **7.23**
18-Octadecanolide, 128, **5.30**
3-Octanol, **6.6**
3-Octanone, **6.6**
Octopamine, 263, **9.7**
2-Octynoic acid, 39, 115, 154, 191
Oenocytes, 83, 88, 97, 123, 124
Oestradiol, 244
Oleandrin, **11.2**
Oleic acid, 45, 46, 72, 73, 120, **4.1**, **4.11**, **5.21**
Ommins, 284, **9.25**
 ommin A, **9.25**
Ommochromes, 284, **9.25**
 as excretory products, 284
Oncopeltus fasciatus, 228, 281, Plate 36
One-carbon fragments, 26–28, **2.17**, **2.18**, **2.19**
Oocytes, 236
Opiliones, 159, **6.16**
Opsin, 249
Oreina beetles, 320
Oribotritia berlesi, 201
Ornithine, 298, 318, **10.5**
Oryzaephilus beetles, 113, **5.15**
 O. mercator, 203, **7.26**
Osmeteria, *see Papilio* butterflies
Oviposition deterrent, 395, **10.12**
Oxidus gracilis, 258, 323
9-Oxo-2-decenoic acid, **5.28**
Oxyluciferin, 308, **10.15**
Oxytrigona bees, 29
Oxytrigona mediorufa, 150

PBAN, 97, 100, 109
Paederus beetles, 57, 172, Plate 19
  bacterial symbionts, 172
  P. fuscipes, 172
Paederidus beetles, 172
Palasonin, 208, **7.30**
Palasoninimide, **7.30**
Palmitic acid, 70, **4.1, 4.6**
Palpatores, 159, **6.16**
Paltothyreus tarsatus, 43
Papilio butterflies, 206
  osmeterial defence, 206,
    Plate 23
  P. helenus, 206
  P. memnon, 206
  P. phorcas, 282
  P. protenor, 206
  P. sarpedon, 283, 284, Plate 37
Papiliochromes, 289, **9.29**
Paratrechina ants, 302
Paropsis atomaria, 324
Pectinophora gossypiella, 247, **5.4**
Pederin, 57–58, 82, 172–175, **6.31, 6.32, 6.33**
Pederone, 172, **6.31**
Pentamethyldocosane, 88, **4.19**
Pentatomid bugs, 159–160
Perillene, 205, **7.28**
Perillus bioculatus, 247
Periplaneta americana, 77, 83, 84, 85, 90, 203, 237, 266, **7.25,** Plate 8
Periplaneta fuliginosa, 86
Periplanones A, B, C, D, 203, **7.25**
Phaedon amoraciae, 198, 199
Phaedon cochleariae, 182,198, 199, 328, Plate 20
Phaeomelanin, 274, **9.16**
Pharmacophagy, 315
Pheidole biconstricta, 295
Phenol, 261
Phenoloxidase, 273, 288

Phenols, 265–269
  m-alkylphenols and resorcinols, 268, **9.11**
  oxidative coupling, 275, **9.17**
  phenolic ketones, 267, **9.10**
Phenylalanine, 255, 256, 261, 323, **9.2**
Phenyl-C_3 compounds, 256–260
  derivatives as pheromones, 60–261
2-Phenylethanol, 260, **9.2**
Phenylpyruvic acid, 256, **9.1, 9.2**
Pheomelanin, 274, **9.16**
Pheromones, 8
  and chirality
  contact, 137, **5.42**
  in pest control, 8–9
  lepidopteran blends, 109–110, **5.5, 5.6. 5.9**
  spacing or epidiectic, 186
Philanthus triangulum, 303, **10.10**
Philonthus beetles, 295
Phorcabilin, 282, 284, **9.24**
Phosphoenol pyruvate, 30, **1.2**
Photinus fireflies, 241–242, Plate 44
  P. pyralis, 308
Photosynthesis, 44, **1.2, 1.4, 3.4**
Photurus fireflies, 242
Phyllium westwoodii, 300
Phyllopertha diversa, 296
Phytosterols, 226–227
  dealkylation by insects, 228–230, 235, **8.9**
Phragmatobia fuliginosa, 100
Phratora laticollis, 199
Pieris brassicae, 278, 281, 282, 285
Pieris rapae, 110, 278
Pigments of insects, 271–286, 289
Pinenes, 184, 188, **7.5, 7.8, 7.23**
Pityogenes chalcographus, 112, 166, **6.24**

Pityogenes hopkinsi, 112
cis-Pityol, 189, 190, 194, **7.11**
Pityophthorus beetles, 190
Plagiodera versicolora, 198, 199
Plagiodial, 199
Plant biosynthesis, 5–7
Plant volatile elicitors, 309
Platyphora beetles, 321
 P. boucardi, Plate 47
 P. kollari, 242, **8.22**
Platypus flavicornis, 189
Platypus mutates, 190
Plectreurys tristis, **10.10**
Poekilocerus bufonius, 316
Polyketides, 58, 146, 147–148, 255, 270
 in insects? 148, 163–165
 synthases, 147–148, 173
 types I, II and III, 148
Polygraphus polygraphus, 189
Polymerase chain reaction, 55–58, 59, 172
Polyommatus icarus, 286
Pogonomyrmex salinus, 153
Ponasterone A, 236, **8.15**
Porphobilinogen, **9.24**
Porphyrins, 281
Poststerone, 238, **8.17**
Precoccinelline, 115, **5.16**
Pregnanes, 245
Prenyl transferase. *see* geranyl diphosphate synthase
Prephenic acid, 256, **9.2**
Prochirality, 22, 33, 46, **2.11, 2.24, 2.25**
Propionic acid, 70, 81, 85, 164
 metabolism by insects, 87, **4.18**
Prorhinotermes canalifrons, 202
Prostaglandins, 77, 79, **4.12**
 effects in insects, 79
Protein data bank, 13
Prothoracic gland, 232–234

Prothoracicotrophic hormone, 232, 236
Protoaphins, 277, **9.19**
Protocatechuic acid, 266, **9.9**
Protoporphyrin IX, 281, **2.4, 9.24**
Protoveratrines, 326, **11.9**
Prunasin, **11.5**
Pseudopederin, 172, **6.31**
Pseudozizeeria maha, 285
Psyllobora vigintiduopunctata, 115
Psylloborine A, 115, **5.17**
Pteleobius vittatus, 189, 190
Pterins, 277–281
Pterobilin, 282, 283, **9.24**
Pterostichus californicus, 47
Pumiliotoxins, **10.8**
Putrescine, 298, 303, **10.5, 10.10**
Pyrazines, alkyl, 299–300, 318, 325, **10.7, 11.1**
Pyrearinus termitilluminans, 308
Pyridoxal phosphate, 25, **2.15, 2.16**
Pyrrhocoris apterus, 247, 281
Pyrrolidines, 295, **10.1**
 alkylpyrrolidines, -piperidines, -pyrrolines, 133–134, **5.35, 5.36, 5.37**
Pyrrolizidines, 297–298, 318–322, **10.4**
 bitter taste, 322
 N-oxides, 319
Pyruvic acid, 24–25, 29–30, 185, **1.2, 2.14, 7.7**

Quinones, 269–271, **9.12**

Radio-labelling, 39–43
 double labelling, 43, **3.3**
 specific labelling,
 uniform labelling,
RasMol, 13
Re face, 34, 109, **2.25**

Reticulitermes termites, 221
  R. santonensis, 224
Retinal, 249, **8.28**
Retronecine, 321, **11.4**
Reverse transcription, 37
Rhadinoceraea nodicornis, 325
Rhagoletis cerasi, 305, **10.12**,
  Plate 43
Rhipicephalus microplus, 237
Rhodanese, 323–324, **11.7**
Rhodopsin, 249, **8.28**
Rhodnius prolixus, 282
Rhodommatin, 284, **9.25**
Rhyacophila fasciata, 150
Rhyssa persuasoria, 189,
  Plate 22
Rhytidoponera aciculate,
  42, 163
Rhytidoponera chalybaea, 164
Rhyzoglyphus robini, 230, **8.11**
Rinderine, 321, **11.4**
Ring gland, 232
Romalea microptera, 266
Royal jelly, 126–127, **5.29**

^{35}S labelling, 39, 43, 284, **3.3**
Salicylaldehyde, 327
Salicylic acid, 261, **9.5**
  alkylsalicylic acids, 268, **9.11**
Saponarin, 285, **9.26**
Sarcosine, **10.12**
Sarpedobilin, 283
Scale insects, 224, 275–276
Scarites subterraneus, 46, 48,
  Plate 3
Schemochromes, 272
Schistocerca gregaria, 237, 247,
  265, 284, 288, Plate 38
Schizura concinna, 29, Plate 1
Sclerotisation, 288, **9.28**
Scolopendra subspinipes mutilans,
  164

Scolytus beetles, 153
  S. multistriatus, 186
  S. scolytus, 186, 193
Semiochemicals, 8
Seneciphylline, 319, **11.3**
Sequestered plant compounds,
  315
  origins of sequestration,
    327–328
Serine, 25, 26, 29, 115, 120
Serotonin, 264–265, **9.8**
Serricornin, 160, **6.18**
Serricorole, **6.18**
Serricorone, **6.18**
Sesquiterpenes, 179, 202–214
  defences, 204–208
  pheromones, 202–204
Sesterterpenes, 224
Shikimic acid, 39, **1.2**
  pathway, 255–256, **9.1**
Shodromantis bioculata, 247
Si face, 34, **2.25**
Signatipennine, **5.24**
Silpha novaboracensis, 245
Sinapus beetles, 220
Sitophilate, 155, **6.10**
Sitophilus weevils, 155, **6.10**
  S. granarius, Plate 16
  S. oryzae, 155, **6.10**
  S. zeamais, 155, **6.10**
Sitophinone, 155, **6.10**
Sitosterol, 226
Skatole, **9.5**
Solenopsins, 133–134, **5.35**
Solenopsis ants, 132–134, 297,
  5.35, 5.36, 5.37, 10.4
  S. geminata, 133, **5.35**
  S. invicta, 133, 161, Plate 10
  S. richteri, 133
Specific incorporation, 40
Spermidine, 303, **10.10**
Spermine, 303, **10.10**

Spiroacetals, 166–172, **6.23, 6.24, 6.25, 6.30**
   biosynthesis, 167–172, **6.26, 6.27.6.28, 6.29**
Spodoptera moths,
   S. eridania, 86, 89
   S. exigua, 309
   S. littoralis, 105, 107, 238, **5.9**
Springene, 201, 220, **8.1**
Squalene, 225, 226, **8.7**
Stegobinone, 160, **6.18**
Stegobium panicetum, 160, 228
Stephanitis pyrioides, 266–267, Plate 28
Stephanitis rhododendri, 267
Sterols, 6,
   dietary and tissue, 228
   defensive, 240
   in insects, 227–242
   as pheromones, 239
   pyrones, 241
Stigmasterol, 226, 230
Stiretrus anchorago, 159, **6.15**
Subcoccinella vigintiquatropunctata, 120–121, **5.22**
Succinic acid, 86, 168, **4.17**
Sulcatol, 189, 190, **7.11**
Sulcatone, 190, **7.11**
Symbiotic bacteria, 31, 172, 228, 244, 266, 275
Synomones, 8
Syringaldehyde, 261, **9.4**

Tenebrio molitor, 155, 229
Termites, 135–137, 201, 221, **5.40**
   terpene defensive compounds, 201, **7.23**
Terpinen-4-ol, 189, **7.9**
Terpineol, **7.5**
Testosterone, 244
11-Tetradecenyl acetate, **5.4**

11-Tetradecenal, **5.4**
Tetrahydrofolic acid, 26–28, 278, **2.17, 9.22**
Tetramorium aculeatum, 131, **5,34**
Tetramorium impurum, 47, 163
Tetraponerines, 297, **10.4, 10.5**
Tetrapyrroles, 281–284
Tetraterpenes, 245–248
Thaumetopoea pityocampa, 105, **5.8**
Thelodesmus armatus, 261
Thiamine diphosphate, 23–25, **2.13, 2.14**
Thioesterase, 68–69
Threonine, 300
Tiglic acid, 46, **3.6**
Transamination, 25, **2.15, 2.16**
Trehalose, 287, **9.27**
Triatoma infestans, 83
Tribolium castaneum, 60, 76, 154, **6.9**, Plate 7
Tribolium confusum, 91, 111, 155, **5.13, 6.9**
Trichoplusia ni, 56, 109, **5.5**
9-Tricosene, 83, 122–123, **5.25**
2-Tridecanone, 41
Triglycerides, 66, **4.1**
4,6,8-Trimethyldecan-2-one, 160, **6.17**
Trimethyldotetracontane, 89
Trimethylhexatriacontane, 89
3,4,7-Trimethyl-2,6-nonadienol, 211, **7.33**
3,4,7-Trimethyl-2,6-octadienol, 210, **7.33**
3,5,7-Trimethyloctanolide, 161, **6.19**
3,5,7-Trimethylnonanolide, 161, **6.19**
6,10,13-Trimethyltetradecanol, 159, **6.15**
Trimethyltetratriacontane, 89

2,6,10-trimethyl-5,9-undecadien-1-ol, 203, **7.25**
Trinervitanes, 222, **8.3**
Tryptophan, 255, 261, 284, 284, **9.5, 9.28**
Triterpenes, 179, 224, 226
   saponins, **8.22**
Trogoderma granarium, 81
Trypodendron domesticus, 188
Trypodendron lineatum, 188
Tryptamine, 264, **9.8**
Tryptophan, 261, 264, 284, **9.5, 9.8, 9.25, 9.28**
Tyrosine, 256–257, 261, 263, 272, 276, 288, **9.2, 9.7, 9.28**

Uric acid, 25, 307
Uroleuconaphins, 277, **9.20**
Uroleucon nigrotuberculatum, 277
Utethesia moths, 319
   U. ornatrix, 319, Plate 46

Valine, 46, 79, 80, 137, 210, 323, 324, **3.5, 4.13, 4.20, 5.3, 11.8**
Vanillin, 261, **9.4**
Veratrum alkaloids, 325–326
cis-Verbenol, 188, **7.8**
Verbenone, 188, **7.8**
Vernolic acid, 129, **5.31**

Vespa wasps
   V. crabro, 281
   V. germanica, 281
   V. vulgaris, 281
Vibidia duodecimguttata, 115
Virginae butanolide A, 50–51, 53, **3.9, 3.10**
Visual spectrum, 249, 272
Vitamins, 28, 228
   vitamin A, 249
   vitamin B_{12}, 86–88, **4.17**
cis-Vittatol, 188, 190, 194, **7.11**
Volicitin, 309, **10.16**
Vonones sayi, 270

Wasps, 129–130

Xanthommatin, 284, **9.25**
Xanthopterin, 278, 281, **9.22**

Y-organ, 236

Zingiberene, **7.1**
Zonocerus variegatus, 321, Plate 48
Zootermopsis angusticollis, 85, 86
Zophobas atratus, 78
Zophobas rugipes, 26
Zygaena moths, 324–325
   secretion from mouth, 325
   Z. filipendula, 324–325
   Z. trifolii, 324, Plate 50

WITHDRAWN
FROM STOCK
QMUL LIBRARY